Applications of Statistical Physics

Outside front cover: Flow through a percolating cluster of sandbodies.
Different colours represent times taken by a tracer to get to that point.
(Courtesy of P.R. King.)

Applications of Statistical Physics

Proceedings of the NATO Advanced Research Workshop
held at Technical University of Budapest
Hungary, 19–22 May 1999

Editors:

Adam Gadomski

Department of Theoretical Physics
Institute of Mathematics and Physics
Technical University of Bydgoszcz
Al. Kaliskiego 7/402
PL-85796 Bydgoszcz
Poland

János Kertész

Institute of Physics
Technical University of Budapest
Budafoki ut 8.
Budapest, H-1111 Hungary

H. Eugene Stanley

Boston University
Center for Polymer Studies
Department of Physics
590 Commonwealth Avenue
Boston, Massachusetts 02215, USA

Nicolas Vandewalle

Charge de recherches FNRS
GRASP, Institut de Physique B5
Université de Liège
B-4000 Liège, Belgium

1999

ELSEVIER

AMSTERDAM - LAUSANNE - NEW YORK - OXFORD - SHANNON - SINGAPORE - TOKYO

ELSEVIER SCIENCE B.V.
Sara Burgerhartstraat 25
P.O. Box 211, 1000 AE Amsterdam, The Netherlands

Reprinted from: *Physica A, vol. 274/1,2, 1999*

Library of Congress Cataloging in Publication Data
A catalog record from the Library of Congress has been applied for.

Transferred to digital printing 2006

ISBN: 0-444-50409-5

⊛   The paper used in this publication meets the requirements of ANSI/NISO Z39.48-1992 (Permanence of Paper).
Printed and bound by Antony Rowe Ltd, Eastbourne

Contents

APPLICATIONS OF STATISTICAL PHYSICS

NATO Advanced Research Workshop, Budapest, May 19–22, 1999

The field of statistical physics has undergone a spectacular development in recent years. The *fundamentals* of the subject have advanced spectacularly with multidisciplinary approaches involving physicists, chemists, and mathematicians. Equally spectacular has been the development of *applications* of statistical mechanics to shed light on a wide range of problems, many of them arising in fields quite distant from traditional physics disciplines. Indeed, a wide range of possible applications opens since one of the main goals of statistical physics is to understand the collective behavior of a large number of interacting units. Recent applications range from such topics as oil recovery from porous rock to protein folding, DNA structure, morphogenesis and the cooperative behavior of living creatures. Concepts and methods of statistical physics have been applied successfully to "exotic" problems that seem to be far from physics, such as vehicular and pedestrian traffic, or economy and finance. Sometimes addressing practical problems motivates fundamental developments in statistical physics, as in for example polymer physics or viscous fingering.

No wonder that the dynamic development and the broadening of the application fields of statistical physics moves rapidly forward. This fact motivated our interest in organizing a NATO Advanced Research Workshop to anticipate the new efforts on applications of statistical physics likely to take place in coming years as we enter a new century. The response to the idea of such a workshop was remarkably positive, so we could gather a good mixture of participants including senior scientists and young students, experimentalists and theoreticians, "Wessies" and "Ossies" (the German nicknames for people coming from Western and Eastern countries). The talks demonstrated indeed the extraordinary power and breadth of the statistical physics discipline. Applications of statistical physics included topics such as DNA migration, wetting, chemical waves, granular media, molecular motors, biological pattern formation and motion, as well as such really practical problems as heart diagnosis, internet traffic jamming, oil recovery and econophysics (a neologism for statistical physicists attempting to shed light on questions of economics and finance). In addition to the keynote lectures, a number of high quality, interesting contributed communications were also presented. This Proceedings contains the keynote invited talks in the order of their presentation and the contributed communications in alphabetical order. This collection of papers is designed to hint at some trends that will become more further elaborated in the coming century.

Special thanks are due to many who made possible this meeting. First and foremost, we thank the Scientific Affairs Division of NATO, who kindly supported our efforts: under their new priorities for supporting "applications" and "East-West interactions". The organization of this workshop reflects well the extremely fast-changing world:

0378-4371 - see front matter © 1999 Elsevier Science B.V. All rights reserved.
PII: S 0 3 7 8 - 4 3 7 1 (9 9) 0 0 4 2 2 - 7

When we submitted our proposal, Hungary, Poland and the Czech Republic were Partner states to NATO – by the time the meeting took place they had become full members. We received supplemental support from the Technical University of Budapest, the Hungarian Academy of Sciences and the Ministry of Education (NEFIM) which is acknowledged hereby with thanks. Marcel Ausloos provided most generous help at every stage of the meeting. We are also immensely grateful to Luuk Holla and Jan Hanraads who facilitated the extremely rapid and efficient publication of this volume, and to Marten Stavenga whose vision led to the simultaneous publication of a special issue of Physica A as this book.

Bydgoszcz-Budapest-Boston-Liège, August 1999   Adam Gadomski[1]
                                                      János Kertész[2]
                                                      H. Eugene Stanley[3]
                                                      Nicolas Vandewalle[4]

[1] Tel.: + 48-52 340 8616; fax: + 48-52 340 8643; e-mail: agad@tower.atr.bydgoszcz.pl
[2] Tel.: + 36-1 463 3568; fax: + 36-1 463 3567; e-mail: kertesz@phy.bme.hu
[3] Tel.: + 1-617 353 2617; fax: + 1-617 353 3783 (on desk) 617/353-9393 (dept. office); e-mail: hes@bu.edu
[4] Tel.: + 32-4 366 3703; fax: + 32-4 366 2990; http: //www.supras.phys.ulg.ac.be/ statphys/nico/nico.html

ELSEVIER

Physica A 274 (1999) 1–7

www.elsevier.com/locate/physa

Problems of DNA entry into a cell

Pierre-Gilles de Gennes *

Collège de France, 11 place Marcelin Berthelot 75231 Paris Cedex 05, France

Abstract

This paper describes some basic processes involved when a DNA molecule faces a pore and gets through. In the simple case considered here, the only driving force is a concentration difference between DNA outside and inside. For plausible conditions, the transfection time (the time required to force *one* molecule in) is expected to be reasonably short. © 1999 Elsevier Science B.V. All rights reserved.

1. Introduction

Gene therapy is one of the major goals of our time. The basic process amounts to transfer a long, double stranded, piece of genetic code from an external medium into the cytoplasm of a target cell. Of course, even if this is achieved, various problems of cellular transfer still arise. But the cytoplasmic barrier is essential.

There are (at least) three major techniques of transfer. The first amounts to construct a certain lipid/DNA package, which should coalesce efficiently with the cytoplasmic membrane [1]. The second technique uses the capside of a natural virus: this is engineered to perform exactly the right function – attaching to the cell wall, opening a hole and discharging its DNA content into the cell. One serious difficulty here is due to the large immunoresponse which most capsides induce in our organism. However, certain viruses (associated adenovirus, or AAV) appear to be non pathogenic and are nearly invisible to the immune system: they may give us a solution [2].

A third technique is based on the opening of a pore in the cell wall. Various opening tricks have been proposed, such as the molecular drill of Ref. [3]. But the main hopes are related to *electroporation*: applying a field over a cell, we may, in some instances, open a transient pore, and use it to channel in the DNA.

The present paper is mainly concerned with this approach. In Section 2, we summarize the electroporation process. In Section 3, we discuss the DNA motions. In

---

* Fax: 33-1-45-35-14-74.

E-mail address: pierre-gilles.degennes@espci.fr (P.G. de Gennes)

0378-4371/99/$ - see front matter © 1999 Elsevier Science B.V. All rights reserved.
PII: S 0378-4371(99)00309-X

Section 4, we estimate the resulting transfection time for a simple vesicle. All our approach is very crude, and holds only at the level of scaling laws. Hopefully, this can be the starting point for certain model experiments. A brief account of the present ideas has already been published [4].

2. Electroporation

A large electric field E_0 ($\sim 10^4$ V/cm) is applied to a cell as a short pulse (microseconds to milliseconds). The cell opens up. In recent experiments [5], one is able to visualize the entry of certain external species (e.g. Ca^{++}), using a calcium-sensitive dye incorporated in the cell. The threshold voltage difference V across the cell wall is of order 1 V.

A classical (naive) argument describing the opening is based on the following free energy for a pore of radius r:

$$f = -\frac{\varepsilon E_0^2}{8\pi} d\pi r^2 + \Im 2\pi r , \tag{1}$$

where ε is the dielectric constant, and d the thickness of the lipid bilayer. The vesicle is assumed to be floppy (i.e. to have optimised the area per polar head). The parameter $\Im$ is a line energy, describing the folding of the monolayers near the pore edge. Eq. (1) leads to an energy barrier with a certain radius r^* and energy f^*. When f^* becomes comparable to the thermal energy kT, the pore opens, and this gives an estimate of the threshold field.

This analysis has been modified by Isambert [6], who considered a spheroidal vesicle (of radius R) and showed how curvature effects are important. The following is a poor man's version of Isambert's ideas.

Without any pore, the field vanishes in the water regions, and is concentrated in the bilayer, with a value:

$$E_m \sim E_0 \frac{R}{d} \sim \frac{V}{d} . \tag{2}$$

The normal stress on the membrane is of order:

$$p_n = \frac{\varepsilon E_m^2}{8\pi} . \tag{3}$$

For a flat membrane, the stresses on both sides would balance out. But for a curved membrane, there is a non compensated stress outwards of order:

$$p_1 \cong p_n \frac{d}{R} . \tag{4}$$

This puts the membrane into tension like an inflated balloon. The two-dimensional tension σ is related to p_1 by the Laplace formula:

$$p_1 = \frac{2\sigma}{R} . \tag{5}$$

Rupture occurs when this tension reaches a certain critical value σ_m. Rearranging Eqs. (2)–(5), one then arrives at a threshold voltage V which is independent of the cell radius, and of order 1 V – rather similar to the estimate from Eq. (1).

There are many interesting questions concerning the fate of the pore after opening. Related observations on pores opened by mechanical tension in vesicles have been performed recently by Sandre and Moreaux, and analysed by Brochard-Wyart [7]. For our present purposes, we shall not enter into this problem: we shall assume (a) that a pore has been opened (b) that there is no macroscopic flow of water inward or outward (c) that the pore is larger than the Debye screening length, so that the negatively charged DNA does not suffer from repulsive interactions with the lipids. The only force driving the DNA is then its concentration gradient.

It is of importance to realise that once the DNA has inserted a significant portion inside, the pore may close on itself, without stopping the entry process: the only difference is that we then have an extra friction between DNA and lipid. But this local friction is often expected to be a minor correction, when the DNA is long and hydrodynamic friction is large: we return to this point in Section 3.

We can estimate the time for passive closure of a pore (of initial size r_p) due to the line tension $\Im$ and the two-dimensional viscosity η_2 of the bilayer: this is of order $r_p \eta_2 / \Im \sim$ ms.

3. The dynamics of passive entry

3.1. The driving force

In the outer compartment, we have a concentration c of DNA (nb of base pairs per unit volume) and a chemical potential $\mu + \Delta\mu$. Inside the vesicle, we take $c = 0$, and we have a potential μ. The scaling structure of $\Delta\mu$ is

$$\Delta\mu \cong kT \frac{c}{c_s} \,, \tag{6}$$

where c_s is the ionic strength (and $c < c_s$).

Eq. (6) may be understood through the following crude argument. Any unit (base pair) along one DNA carries a charge $\sim e$ and is exposed to electrostatic repulsions from other chains, within distances r comparable to the screening radius κ^{-1}. Thus:

$$\Delta\mu \sim \int \frac{e^2}{\varepsilon r} e^{-\kappa r} c 4\pi r^2 \, dr = \frac{4\pi c e^2}{\varepsilon \kappa^2} \tag{7}$$

while (for monovalent salts):

$$\kappa^2 = \frac{4\pi e^2 c_s}{\varepsilon kT} \quad (c < c_s). \tag{8}$$

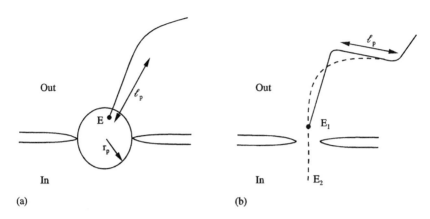

Fig. 1. A schematic picture for the entry process: a chain end E moves from E_1 to E_2 in the vicinity of the pore. The chain has a large persistence length ℓ_p. The penetration process implies motions over a few persistence lengths.

3.2. The entry process

Our scheme for entry is displayed in Fig. 1. A chain end initially lies close to the pore: then the semi rigid chain unfolds (keeping its persistence length ℓ_p) under the concentration gradient. As seen in the figure, this is a local process, involving only a few persistence lengths. The corresponding friction coefficient is

$$\zeta_\ell \sim \eta \ell_p \,, \tag{9}$$

where η is the viscosity of water. The driving force is $\Delta\mu/a = f$, where a is the distance between base pairs along one DNA ($a \sim 3$ Å). The velocity of entry is f/ζ_ℓ, and the time for one chain end, to go from "out" to "in" is

$$\tau_e = \frac{r_p f}{\zeta_\ell} = \frac{\eta r_p \ell_p a}{\Delta\mu} \,. \tag{10}$$

Taking $c/c_s = 10^{-3}$, $\ell_p = 100$ nm, $r_p = 10$ nm, we arrive at $\tau_e \sim 10^{-4}$ s.

3.3. Sliding

After entry, most of the chain is still outside and has to be pushed in by the force f. This sliding process has been studied experimentally [8] and theoretically [9] on a somewhat related system: a single strand polyuracil entering via a proteic pore (S. Aureus α hemolysin) of inner diameter ~ 15 Å. Here, the driving force is an electric field. Chain passage is detected by a patch clamp technique, observing pulses in the channel conductance. The DNA is short ($\sim$200 base pairs). Thus fluctuations in the speed are superimposed to an average drift [9].

Our system differs in some respects: (a) our DNA is double stranded, and locally rigid (b) the chains are long ($\sim$50 kbase pairs): drift dominates over fluctuations

(c) the pore can be open or can have closed back, giving different friction regimes:

(i) If the pore is relatively large, we deal with hydrodynamic friction. This may be of two types:

(α) If the DNA molecules (in the outer compartment) overlap significantly, they move by reptation, and the friction coefficient ζ is of the form:

$$\zeta \cong \eta L, \tag{11}$$

where L is the contour length of the chain.

(β) If we are talking about a very dilute DNA system, we face the problem of a flexible (at large scales) coil being force through a pinhole. Here, the hydrodynamic interactions are dominant, and we expect [3]:

$$\zeta = \zeta_\beta \sim \eta R, \tag{12}$$

where R is the coil size of our DNA. $R \sim (L\ell_p)^{1/2}$.

This second estimate is smaller than the first (roughly by one decade).

(ii) If the pore has closed back, we should add on to this, a contribution ζ_γ describing the friction between DNA and lipid. This ζ_γ may be large, but is independent of chain length. Thus we expect processes α or β to dominate for long chains. This leads to a sliding time (for case α):

$$\tau_s \sim \frac{\eta L^2 a}{\Delta \mu} \quad \text{(case } \alpha \text{)}. \tag{13}$$

which may be of order 1 s.

4. The transfection time

Entry is the bottleneck of our process. But we must now construct the statistics of entry. Two distinct questions emerge:

(a) What is the time required to be sure that one DNA has entered? We call this the transfection time τ_t.

(b) What is the time required for equilibration of concentrations between the two compartments. We call this the final time τ_f.

The time τ_t is clearly related to the entry time τ_e, but two extra features must be incorporated:

(1) In Fig. 1, we assumed that one chain end was initially near the pore (of radius r_p). In fact, the number of chain ends available in this region (of volume r_p^3) is of order:

$$n = 2\frac{c}{N}r_p^3 \quad (\ll 1), \tag{14}$$

where $N = L/a$ is the number of base pairs per chain, and $2c/N$ the concentration of chain ends.

(2) In most electroporation processes, the field E_0 cannot be applied continuously, because of the resulting Joule heat in water. The field is present only for a fraction ψ of the time.

Thus we are led to write:

$$\tau_t = \frac{1}{n\psi}\tau_e \,,$$ (15)

where τ_e is given by Eq. (10).

This leads to transfection times τ_t, which are in the range of hours for $\psi = 10^{-3}$.

How does this relate to the full equilibrium time τ_f? The time τ_f can be examined via a rate equation for the internal concentration c_i, inserting the appropriately corrected driving force $\Delta\tilde{\mu} = kT(c - c_i)/c_s$. But the scaling result is obvious: to equilibrate, we need to pass a number $\sim c/N\, R^3$ chains in the vesicle of volume R^2. Thus:

$$\tau_f = \frac{c}{N}R^3\tau_t$$ (16)

and τ_f is very large.

Fortunately, for our purposes, the most relevant time is the transfection time, and this time appears not exceedingly large for medical purposes.

Of course, our discussion is very schematic; many extra features should be incorporated.

(α) *Role of electrical charges*: when we open a pore, we may have a drift of the DNA in the residual electric fields, and a flow through the pore induced by electroosmosis. But we expect these features to die out fast (within a dielectric relaxation time) when the pore opens.

(β) *Overall flows*: if, for some reason, the inside is not in full osmotic equilibrium with the outside, water will flow [6]. Depending on signs, this may stop, or may facilitate the entry of DNA.

(γ) *Multiple entry*: if there is more than one pore, we may find cases where one DNA chain enters into two of them (p_1 and p_2). If we deal with a single vesicle, the driving force will then pull p_1 towards p_2, and they should ultimately coalesce. But if we deal with a real cell, the cytoskeleton will prevent these motions, and our DNA is stuck.

5. Summary

An electroporation process must generate a pore larger than the screening length. The electrical field pulse must last longer than the entry time τ_c (Eq. (10)). If these conditions are satisfied, passive entry should occur, and lead to transfection times which are not too large.

Acknowledgements

I have benefited from discussions with F. Brochard-Wyart on the dynamics of pore closure (with or without surface tensions).

References

[1] R.G. Crystal, Science 270 (1995) 404.
[2] R.J. Samulski, M. Sally, M. Muzyksa, Development of Human Gene Therapy, Cold Spring Harbor Press, 1999 (Chapter 7).
[3] M.A. Guedeau-Boudeville, L. Jullien, J.M. di Meglio, Proc. Nat. Acad. Sci. (USA) 92 (1995) 1.
[4] P.G. de Gennes, Proc. Nat. Acad. Sci. (USA) Vol. 96 (1999), pp. 7262–7264.
[5] B. Gabriel, J. Teyssié, Biophys. J. 76 (1999), in press.
[6] H. Isambert, Phys. Rev. Lett. 80 (1998) 3404.
[7] O. Sandre, F. Brochard-Wyart, Proc. Nat. Acad. Sci. (USA) Vol. 96 (1999), in press.
[8] J. Kasianowicz, E. Branoin, D. Branton, D. Deamer, Proc. Nat. Acad. Sci. (USA) 93 (1996) 13 770.
[9] D. Lubensky, D. Nelson, Phys. Rev. E, to be published.

ELSEVIER

Physica A 274 (1999) 8–18

www.elsevier.com/locate/physa

Charge inversion in DNA–amphiphile complexes: possible application to gene therapy

Paulo S. Kuhn, Yan Levin*, Marcia C. Barbosa

Instituto de Física, Universidade Federal do Rio Grande do Sul, Caixa Postal 15051, CEP 91501-970, Porto Alegre, RS, Brazil

Abstract

We study complex formation between the DNA and cationic amphiphilic molecules. As the amphiphile is added to the solution containing DNA, a cooperative binding of surfactants to the DNA molecules is found. This binding transition occurs at a specific density of amphiphile, which is strongly dependent on the concentration of the salt and on the hydrophobicity of the surfactant molecules. We find that for amphiphiles which are sufficiently hydrophobic, a charge neutralization, or even charge inversion of the complex is possible. This is of particular importance in applications to gene therapy, for which the functional delivery of specific base sequence into living cells remains an outstanding problem. The charge inversion could, in principle, allow the DNA–surfactant complexes to approach the negatively charged cell membranes permitting the transfection to take place. © 1999 Elsevier Science B.V. All rights reserved.

PACS: 87.14.Gg; 87.15.Nn

1. Introduction

In the last few years gene therapy has received significant attention both from the scientific community and from the general public. The development of new techniques for transferring genes into living cells allowed for the potential treatment of several diseases of genetic origin [1–11]. The central problem of gene therapy lies in the development of safe and efficient gene delivery system. Since both the DNA and the cell membranes are negatively charged, the naked polynucleotides are electrostatically prevented from entering the cells. Furthermore, the unprotected DNA is rapidly degraded by nucleases present in plasma [11].

---

* Corresponding author. Fax: 55-51 319 1762.
E-mail address: levin@if.ufrgs.br (Y. Levin)

Although, much effort has concentrated on viral transfection, non-viral methods have received increased attention. This is mostly due to the possible complications which can arise from recombinant viral structures, and the consequent risk of cancer. In the non-viral category, the DNA–liposome complexes have shown the most promise. Cationic liposomes can associate with the DNA segments, neutralizing or even inverting the electric charge of nucleotides, thus significantly increasing the efficiency of gene adsorption and transfection by cells.

In this paper we present a model of DNA–amphiphile solutions. We find that in equilibrium, solution consists of complexes composed of DNA and associated counterions and amphiphiles. As more amphiphiles are added to solution, a cooperative binding transition is found. At the transition point, a large fraction of the DNA's charge is neutralized by the condensed surfactants. If the density of surfactant is increased beyond this point, a charge inversion of the DNA becomes possible. The necessary density of amphiphile needed to reach the charge inversion is strongly dependent on the characteristic hydrophobicity of surfactant molecules. In particular, we find that for sufficiently hydrophobic amphiphiles, such as for example some cationic lipids, the charge inversion can happen at extremely low densities.

2. The model

Our system consists of an aqueous solution of DNA segments, cationic surfactants, and monovalent salt. Water is modeled as a uniform medium of dielectric constant D. In an aqueous solution, the phosphate groups of the DNA molecules become ionized resulting in a net negative charge. The salt is completely ionized, forming an equal number of cations and anions. Similarly, the surfactant molecules are assumed to be fully dissociated producing negative anions and polymeric chains with cationic head groups.

Following the usual nomenclature, we shall call the ionized DNA molecules the "polyions", the positively charged ions the "counterions", and the negatively charged anions the "coions". To simplify the calculations, all the counterions and coions will be treated as identical, independent of the molecules from which they were derived. The DNA strands will be modeled as long rigid cylinders of length L and diameter a_p, with the charge $-Zq$ distributed uniformly, with separation $b \equiv L/Z$, along the major axis. The cations and anions will be depicted as hard spheres of diameter a_c and charge $\pm q$. For simplicity we shall also suppose that each one of the s surfactant monomers is a rigid sphere of diameter a_c with the "head" monomer carrying the charge $+q$. The interaction between the hydrophobic tails is short ranged and characterized by the hydrophobicity parameter χ (see Fig. 1). The density of DNA segments is $\rho_p = N_p/V$, the density of monovalent salt is $\rho_m = N_m/V$, and the density of amphiphile is $\rho_s = N_s/V$, where N_i is the number of molecules of specie i and V is the volume of the system.

The strong electrostatic attraction between the polyions, counterions, and amphiphiles, leads to the formation of complexes consisting of *one* polyion, n_c counterions, and n_s

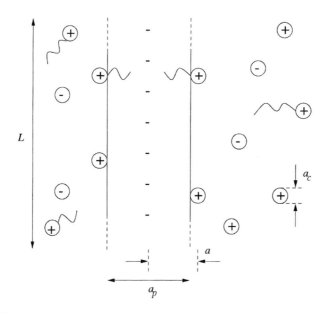

Fig. 1. A cylindrical polyion of diameter a_p, length L, and charge $-Zq$, surrounded by spherical ions of radius a_c and amphiphilic molecules of s monomers. Each monomer of a macroion is free or has *one* counterion, *or* a ring made of l amphiphilic molecules associated with it.

$$- \; C \; - \; S_2 - \; - \; S_3 - \; - \; C \; - \; -$$

Fig. 2. Schematic representation of a complex. Empty sites (monomers) $(-)$, sites with associated counterion (c), sites with l associated amphiphiles (s_l).

amphiphilic molecules. We shall assume that to each phosphate group of the DNA molecule can be associated at most *one* counterion or $l \leqslant l_{max}$ surfactants. This assumption seems to be quite reasonable in view of the fact that the electrostatic repulsion between the counterions will prevent more than one counterion from condensing onto a given monomer. On the other hand, the gain in hydrophobic energy resulting from the close packing of the surfactant molecules might be able to overcome the repulsive electrostatic interaction between the surfactant head groups, favoring condensation of more than one surfactant on a given monomer (see Fig. 2). The l amphiphilic molecules form a "ring" of radius a around the central negative monomer of the DNA (see Fig. 3). If we assume that most of the hydrocarbon chain of the associated surfactants is hidden inside the DNA molecule, the maximum number of surfactants in a ring can be estimated from the excluded volume considerations, $l_{max} = 2\pi a/a_c$, where $a \equiv (a_p + a_c)/2$ is the radius of the exclusion cylinder around a polyion.

At equilibrium, each site (monomer) of a polyion can be free or have one counterion *or* a ring of $l = 1, \ldots, l_{max}$ surfactants associated to it. We define the surface coverage of counterions as $p_c = n_c/Z$, and the surface coverage of surfactant rings as $p_l = n_l/Z$, where

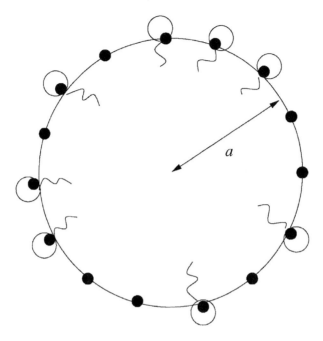

Fig. 3. Ring composed of l surfactant molecules, $l_{\max} = 15$.

n_c is the number of condensed counterions and n_l is the number of rings containing l surfactants. Each polyion has a distribution of rings containing from one to $l_{\max}$ surfactants. We shall neglect the polydispersity in the size of the complexes, assuming that all the complexes have n_c counterions and n_s amphiphilic molecules — in rings of $\{p_l\}$ — with

$$n_s = \sum_{l=1}^{l_{\max}} Zl\, p_l \,. \tag{1}$$

The total charge of each polyion is, therefore, renormalized from $-Zq$ to $-Z_{\mathit{eff}}q$, with $Z_{\mathit{eff}} \equiv Z - n_c - n_s$ [12–16]. From overall charge neutrality, the density of free cations is $\rho_+ = \rho_m + (Z - n_c)\rho_p$, the density of free anions is $\rho_- = \rho_m + \rho_s$, and the density of free surfactants is $\rho_s^f = \rho_s - n_s\rho_p$. We shall restrict our attention to the limit of low surfactant densities, so as to prevent micellar formation in the bulk.

The aim of the theory is to determine the characteristic values of n_c, n_s, and the surface coverage by rings $\{p_l\}$. To accomplish this, the free energy of the DNA–surfactant solution will be constructed and minimized.

3. The Helmholtz free energy

The free energy is composed of three contributions,

$$F = F_{\text{complex}} + F_{\text{electrostatic}} + F_{\text{mixing}} \,. \tag{2}$$

The first term is the free energy needed to form the isolated complexes. The second term accounts for the electrostatic interaction between the counterions, coions, surfactants and complexes. Finally, the third term is the result of entropic mixing of various species.

To calculate the free energy required to construct an isolated complex composed of one polyion, n_c condensed counterions, and n_s condensed surfactants, we employ the following simplified model. Each monomer of a polyion can be free or occupied by a counterion, or by $1 \leqslant l \leqslant l_{max}$ amphiphiles (see Fig. 2). Therefore, to each monomer i we associate occupation variables $\sigma_c(i)$ and $\{\sigma_l(i)\}$, which are nonzero if that particular monomer is occupied by a condensed counterion or a ring with l surfactants, respectively. The free energy of N_p isolated complexes can then be written as

$$\beta F_{complex} = -N_p \ln \sum_v e^{-\beta E_v} , \tag{3}$$

where the sum is over all possible configurations of counterions and surfactants along a complex. For a particular configuration v, the energy can be expressed as the sum of three terms, $E_v = E_1 + E_2 + E_3$. The first one is the electrostatic contribution arising from the Coulombic interactions between all charged sites of a complex,

$$E_1 = \frac{q^2}{2} \sum_{\substack{i \neq j}}^{Z} \frac{[-1 + \sigma_c(i) + \sum_l^{l_{max}} l\sigma_l(i)][-1 + \sigma_c(j) + \sum_l^{l_{max}} l\sigma_l(j)]}{D|r(i) - r(j)|} , \tag{4}$$

where we have assumed that the only effect of association is the renormalization of the effective charge of each monomer. The second term E_2, is due to hydrophobic interactions between the surfactant molecules,

$$E_2 = \frac{\chi}{2} \sum_{\langle i,j \rangle}^{Z} \sum_{l,l'=1}^{l_{max}} \frac{(l+l')}{2} \sigma_l(i)\sigma_{l'}(j) , \tag{5}$$

where in order to simulate the short-ranged nature of hydrophobic interactions, the first sum is constrained to run over the nearest neighbors. The hydrophobicity parameter χ is negative, representing the tendency of the two adjacent surfactant molecules to expel water. We can estimate its value from the experimental measurement of the energy necessary to remove an amphiphile from a monolayer and place it in the bulk [17].

The third contribution E_3, accounts for the internal energy of each ring,

$$E_3 = \sum_i^{Z} \sum_{l=2}^{l_{max}} \sigma_l(i)E_l . \tag{6}$$

E_l is the interaction energy between l surfactants forming a ring. Each ring contains a maximum of l_{max} sites, which can be occupied by surfactants. To each one of these sites we associate an occupation variable $\tau(j)$, which is zero if site j is unoccupied by a surfactant and is one if it is occupied (see Fig. 3). The interaction energy of surfactants forming a ring can then be written as

$$E_l = \frac{q^2}{2D} \sum_{\substack{i \neq j}}^{l_{max}} \frac{\tau(i)\tau(j)}{2a \sin(\pi|i - j|/l_{max})} + \frac{\chi}{2} \sum_{\langle i,j \rangle}^{l_{max}} \tau(i)\tau(j) . \tag{7}$$

The first term of Eq. (7) is due to electrostatic repulsion between the surfactant head groups, while the second is the result of attraction between the adjacent hydrocarbon tails.

The exact solution of even this simpler sub-problem (i.e. evaluation of the sum in Eq. (3)) is very difficult due to the long ranged electrostatic interactions. We shall, therefore, resort to mean-field theory, which works particularly well for long-ranged potentials. Evaluating the upper bound for the free energy, given by the Gibbs–Bogoliubov inequality, and neglecting the end effects we obtain,

$$\beta F_{complex} = \beta N_p[f_{el} + f_{hyd} + f_{ring} + f_{mix}] \,. \tag{8}$$

The first term,

$$\beta f_{el} = \xi S \left[-1 + p_c + \sum_{l=1}^{l_{max}} l p_l \right]^2 - \xi S N_p \,, \tag{9}$$

is the electrostatic interaction between the sites along one rod and is related to E_1. S is expressed in terms of the digamma function [18],

$$S = Z[\Psi(Z) - \Psi(1)] - Z + 1 \,, \tag{10}$$

and $\xi \equiv \beta q^2/Db$ is the Manning parameter [19,20]. The second term in Eq. (8),

$$\beta f_{hyd} = \beta \chi (Z - 1) \sum_{n,m}^{l_{max}} \frac{(n+m)}{2} p_m p_n \,, \tag{11}$$

is the hydrophobic attraction between the rings inside a complex.[1] The third term is the free energy due to the electrostatic and hydrophobic interactions between the surfactants forming a ring,

$$\beta f_{ring} = \frac{2 \ln l_{max} + v_0}{4\pi T^*} \sum_{l=2}^{l_{max}} Z p_l l^2 + \frac{\beta \chi}{l_{max}} \sum_{l=2}^{l_{max}} Z p_l l^2$$

$$+ \sum_{l=1}^{l_{max}} Z p_l l \ln\left(\frac{l}{l_{max}}\right) + \sum_{l=1}^{l_{max}} Z p_l l_{max} \left(1 - \frac{l}{l_{max}}\right) \ln\left(1 - \frac{l}{l_{max}}\right) \,, \tag{12}$$

where $v_0 \approx 0.25126591$, and the reduced temperature is $T^* = k_B T Da/q^2$. Finally, the free energy of mixing for rings and counterions of a complex is,

$$\beta f_{mix} = Z \left(1 - p_c - \sum_{l}^{l_{max}} p_l\right) + \ln\left(1 - p_c - \sum_{l=1}^{l_{max}} p_l\right) + Z p_c \ln p_c$$

$$+ Z \sum_{l=1}^{l_{max}} p_l \ln p_l - Z p \ln l_{max} + Z p l_{max} \left(1 - \frac{1}{l_{max}}\right) \ln\left(1 - \frac{1}{l_{max}}\right) \,, \tag{13}$$

---

[1] For the present calculation we shall neglect the additional hydrophobic contribution which arises from the interaction of amphiphiles with the backbone of the DNA.

where to be consistent with expression (12), we have included a contribution to the free energy arising from the azimuthal motion of condensed counterions around the polyion, i.e. the last two terms of Eq. (13).

Once a cluster, constructed in isolation, is introduced into solution, it gains an additional solvation energy due to its interaction with other clusters, free counterions, free coions, and free surfactants. The electrostatic repulsion between the complexes is screened by the ionic atmosphere, producing an effective short-ranged potential of DLVO form [21–25]. The electrostatic free energy due to interactions between various clusters can be estimated from the second virial coefficient,

$$\beta F^{cc} = (Z - n_c - n_s)^2 \frac{2\pi N_p^2 a^3 e^{-2\kappa a}}{V T^* (\kappa a)^4 K_1^2(\kappa a)}, \tag{14}$$

where $(\kappa a)^2 \equiv 4\pi \rho_1^*/T^*$ and $\rho_1^* \equiv a^3[\rho_p(Z - n_s - n_c) + 2\rho_m + 2\rho_s]$ is the reduced density of free ions. The free energy due to interaction between the complexes and free ions and surfactants can be obtained following the general methodology of the Debye–Hückel–Bjerrum theory [13,14,26–32],

$$\beta F^{ci} = N_p (Z - n_c - n_s)^2 \frac{(a/L)}{T^* (\kappa a)^2} \left[-2\ln(\kappa a K_1(\kappa a)) + I(\kappa a) - \frac{(\kappa a)^2}{2} \right] \tag{15}$$

with

$$I(\kappa a) = \int_0^{\kappa a} \frac{x K_0^2(x)}{K_1^2(x)} \, dx, \tag{16}$$

where K_n is the modified Bessel function of order n. The contribution to the total free energy arising from the interactions between the free ions and surfactants is given by the usual Debye–Hückel expression [26,27]

$$\beta F^{ii} = -\frac{V}{4\pi a_c^3} \left[\ln(1 + \kappa a_c) - \kappa a_c + \frac{(\kappa a_c)^2}{2} \right]. \tag{17}$$

This term is very small and is included only for completeness.

The last contribution to the total free energy, Eq. (2), results from the entropic mixing of the counterions, coions, surfactant and complexes,

$$F_{\text{mixing}} = F_{m+} + F_{m-} + F_s + F_c. \tag{18}$$

The free energy of mixing is obtained following the general ideas introduced by Flory [33],

$$\beta F_{m+} = N_{m+} \ln \phi_{m+} - N_{m+},$$

$$\beta F_{m-} = N_{m-} \ln \phi_{m-} - N_{m-},$$

$$\beta F_s = N_s \ln (\phi_s/n_s) - N_s,$$

$$\beta F_c = N_p \ln \left(\frac{(Z + n_c + n_s)\phi_c}{Z + n_c + n_s s} \right) - N_p. \tag{19}$$

In the above expression $m+$ denotes free counterions, $m-$ free coions, s free surfactant molecules, and c complexes. The

$$\phi_{m+} = \frac{\pi\rho_+^*}{6}\left(\frac{a_c}{a}\right)^3,$$

$$\phi_{m-} = \frac{\pi\rho_-^*}{6}\left(\frac{a_c}{a}\right)^3,$$

$$\phi_s = \frac{s\pi\rho_s^{f*}}{6}\left(\frac{a_c}{a}\right)^3,$$

$$\phi_c = \pi\rho_p^*\left[\frac{1}{4(a/L)}\left(\frac{a_p}{a}\right)^2 + \frac{1}{6}(n_c + n_s s)\left(\frac{a_c}{a}\right)^3\right] \tag{20}$$

are the volume fractions occupied by the free counterions, coions, surfactants, and complexes, respectively.

4. Results and conclusions

The equilibrium configuration of the polyelectrolyte–surfactant solution is determined by the requirement that the Helmholtz free energy be minimum. Since F is the function of n_s, n_c, and the surface coverage by rings $\{p_l\}$, minimization of F implies that

$$\delta F = \frac{\partial F}{\partial n_s}\delta n_s + \frac{\partial F}{\partial n_c}\delta n_c + \sum_{l=1}^{l_{\max}}\frac{\partial F}{\partial p_l}\delta p_l = 0. \tag{21}$$

Using the constraint Eqs. (1) and (2) can be separated into $l_{\max} + 1$ equations,

$$\frac{\partial F}{\partial n_c} = 0 \tag{22}$$

and

$$\frac{\partial F}{\partial n_s}Zl + \frac{\partial F}{\partial p_l} = 0, \quad l = 1, \ldots, l_{\max}. \tag{23}$$

The system of Eqs. (22) and (23) can, in principle, be solved numerically. However, for reasonable values of $l_{\max}$ this requires a significant numerical effort. Instead of pursuing this brute force method, we note that to a reasonable accuracy, the surface coverage by rings, $\{p_l\}$, can be approximated by an exponential distribution [34],

$$p_l = \frac{n_s e^{\alpha l}}{Z\sum_{l=1}^{l_{\max}} l e^{\alpha l}}. \tag{24}$$

We have checked that this is, indeed, a good approximation by numerically solving Eq. (23) for an isolated complex. Using ansatz (24), the total free energy becomes a function of n_c, n_s, and α. For a fixed volume and number of particles, the equilibrium

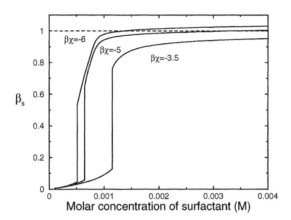

Fig. 4. Effective binding fraction of amphiphiles $\beta_s \equiv n_s/Z$, as a function of amphiphile concentration ρ_s. The concentrations of DNA and of added salt is 2×10^{-6} M and 18 mM, respectively. The length of the DNA segments is 220 base pairs. The solvent is water at room temperature, so that $\xi = 4.17$.

corresponds to the minimum of Helmholtz free energy,

$$\frac{\partial F}{\partial n_c} = 0 \,, \tag{25}$$

$$\frac{\partial F}{\partial n_s} = 0 \,, \tag{26}$$

$$\frac{\partial F}{\partial \alpha} = 0 \,. \tag{27}$$

These are three coupled algebraic equations, which can be easily solved numerically to yield the characteristic number of condensed counterions, surfactants, as well as the shape of the distribution of ring sizes (α). In Figs. 4 and 5 we present a numerical solution of these equations. As a specific example we consider a cationic surfactant with an alkyl chain of $s = 12$ groups. In this case the hydrophobicity parameter can be estimated [30] to be in the range of $\chi \approx -3,5 k_B T$. To explore the dependence of condensation on the hydrophobicity of surfactant, we shall vary this value within reason. The density of monovalent salt and the DNA is taken to be 18 and 2×10^{-3} mM, respectively.

The resulting binding isotherms are illustrated in Fig. 4. The fraction of associated amphiphilic molecules $\beta_s = n_s/Z$, is plotted against the density of surfactant for a fixed amount of monovalent salt, ρ_m. For small concentrations of cationic surfactant, few amphiphilic molecules associate with the DNA segments. At a certain critical concentration, however, the system forms surfoplexes [30,32] — complexes in which the charge of the DNA is almost completely neutralized by the associated amphiphiles. If the density is increased further, on average, more than one surfactant molecule will associate to each phosphate group, leading to charge inversion of the surfoplexes. For highly hydrophobic surfactants the charge inversion can happen very close to the

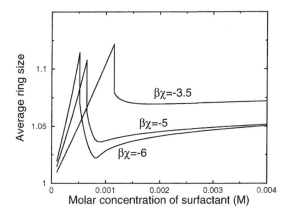

Fig. 5. Average size of rings in a complex (parameters are the same as in Fig. 4).

cooperative binding transition. We note that our theory predicts the binding transition to be discontinuous, this, most likely, is an artifact of the mean-field approximation [32].

We have presented a simple theory of DNA–surfactant solutions. Our results should be of direct interest to researchers working on the design of improved gene delivery systems. In particular, we find that addition of cationic surfactants leads to a strong cooperative binding transition. This transition happens far below the critical micell concentration. A further increase of amphiphile density can result in charge inversion of the DNA–surfactant complexes. This regime should be particularly useful in designing gene or oligonucleotide delivery systems. Until now most of nonviral gene–delivery systems were in the form of lipoplexes — complexes formed by DNA and cationic liposomes. To form the liposomes, however, is required a significant concentration of cationic lipid. Unfortunately, at high concentrations both lipids and surfactants are toxic to organism. Our model suggests that the charge inversion can be achieved with quite a small concentration of cationic amphiphile, *if* it is sufficiently hydrophobic. This should reduce the risk of unnecessary medical complications.

Acknowledgements

This work was supported in part by Conselho Nacional de Desenvolvimento Científico e Tecnológico and Financiadora de Estudos e Projetos, Brazil.

References

[1] T. Friedmann, Sci. Am. 276 (1997) 80.
[2] P.L. Felgner, Sci. Am. 276 (1997) 86.
[3] P.L. Felgner, G.M. Ringold, Nature 337 (1989) 387–388.
[4] P.L. Felgner, G. Rhodes, Nature 349 (1991) 351.

[5] I.M. Verma, N. Somia, Nature 389 (1997) 239.
[6] J.O. Rädler et al., Science 275 (1997) 810.
[7] D. Harries et al., Biophys. J. 75 (1998) 159.
[8] W.F. Anderson, Nature 392 (Suppl.) (1998) 25.
[9] A.V. Gorelov et al., Physica A 249 (1998) 216.
[10] K. Shirahama et al., Bull. Chem. Soc. Japan 60 (1987) 43.
[11] M.J. Hope, B. Mui, S. Ansell, Q.F. Ahkong, Mol. Membr. Biol. 15 (1998) 1.
[12] S. Alexander et al., J. Chem. Phys. 80 (1984) 5776.
[13] M.E. Fisher, Y. Levin, Phys. Rev. Lett. 71 (1993) 3826.
[14] Y. Levin, M.E. Fisher, Physica A 225 (1996) 164.
[15] Y. Levin, Europhys. Lett. 34 (1996) 405.
[16] Y. Levin, M.C. Barbosa, J. Phys. II (France) 7 (1997) 37.
[17] J.N. Israelachvili, D. Mitchell, B.W. Ninham, J. Chem. Soc. Faraday Trans. 72 (1976) 1525.
[18] I.S. Gradshteyn, I.M. Ryzhik, Table of Integrals Series and Products, Academic Press, New York, 1965.
[19] G.S. Manning, J. Chem. Phys. 51 (1969) 924.
[20] J.L. Barrat, J.F. Joanny, Adv. Chem. Phys. 94 (1996) 1.
[21] B.V. Derjaguin, L. Landau, Acta Phys. (USSR) 14 (1941) 633.
[22] E.J.W. Verwey, J.Th.G. Overbeek, Theory of the Stability of Lyophobic Colloids, Elsevier, Amsterdam, 1948.
[23] M. Medina-Noyola, D.A. McQuarrie, J. Chem. Phys. 73 (1980) 6279.
[24] X.-J. Li, Y. Levin, M.E. Fisher, Europhys. Lett. 26 (1994) 683.
[25] M.E. Fisher, Y. Levin, X.-J. Li, J. Chem. Phys. 101 (1994) 2273.
[26] P.W. Debye, E. Hückel, Phys. Z. 24 (1923) 185.
[27] D.A. McQuarrie, Statistical Mechanics, Harper and Row, New York, 1976.
[28] N. Bjerrum, Kgl. Dan. Vidensk. Selsk. Mat.-Fys. Medd. 7 (1926).
[29] P.S. Kuhn, Y. Levin, M.C. Barbosa, Macromolecules 31 (1998) 8347.
[30] P.S. Kuhn, Y. Levin, M.C. Barbosa, Chem. Phys. Lett. 298 (1998) 51.
[31] P.S. Kuhn, Y. Levin, M.C. Barbosa, Physica A 266 (1999) 413.
[32] P.S Kuhn, M.C. Barbosa, Y. Levin, Physica A 269 (1999) 278.
[33] P. Flory, Principles of Polymer Chemistry, Cornell University Press, Ithaca, New York, 1971.
[34] Y. Levin, Phys. Rev. Lett. 83 (1999) 1159.

ELSEVIER

Physica A 274 (1999) 19–29

www.elsevier.com/locate/physa

Application of statistical mechanics to the wetting of complex liquids

R. Fondecave, F. Brochard-Wyart*

*P.C.C., UMR 168, Institut Curie, Section de Recherche, 11 rue Pierre et Marie Curie,
75231, Paris Cedex 05, France*

Abstract

We study the wetting laws for binary mixtures with antagonist components: the solvent wets but the solute does not. Naïvely, we would expect a wetting transition at a composition $\phi = \phi_W$. We measure the contact angle $\theta(\phi)$ which decreases from $\theta(1)$ up to a plateau value θ_L for $\phi \leqslant \phi_L$. In the plateau regime, the solution droplet is in equilibrium with a precursor film of pure solvent. At ϕ_L, we have a "leak out transition", which results from the frustration of the solvent attracted by both the polymer and the solid. Because the contact angle is finite at all composition, films of solution dewet below a critical thickness $e_c(\phi)$. We observe two regimes of dewetting (i) dry dewetting at composition $\phi > \phi_L$. The final state after dewetting are multitude of droplets on a dry solid, (ii) "wet" dewetting for $\phi < \phi_L$, where the final droplets coexist with a film of pure solvent. All these results can be extended to other complex systems, where wetting processes and phase separation are coupled. © 1999 Elsevier Science B.V. All rights reserved.

1. Introduction

The wetting and the dewetting of pure liquids have been intensively studied in the recent past. A number of practical problems are controlled by these processes: the deposition of liquid films (paints, lubricants, fungicides or insecticides, etc.), the fast drying induced by chemical agents, the hydroplaning of motorcars, etc.

However, in all practical applications, the formulation of a liquid implies a complex mixture of additives. Our aim here is to extend the laws of wetting obtained with pure liquids to binary mixtures. We study a challenging case: a liquid mixture with antagonist components: the solvent wets (the spreading parameter $S_0 > 0$), but the solute does not ($S_1 < 0$) (Fig. 1).

---
* Corresponding author.
E-mail address: brochard@curie.fr (F. Brochard-Wyart)

0378-4371/99/$ - see front matter © 1999 Elsevier Science B.V. All rights reserved.
PII: S 0378-4371(99)00323-4

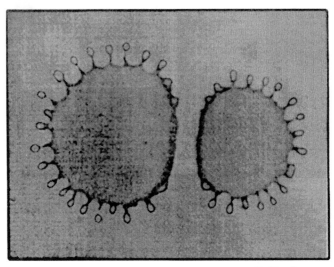

R. Fondecave

Fig. 1. Spreading of two adjacent drops of complex liquids.

2. The system: partial and complete wetting

The wetting by a nonvolatile liquid of a smooth chemically homogeneous solid substrate is well understood [1]. The control parameter is the spreading coefficient S [2,3], giving the energy difference between bare and wet substrate

$$S = \gamma_{S0} - (\gamma_{SL} + \gamma),\tag{1}$$

where the γ_{ij} are, respectively, the solid/air, solid/liquid and liquid/air interfacial tensions.

The sign of S separates two regimes:

(A) if $S > 0$, *complete wetting*, a liquid drop deposited onto the solid surface spreads completely and ultimately becomes a flat pancake, of thickness e_0 [4,5]. e_0 results from a competition between long-range forces which tend to thicken the film and S, written as

$$S = e\Pi + P,\tag{2}$$

where $P(e)$ is the contribution of long-range van der Waals forces ($P(e) = A//12\pi e^2$, where A is the solid/liquid Hamaker constant assumed to be positive), and $\Pi(e) = -\mathrm{d}P//\mathrm{d}e$ is the disjoining pressure [6].

Eq. (2) gives a thickness $e_0 = a\sqrt{3\gamma//2S}$, where a is a nanoscopic length defined by $A//6\pi = \gamma a^2$.

(B) If $S < 0$, *partial wetting*, a liquid drop remains localised on a small area. A small drop achieved a spherical cap, with a contact angle θ defined by the Young

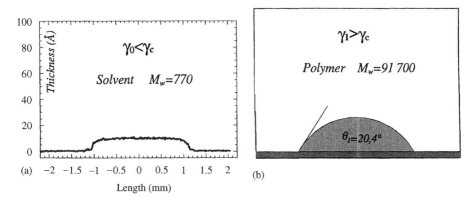

Fig. 2. Wetting of the two components: (a) complete wetting of the pure solvent ($M_w=770$). The spreading of a sessile micro drop is monitored by ellipsometry. The final state is the "van der Waals pancake" of thickness e_0 ($e_0 = 10$ Å), (b) *partial wetting of the polymer*: a droplet of pure polymer achieves at equilibrium a spherical cap, with a contact angle θ_1.

relation

$$\gamma_{SO} = \gamma_{SL} + \gamma \cos \theta_E .\tag{3}$$

A large drop is flattened by gravity. Its thickness e_c results from a balance between gravity forces and S, written as [14]

$$-S = \tfrac{1}{2}\rho g e_c^2 ,\tag{4}$$

where ρ is the liquid density and g the acceleration due to gravity. e_c is also the critical thickness below which thin films are unstable and they dewet.

We have built a system where the pure solvent wets (regime A), and the solute does not wet (regime B):

The *substrate* is an oxidised silicon wafer, grafted with a monolayer of hexa (HTS) or octadecyltrychlorosilanes (OTS), prepared by the method of Sagiv [7]. The critical surface tension, measured by using alcane series, is $\gamma_c = 20, 2 \pm 0.3$. The hysteris measured from the difference between advancing and receding contact angles is extremely low ($\theta_a - \theta_r \sim 2°$ for $\theta \sim 10°$).

The *solvent* is an oligomer "P dimethylsiloxane" of P monomers. We have studied three molecular weights $M_w = 770$, 950 and 1250 corresponding to $P = 10, 12, 17$. For these three oligomers, $\gamma < \gamma_c$, and they wet the substrate. We show in Fig. 2a the profile of a spreading micro droplet. The final stage is a "pancake" of thickness 10 Å for $P = 10$. The thickness e_0 (Table 1) increases with P, as expected from Eq. (2) with S decreasing.

The *polymer* is a N dimethylsiloxane ($M_w = 91700$), with $N = 1230$ large. The surface tension of the polymer is larger than γ_c. A small droplet achieved a spherical cap, with a contact angle $\theta = 20.4°$ (Fig. 2b). From θ, one can deduce the value of the polymer spreading coefficient $S_1 = -1, 30 \pm 0, 02$ mNm^{-1}. The critical thickness below which a film dewets, deduced from Eq. (4), is 520 µm.

Table 1
Summary of experimental results

M_w	P	N	$\gamma(\text{mN m}^{-1})$	e_0 (Å)	θ°	ϕ_L	ϕ_w
770	10	—	19.2 ± 0.2	10 ± 1	—	0.54	0.47
950	12	—	19.3 ± 0.2	12 ± 1	—	0.47	0.41
1250	17	—	19.6 ± 0.2	23 ± 1	—	0.29	0.26
2000	27	—	20 ± 0.2	—	7°	—	—
91 700	—	1239	20.7 ± 0.2	—	20.4°	—	—

The solvent wets, the polymer dewets: what happens if we mix these two components?

3. Sessile droplets of solutions: the leak out transition

Micro droplets (volume $\Omega \sim$ nl) are gently deposited on the silanated silicon wafer with micro glass capillaries. The contact angles are measured optically by two methods:

(1) *Reflection of a laser beam*. The droplet acts like a convex mirror for a parallel light beam from a He–Ne laser, incident perpendicularly to the solid substrate. The divergence of the reflected beam is proportional to the contact angle θ at the contact line. This method [8] has an accuracy of $\pm 0.5^\circ$ and is used for $\theta > 4^\circ$.

(2) *Interferential reflection contrast microscopy* (RCIM). From the interference fringes, one can construct the complete profile of the drop and extract the contact angle and the volume. This technic allows to measure with high accuracy (± 0.01) small contact angles $1 < \theta < 4^\circ$.

The pure solvent has a positive spreading coefficient S_0,

$$S_0 = \gamma_{S0} - (\gamma_{SL0} - \gamma_0), \tag{5}$$

where γ_{SL0} and γ_0 are the solvent/solid and solvent/air interfacial tensions.

Adding polymer, we form a solution of polymer volume fraction ϕ. The spreading parameter $S(\phi)$ is positive for $\phi=0$, and negative for $\phi=1$. For intermediate ϕ values, we can write

$$S(\phi) = \gamma_{S0} - [(\gamma_{SL}(\phi) + \gamma(\phi))] = S(0) - \Delta\gamma(\phi). \tag{6}$$

The polymer is depleted at both interfaces: (a) at the liquid/air interface because $\gamma_1 > \gamma_0$. (b) At the passive solid/liquid interface, because small chains are entropically favoured. The polymer is in the semi-dilute regime, and can be pictured as a transient network of mesh size ξ [9,10], related to ϕ by

$$\xi = a_0 \phi^{-3/4}. \tag{7}$$

The osmotic pressure of the polymer solution $\Pi_{OSM}(\phi)$ scales like [9]

$$\Pi_{OSM} = \frac{kT}{\xi^3}. \tag{8}$$

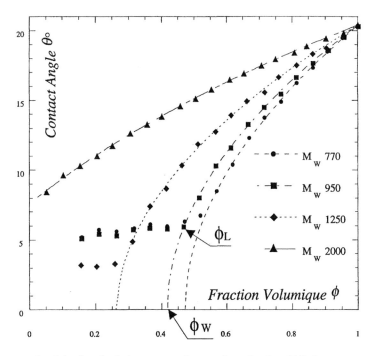

Fig. 3. Contact angle of droplet of solution versus polymer volume fraction. $\theta(\phi)$ decreases up to a plateau value θ_L for $\phi < \phi_L$. ϕ_L and ϕ_w are the composition of the "leak out" and wetting (masked) transition, respectively. Configuration of droplets above and below ϕ_L are shown in Fig. 6.

The increase in surface tension $\Delta\gamma$ at both interface is the work required to expel the polymer from the depletion layer of thickness ξ. We then have [11]

$$\Delta\gamma \sim \frac{kT}{\xi^2} \, . \tag{9}$$

Clearly, the contact angle $\theta(\phi)$ should decrease when the polymer concentration decreases. From the Young equation, $\theta(\phi)$ and $S(\phi)$ are related by

$$\sin^2 \frac{\theta}{2}(\phi) = -\frac{S(\phi)}{2\gamma(\phi)} \, . \tag{10}$$

Normally, we would expect a wetting transition [1] $\theta(\phi_w) = 0$ at a volume fraction ϕ_w such as $S(\phi_w) = 0$. The plot of $\theta(\phi)$ is different and is shown in Fig. 3.

We observe two domains, separated by a critical composition ϕ_L (listed for various oligomers). Above ϕ_L, the contact angle increases monotonously up to θ_L. Below ϕ_L the contact angle remains constant, and equal to θ_L. Using an ellipsometer, we observe (Fig. 4) that the droplet is surrounded by a halo, a nanoscopic film of thickness e_0, equal to the thickness of the van der Waals pancake of the pure solvent. We call it the "fried egg" drop (Fig. 5). By extrapolation of the curve $\theta(\phi)$, one can determinate ϕ_w, but we do not reach a wetting transition at ϕ_w.

We conclude that at and below ϕ_L, the solvent escapes from the droplet. That is why we named this new transition the "leak out transition" [12]. Below ϕ_L, the "fried

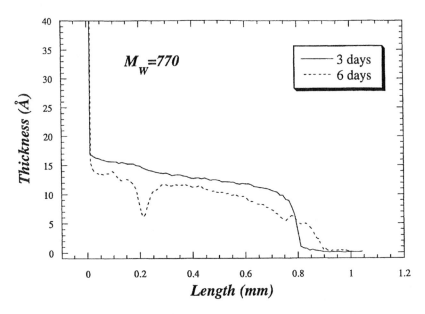

Fig. 4. Ellipsometric profile of a precursor film emanating from the drop when $\phi < \phi_L$ on HTS wafer at (———) three days and (- - - - -) six days after the deposition.

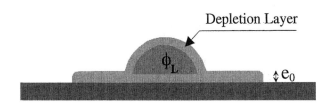

Fig. 5. "Fried egg" configuration of a drop of polymer solution for $\phi < \phi_L$. The final state of a sessile droplet is a spherical cap of polymer solution at fixed composition ϕ_L, surrounded by a halo of pure solvent. At equilibrium, the osmotic pressure in the drop balances the disjoining pressure in the film.

egg" droplet represented schematically in Fig. 5 behaves like a micro osmometer: the polymer in semi-dilute solution in the drop is excluded from the surrounding film because $e_0 < \xi(\phi_L)$, the correlation length at $\phi = \phi_L$. The drop/film equilibrium is ruled by the equality of the solvent chemical potential in the two phases $\Pi_{0,\text{drop}} = \Pi_{S,\text{film}}$, written as

$$\frac{kT}{\xi_L^3} = \frac{\gamma a^2}{e^3} \,. \tag{11}$$

The halo of solvent coexists with the solid. This imposes $S = e\Pi + P$, i.e., $e = e_0$. Then:

$$\xi_L = e_0 \left(\frac{kT6\pi}{A} \right)^{1/3} \,. \tag{12}$$

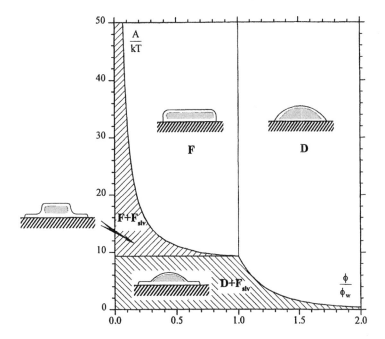

Fig. 6. Wetting diagram of a polymer solution. D = droplet, with contact angle $\theta(\phi)$. F = Film of polymer solution. F_{slv} = Film of pure solvent. The biphasic regions are shaded.

Conclusion. Above ϕ_L we observe a spherical cap, with a composition ϕ in polymer. Below ϕ_L we have been able to observe "fried egg" droplets, where a spherical cap containing the solution at composition ϕ_L is surrounded by a halo of pure solvent, as shown in Fig. 5.

4. The wetting diagram of a polymer solution

We have observed a "leak out transition" predicted first by Boudoussier [13] in a mean field approximation. This transition results from a competition: the solvent has an affimity for the polymer, but also for the wettable substrate. For $\phi > \phi_L$, the droplet keeps it solvent, and is in contact with the dry solid. For $\phi < \phi_L$, attraction by the solid wins, the solvent escapes and the droplet of solution is surrounded by a precursor film of pure solvent ("fried egg" configuration). The leak out transition masks the wetting transition and the contact angle remains finite at all compositions.

We have constructed the full diagram of wetting of a polymer solution starting from the correct scaling laws to describe the polymer solution shown in Fig. 6. There is a single dimensionless parameter $\tilde{A} = A//kT$.

(a) If $\tilde{A} < A_c$, $\phi_L > \phi_w$: the leak out transition arises in the condition of partial wetting. The solvent escapes from the droplet. The final state, named "fried egg drop", is a droplet of polymer solution at composition ϕ_L coexisting with a nanoscopic aureola of pure solvent.

(b) If $\tilde{A} > A_c$, $\phi_L < \phi_w$, the leak out transition arises in the condition of complete wetting. If droplets of decreasing concentration ϕ are deposited, the contact angle will decrease up to $\theta(\phi_W) = 0$. At ϕ_L, a film of pure solvent will escape from the solution pancake. All our experiments are performed in the regime $\tilde{A} < A_c$.

5. The dewetting of polymer solutions

The dewetting of pure polymer PDMS films has been studied first by Redon [14]. A film dewets below a critical thickness

$$e_c(1) = 2\kappa^{-1} \sin \frac{\theta_1}{2},$$

where e_c is the thickness of the large drops flattened by gravity. Thick films are metastable: one must nucleate a dry patch, which grows if its radius R is larger than R_c, the critical nucleation radius, comparable to the initial film thickness e. Microscopic films ($e < \mu m$) are unstable, and break spontaneously in a multitude of droplets arranged in polygons [15–17]: one names this regime "spinodal decomposition" by analogy with phase transitions.

We have observed two regimes of dewetting [18]:

(a) $\phi \geqslant \phi_L$: "*dry dewetting*" (Fig. 7a). When $\phi > \phi_L$, the solution droplets are in condition of classical "partial wetting", i.e., deposited on a dry solid. These solutions behave like simple liquids, with a contact angle $\theta(\phi)$. $e_c(\phi)$ can be deduced from $\theta(\phi)$ by the classical relationship $e_c(\phi) = 2\kappa^{-1} \sin(\theta//2)(\phi)$. We show in Fig. 7a the final state after the rupture of a thin PDMS film ($e \sim \mu m$): droplets are arranged in polygons.

(b) $\phi \leqslant \phi_L$ "*wet dewetting*" (Fig. 7b). The particularity of this situation is that after the film's rupture, the substrate is covered by a nanoscopic film of thickness $e_c(\phi)$ in coexistence with droplets of polymer solution.

We show in Fig. 7b a typical picture of spontaneous dewetting. One sees a multitude of droplets arranged in polygons, but the polygons are now full of droplets. Between the droplets, we have confirmed the existence of a nanoscopic film by ellipsometry.

In the case of the dewetting of pure PDMS, shown for comparison in Fig. 5a, the droplets are arranged in polygons, which are empty. When holes are formed at random, they grow and are surrounded by a rim. When the rims of adjacent holes meet, they break into droplets, which form the polygons.

In the case of dewetting of mixtures, the moving rims exhibit a fingering instability and break itself into droplets, grow again and break again, before adjacent rims meet to form the polygons. The net result of this cascade of Rayleigh instability and fingering are droplets, which are relatively monodisperse in size. The origin of this difference between "dry" and "wet" dewetting must be due to a Marangoni effect, also observed in the spreading of drops [18].

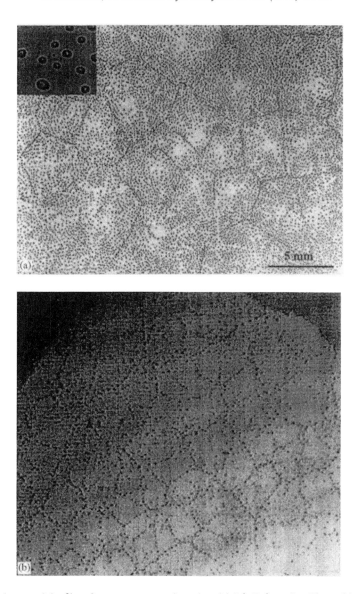

Fig. 7. Final stage of the film after a spontaneous dewetting: (a) *"dry" dewetting.* The multitude of droplets are arranged in polygons on a dry solid, (b) *"wet" dewetting* ($\phi = 0,42$, $M_w = 770$). The polygons are now full of droplets, uniform in size, deposited on a *wet* solid. The thickness e_ϕ of the film between the droplets is 14 Å.

We have also studied the critical thickness $e_c(\phi)$ below which films are unstable, for the two situations of dry/wet dewetting. Film are expected to dewet at all compositions, but $e_c(\phi)$ has to tend to zero with ϕ, because films of pure solvent are stable, whatever their thickness: $(e_c(0) = 0)$. We have found $e_c(\phi) \approx a\phi^{3/4}$ at small composition.

6. Concluding remarks

We have studied the wetting and the dewetting of a binary mixture, with the following properties: (i) *antagonist wetting* properties: the solvent wets totally the substrate, the polymer does not, (ii) the polymer is in *good solvent* conditions, (iii) the polymer is *depleted* at both solid/liquid interfaces.

This system presents two remarkable features:

Phase separation: The spreading induces a *phase separation* of two miscible liquids. We name this transition the "leak out" transition. The solvent is attracted by both the polymer and the solid substrate. At high concentration ($\phi > \phi_L$), the polymer wins and keeps its solvent. Below ϕ_L, the attraction by the solid dominates and the solvent escapes from the droplet and forms a precursor film. The long polymer chains are excluded from the nanoscopic film of solvent, because the confinement energy is huge.

Spreading droplets achieve "micro osmometers". The droplet/film equilibrium imposes that the solvent chemical potential is uniform. This corresponds to the equality of the van der Waals *disjoining pressure* in the film, and the *osmotic pressure* in the droplet.

Dewetting at all polymer concentrations: The film of pure solvent is stable, because the solvent wets totally the substrate. On the other hand, film of polymer solution are always unstable below a critical thickness $e_c(\phi)$.

We observe two regimes: (1) *dry wetting* ($\phi > \phi_L$): after dewetting, the solid is bare, covered by droplets of solution, (2) *wet dewetting* ($\phi < \phi_L$): where droplets coexist with a film of pure solvent. $e_c(\phi)$ decreases to zero with a power law $e_c(\phi) \sim \phi^{0.80}$.

All these features observed here with PDMS polymers in solution in its oligomers can be extended to other systems, if the solute is depleted at both interface, and if the thickness ξ of the depletion layer is larger than the thickness of the wetting film of pure solvent. Large depletion thickness are observed with polymer. One can think of other systems (i) *charged solutions*. Ions are repulsed at the water/air interface, and the thickness of the depletion layer is the Debye length κ_e^{-1}, which becomes large at weak ionic strength. Experiments with polyelectrolytes are under way. Because in these systems, the osmotic pressure is very high, we expect a leak out transition at very low volume fraction. (ii) *colloidal suspensions* — Nanoparticules or emulsions are stabilised by ionic or steric repulsion. In the case of repulsive interfaces, the depletion layer is comparable to the particle size and is large, compared to the wetting film thickness, (iii) *binary mixtures* near the critical point. Compatible mixtures of small AB molecules with antagonist wetting properties may also phase separately if the concentration correlation length ξ becomes larger than the thickness of the wettable component.

The dynamic features not reported here are equally interesting. The spreading of dilute droplets give rise to Marangoni instabilities [18], because the leak out of the solvent builds a gradient of surface tension. The early stage of wet dewetting, where two instabilities are present (amplification of capillary waves coupled to phase transition) are at the moment unknown.

Acknowledgements

We thank L. Vovelle, G. Schorsh for numerous discussions and Rhône-Poulenc for its financial support.

References

[1] P.G. de Gennes, Rev. Mod. Phys. 57 (1985) 827.
[2] W. Cooper, W. Nuttal, J. Agric. Sci. 7 (1915) 219.
[3] F. Brochard-Wyart, J.M. di Meglio, M. Quéré, P.G. de Gennes, Langmuir 7 (1991) 335.
[4] J.F. Joanny, P.G. de Gennes, C.R. Acad. Sci. 299 II (1984) 279.
[5] F. Heslot, A.M. Cazabat, P. Levinson, Phys. Rev. Lett. 62 (1989) 1286.
[6] B. Derjagin, J. Phys. Chim., URSS 14 (1940) 137.
[7] J. Sagiv, J. Am. Chem. Soc. 102 (1980) 92.
[8] C. Allain, D. Ausserré, F. Rondelez, J. Colloid Interface Sci. 107 (1985) 5.
[9] P.G. de Gennes, Scaling Concepts in Polymer Physics, Cornell University Press, Ithaca, NY, 1979.
[10] M. Daoud et al., Macromolecules 8 (1975) 804.
[11] J.F. Joanny, P. Grant, L.A. Turkevich, P. Pincus, J. Phys. 42 (1981) 1045.
[12] R. Fondecave, F. Brochard-Wyart, Europhys. Lett. 37 (1997) 115.
[13] M. Boudoussier, J. Phys. Paris 48 (1987) 445.
[14] C.C. Redon, F. Brochard-Wyart, F. Rondelez, J. Phys. II France 2 (1992) 1671.
[15] G. Reiter, Phys. Rev. Lett. 68 (1992) 75.
[16] F. Reiter, J. Schultz, P. Auroy, L. Auvray, Europhys. Lett. 33 (1996) 29.
[17] R. Xie, A. Karim, J.F. Douglas, C.C. Ham, R.A. Weiss, Phys. Rev. Lett. 81 (1998) 1251.
[18] R. Fondecave, F. Brochard-Wyart, Macromolecules 31 (1998) 9305.

ELSEVIER

Physica A 274 (1999) 30–49

www.elsevier.com/locate/physa

Rotating chemical waves: theory and experiments

András Volford, Péter L. Simon, Henrik Farkas, Zoltán Noszteczius*

Department of Chemical Physics, Technical University, Budapest H-1521, Hungary

Abstract

After a brief introduction in the first theoretic part of this work the geometrical wave theory and its application for rotating waves are discussed. Here the waves are rotating around a circular obstacle which is surrounded by two homogeneous wave conducting regions with different wave velocities. The interface of the inner slow and the outer fast region is also a circle but the two circles (the obstacle and interface) are not concentric. The various asymmetric cases are classified and described theoretically. In the second experimental part chemical waves rotating in a so-called moderately asymmetric reactor are studied. A piecewise homogeneous wave conducting medium is created applying a novel reactor design. All the three theoretical cases of the moderately asymmetric arrangement are realized experimentally and qualitative and quantitative comparison of these results with the theoretical predictions show a good agreement. © 1999 Elsevier Science B.V. All rights reserved.

1. Introduction

1.1. Waves in active media: a possibility for geometrical description

Chemical [1–3] and biological [4,5] wave phenomena became well known and are widely studied in the recent years. While these waves show many similarities with the classical mechanical and electromagnetic waves there are important differences as well. For example, when electromagnetic or mechanical waves travel through a dissipative medium their amplitude decreases inevitably due to energy losses. An amplifier should be applied if we want to restore the original amplitude of a damped wave. Even in the absence of any dissipation, only planar electromagnetic or mechanical waves can propagate with a constant amplitude; their amplitude changes in all other cases. Biological and chemical waves are different. They propagate in a so-called excitable or active media containing evenly distributed energy sources which can be regarded

---

* Corresponding author.
E-mail address: noszti@phy.bme.hu (Z. Noszteczius)

0378-4371/99/$ - see front matter © 1999 Elsevier Science B.V. All rights reserved.
PII: S 0378-4371(99)00331-3

as local amplifiers. As a result the amplitude of chemical and biological waves are preserved while they are propagating in a homogeneous active medium. Such waves can be described simply by evolving fronts and orthogonal rays and this is the basis of the geometric wave theory. Few years ago a systematic research program was started in our laboratory to compare the predictions of the geometric wave theory with the results of chemical wave experiments performed on various membrane rings [6–10] and the present work is a part of that program.

1.2. Rotating waves on concentric membrane rings: motivation and previous results

The biological significance of excitation waves rotating in a two-dimensional annular region was realized first by Wiener and Rosenblueth [11] long time ago. They suggested that a circus movement of waves around an obstacle in the heart is responsible for the arrhyhmia called atrial flutter [12,13]. Moreover, according to their simple geometrical theory, the shape of wavefronts should be involutes of this obstacle in an otherwise homogeneous medium. We could verify this prediction for various obstacles [7] applying a technique developed in our laboratory [6]: after immobilizing a catalyst of the Belousov–Zhabotinsky (BZ) reaction, the membrane became an effective 2D excitable medium supporting chemical waves when other components of the BZ reaction were supplied from a gel slab under the catalytic membrane. A central hole cut into the membrane played the role of the obstacle and a "chemical pinwheel" [14], a system of wavefronts rotating around the hole, was created. After a transient period involute-shaped wavefronts were observed in the annular membrane this verifying the WR theory.

This example also illustrates a special advantage of ring reactors to study chemical waves: an asymptotic state can be achieved in a finite reactor without generating new wavefronts. Consequently, all transient phenomena and irreproducibilities associated with wave generation can be neglected after a transient period.

To meet the conditions of the WR theory the first series of experiments was conducted applying a homogeneous excitable medium. However, most biologically important excitable media are not uniform. For example in recent models of atrial flutter, it is assumed that the obstacle is a 'combined functional-anatomical conduction block' [13], which is surrounded by a region where the wave propagation is much slower than in the other parts of the atrium. To give a theoretical description of the biologically important inhomogeneous problems the geometrical wave theory was generalized for nonuniform excitable media; then this theory was applied for a special case where the central circular obstacle is surrounded by an inner slow and an outer fast region. To check the theoretical predictions experimentally a novel technique [8] to create velocity inhomogeneities was applied. Two concentric rings, a slow inner and a fast outer one, were established in the membrane around the hole. The generalized theory gave the shape of the wavefronts applying descriptive geometrical concepts like caustics [15], minimal loops, involutes and reverse involutes [9] and the experimental results were in complete agreement with these expectations [9].

All the previous theoretical considerations and experimental results are valid only for symmetric arrangements; however, where the slow and fast regions form concentric rings. It will be shown here that a more complex dynamics may appear when these rings are not concentric.

1.3. Aims of this work

In this paper we present a theoretical and experimental study of the so-called moderately asymmetric ring reactor.

In the first theoretical part, after a brief introduction, the foundations of the geometrical theory of waves are outlined together with a classification of the possible cases for a piecewise homogeneous medium [16] as "weak", "moderate" or "strong" asymmetry. In a piecewise homogeneous medium there are always one or more break points in the wavefront. In symmetric arrangements and in the asymptotic state this break point always moves along the borderline of the fast and slow regions. This is not the case with the asymmetric arrangement where a nontrivial break point dynamics can be observed even in the asymptotic state.

In the second part of the paper the theoretical predictions are compared with experimental results obtained with a new reactor. This new reactor gives a better approximation of a piecewise homogeneous medium and reaches its asymptotic state much faster than its predecessor.

2. Theory

2.1. Travelling waves – different approaches

Mathematically, a travelling wave can be described by a wave function u of the type

$$u(r,t) = A(r)G(t - S(r)) \,,$$

where r is the space vector, t the time, A the amplitude, G the phase, and S the eikonal. Wavefronts are defined here as the level surfaces of the eikonal S, because at a given time t the phase G is constant along these surfaces. In the course of wave propagation wavefronts are transformed into subsequent ones: the front at an instant t_1 (belonging to the eikonal value S_1) at a later time t_2 is transformed into the front belonging to the eikonal value

$$S_2 = S_1 + t_2 - t_1 \,.$$

Hence, the waves propagate in the direction perpendicular to the level surface of S with the velocity

$$1/v = |\text{grad}\, S|$$

(eikonal equation). Note that the eikonal in optics [17] is defined in a similar way except for the constant factor c (c is the velocity of light in vacuum).

Travelling waves are obtained in two different ways:
- as a special solution of a differential equation or
- directly by giving the eikonal function S.

Chemical waves [1–3] are fundamentally concentration waves: the wavefronts correspond to equiconcentrational surfaces, or, in 2D situations equiconcentrational curves. The wave function $u(r,t)$ should represent the local state of the system; in most cases u and A are n-component concentration vectors. Vast majority of the theoretical works treat chemical waves via reaction–diffusion equations, since the system of reaction–diffusion equations may have travelling wave solutions. While this approach is an adequate description of all the underlying local processes, many experimentally observed phenomena can be described satisfactorily applying only the more simple geometrical wave theory [7,9,16,18], which directly defines the eikonal.

2.2. The geometrical theory of waves

The foundations of the geometrical theory of waves are very old: Fermat's principle, Huygens' principle. These principles permit us to determine the evolution of wavefronts uniquely if we know the initial front and the propagation velocity v.

The first profound extension of this approach to 2D excitable media is due to Wiener and Rosenblueth [11]. They distinguished three states of points: (i) excitable state, (ii) active state, and (iii) rest state.

At a given instant t, the wavefront is the set of the active points. The wave propagates into the excitable region. The active points become rest points immediately, and after a certain period (T) the rest points will be excitable automatically. This simple kinematic theory was applied to the obstacle problem in homogeneous medium. It turned out that the wavefronts are involutes of the obstacle [11].

2.2.1. Basic concepts

The geometric theory of waves is based on Fermat's principle of the least propagation time. Given two points P and Q, and a curve g lying in the excitable region connecting them, the propagation time belonging to this curve is

$$t(g) = \int \frac{\mathrm{d}s}{v},$$

where s is the arc length, and v the velocity. Fermat's principle selects the family of extremals, that is the curves satisfying the requirement

$$t(g) = \min$$

with fixed endpoints. It is possible to define "distance" between excitable points, and a metric space [18]. Furthermore, given an initial wavefront F_0, we can define rays emanating from the points of F_0 [18] (Fig. 1). The orthogonal trajectories of rays will be the subsequent wavefronts belonging to fixed different "distances" from the initial front (propagation times along the rays). It was proved [18] that the system of evolving wavefronts can be qualified as a dynamical system.

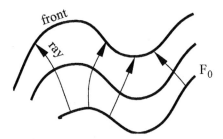

Fig. 1. Fronts are orthogonal to the rays.

According to Fermat's principle, the "true" rays belong to the minimal propagation time. In the generic case, there is only one "true" ray connecting two given points; we will refer to it as the Fermat ray. However, in some critical situations, there may exist several paths belonging to the same minimal value of propagation time. These critical cases are treated in the theory of singularities [15,19]. Another related concept is the "conflict set": this is a set whose points are equally distant from two given sets [20]. To find these critical rays is crucial in the study of the wave portraits.

2.2.2. Factors affecting local wave velocity

The extension to inhomogeneous case where v is a spatial function is straightforward. Using the tools of optimal control, the differential equations of rays and the solution of the obstacle problem can be derived even in anisotropic media, where v depends on the direction too [21].

The local velocity v also depends on the time elapsed since a previous wavefront passed through (dispersion relation). This complex problem was solved exactly in a very simple case only, namely a thin ring (1D) reactor. In this case the problem led to delayed ordinary differential equation [22]. The asymptotic state is very simple, however, because there is only one delay time for the whole ring. Similarly in 2D, this effect has an influence only on the velocity of the stationary wavefronts but not their shape, because the time period is the same for every point of the stationary front.

Finally, v may depend on the curvature of the front. This effect was calculated on the basis of reaction–diffusion equations [23]. A special algorithm was developed to describe the evolution of fronts using the curvature as the parameter of the front line [24,25]. However, the curvature effect is important only when the radius of the curved front is very small, and it often can be neglected. This was the case in our experiments.

2.3. Application of the geometric wave theory to piecewise homogeneous regions. Junction rules of rays at an interfacial boundary

The aim of the present work is to describe the wave propagation in a piecewise homogeneous region, specifically if an obstacle is surrounded by an inner slower and outer faster region. In the piecewise homogeneous case the rays consist of various segments

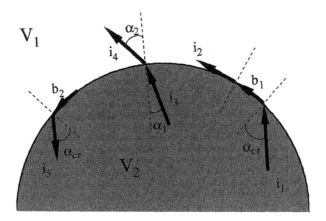

Fig. 2. Junction rules of rays at an interfacial boundary.

which will be called "arcs" whose characteristics and the rules of their connections can be derived from Fermat's principle. There are two classes of arcs:

- Interior arcs; lying inside a homogeneous part, see i_1, i_2, i_3, i_4, i_5 in Fig 2. They are straight segments as a consequence of Fermat's principle.
- Boundary arcs; lying on the interfacial boundary, see b_1, b_2. in Fig 2. This type occurs in the case when the boundary bends toward the region where the velocity is smaller, and the propagation time along the arc is smaller than along the straight short-cut belonging to this arc. If the boundary bends toward the faster region, then the straight short cut yields the minimum propagation time.

Two adjacent arcs of a given ray are joined together according to the law of refraction. According to this law, for the connection of two interior arcs (i_3, i_4) $\sin(\alpha_1)/\sin(\alpha_2) = v_1/v_2$ (Snell's law). For the connection of a boundary arc and an interior arc, there are two possibilities: tangential connection b_1, i_2 (the tangential interior arc is in the faster region), and connection with the critical angle of total reflection α_{cr} ($\sin(\alpha_{cr}) = v_1/v_2$): rays i_1, b_1, b_2, i_5 (the interior arc is in the slower region).

2.4. Results for simple annular regions

Our aim is to determine the front evolution for a special heterogeneous situation, (see Fig. 3). A circular obstacle with radius R_O is surrounded by a slower inner region (wave velocity v_1) and an outer faster region (wave velocity $v_2 > v_1$). The interfacial boundary (separating the two regions) is also assumed to be circular with radius R. A simple initial condition is assumed: a straight wavefront connecting the perimeter of the obstacle and the outer boundary of the reactor. Special cases have already been treated: homogeneous case ($v_1 = v_2$) [7,18], symmetric case (the obstacle and the interface are concentric circles) [9]. In order to deal with the general (asymmetric) case we briefly recall the main points of the previous works.

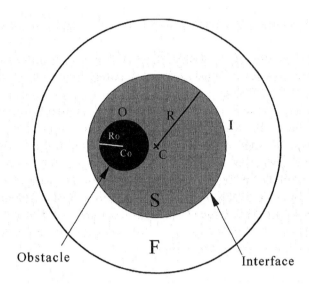

Fig. 3. The geometry of the reactor.

2.4.1. Homogeneous case

In this case there are two relevant types (I and II) of rays:

I. Rays emanating by the initial front. These are straight and perpendicular to the initial front.

II. Rays emanating by a "leading point" running around the boundary of the obstacle. These rays depart tangentially from the obstacle.

Evolution of a pinwheel from the initial straight wavefront takes place in two stages:

- Initial (transient) stage: In this stage both types of rays coexist; although the zone belonging to the rays of type II continuously grows, the wave zone determined by the rays of type I is decreasing.

- Asymptotic stage: Only the rays of type II exist, and the whole field outside the obstacle is occupied by wavefronts generated by the leading point. These wavefronts are involutes of the obstacle's boundary.

2.4.2. Heterogeneous case – symmetric arrangement

Starting from an initial front, two leading points appear at the obstacle boundary (O) and at the interface (I). These points running along O or I, respectively, emanate the rays determining a part of the front. In the first, transient stage, the front consists of parts generated by the initial straight front and parts generated by the leading points. In the second stage of the process the front will consist of parts generated by the leading points only. During the second stage the two leading points compete. The domain of the "best runner" (which has greater angular velocity $\omega = v/R$) will grow at the expense of that of the other. As a result of this competition one leading point will win, that is, the whole front will be determined by it at the third stage of the process (asymptotic wavefront). The asymptotic wavefront rotates with the annular velocity of the winner leading point.

If the angular velocity of the leading point at O is greater than that at I, then the asymptotic wavefronts are generated by rays departing from O tangentially. Reaching I these rays will be refracted according to Snell's law and therefore a break point will appear on the front at I. It turns out that the fronts (orthogonal trajectories of the rays) are involutes of the obstacle in the slower region, and involutes of the relevant caustic (a concentric circle with radius $R_O * v_2/v_1$) in the faster region. In the other case where the leading point at I is the winner, in the outer region the wavefronts are involutes of I while in the inner region they are reverse involutes of a certain caustic C. These front parts are orthogonal to the rays departing from I with the critical angle of total reflection α_{cr}. The envelope of these rays (called caustic) is a concentric circle C inside the obstacle with radius $R * v_1/v_2$. We introduced the term reverse involute to indicate that it bends in the opposite direction.

2.5. Application of the theory for asymmetric arrangements

When the obstacle is located eccentrically, then the asymptotic wave portrait depends not only on the angular velocity ratio but also on the relative position of the obstacle and the above caustic C [16]. According to that we can distinguish three classes: weak, moderate and strong asymmetry. The concept of the minimal loop plays an important role here. The minimal loop is a closed curve encircling the obstacle characterized by the shortest propagation time.

2.5.1. Weak asymmetry
The arrangement is called weakly asymmetric, if either the caustic is inside the obstacle or the obstacle is inside the caustic. In this case the qualitative behavior is identical to that of the symmetric case. That is, the competition between the two leading points is decisive. If the minimal loop is O, then the rays depart tangentially and refracts at I. However, in this case the relevant caustic is not simply a circle. If the minimal loop is I, then the shape of the fronts are identical with those in the symmetric case.

2.5.2. Moderate asymmetry
New peculiarities appear when the asymmetry is greater, namely when the caustic C and the obstacle O intersect each other, this arrangement is called moderately asymmetric. Contrary to the symmetric case, now there is no absolute winner: the loser leading point reappears in each turn and builds a certain "enclave". The other peculiarity is the possibility of the third-type minimal loop, the so-called mixed type minimal loop. This one is composed of arcs on O and on I and common tangent segments connecting them (Fig. 4).

Similar to the symmetric case, after the first transient stage the fronts are determined by two leading points. We introduce the zones Z_O and Z_I; these are the regions where

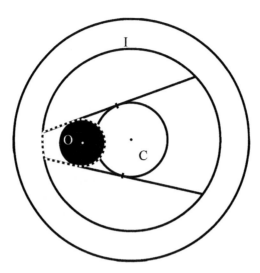

Fig. 4. The mixed loop shown with dashed line for a moderately asymmetric arrangement.

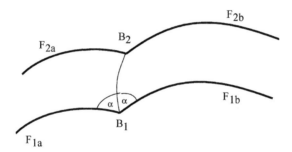

Fig. 5. Motion of the break point: bisector rule.

the fronts are generated by the rays starting from O and I, respectively. Both of these zones may consist of two parts: Z_O has an inner part S_O in the slower region and an outer part F_O in the faster region, and similarly for Z_I. In order to give the asymptotic wave portrait it is sufficient to determine the border of the zones. At any point of the border the front has a break point. During the evolution of the front the break point will move along the border of the zones. So the borders can be obtained as certain break point orbits.

Break point dynamics: Let us consider a front F_1 which consists of two smooth parts F_{1a}, F_{1b} joining at a break point B_1, as shown in Fig. 5. After a short time Δt this front is transformed to F_2 which consists of two corresponding parts F_{2a}, F_{2b} joining at a new break point B_2. This new break point is obtained as a point reached by the two front parts (indexed by a and b) simultaneously, namely after time duration Δt. It is obvious that the break point moves locally in the direction of the actual bisector provided that the velocity depends on space continuously.

Applying these rules of break point motion to the present moderately asymmetric annular arrangement, a direction field can be established for the virtual break point

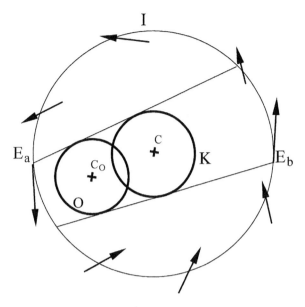

Fig. 6. Direction field for the break point at the interface.

motion. After the first transient stage, at the break point an involute of O (corresponding to F_{1a}) and a reverse involute of C (corresponding to F_{1b}) join. Consequently, at a given position the direction of the break point motion is the bisector of the tangents to O and to C, respectively. So the direction field of the break point motion can be constructed uniquely, and consequently so do the virtual break point orbits.

In the present case, it is easy to see that the common tangents divide I (or an arbitrary circle concentric with it) into two parts according to which the direction field points outward or inward (Fig. 6). Similar division is given for circles concentric with O [16].

Classification of cases of moderate asymmetry: Applying the above detailed technique, it was possible to determine three different classes of asymptotic wave portraits.

- If the minimal loop is O (Fig. 7), then the zone S_O engulfs O entirely; here the fronts are involutes of O. The leading point is rotating periodically on O. However, a break point appears at the point Q_2 (intersection point of I and the common tangent), and from the other leading point (running on I) generates the reverse involute shaped fronts in the "enclave" S_I. In one part of the outer region (denoted by F_I) the fronts are involutes of I while in the other part (F_O) the fronts are generated by the refracted rays coming from O.

- If the minimal loop is I (Fig. 8), then the leading point will rotate in I without any trouble. Thus in the outer region, the wavefronts are involutes of I. In the inner region the wavefronts are reverse involutes of C except in an enclave S_O, where the fronts are involutes of O. The initial point Q_1 of S_O is on the common tangent since this point is the first point on O where the direction field points outward. The enclave S_O partly but not entirely encircles O.

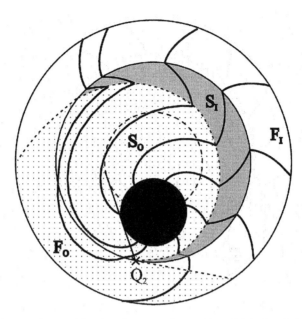

Fig. 7. Stationary fronts and zones. The obstacle generated zone $(S_O \cup F_O)$ is dotted. Case:minimal loop is the obstacle.

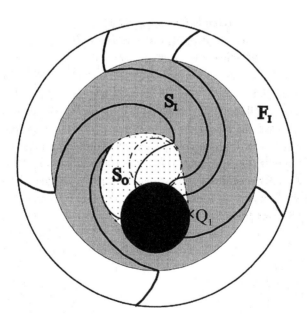

Fig. 8. Stationary fronts and zones. The obstacle generated zone (S_O) is dotted. Case:minimal loop is the interface.

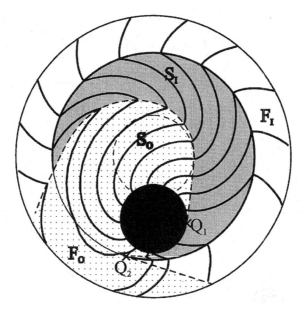

Fig. 9. Stationary fronts and zones. The obstacle generated zone ($S_O \cup F_O$) is dotted. Case:minimal loop is the mixed loop.

- If the minimal loop is mixed (Fig 9), then both S_O and S_I are nonempty "enclaves": the leading points in O and I are not able to make entire rounds because the other leading point ignites somewhere before along its path. In one part of the outer region (denoted by F_I) the fronts are involutes of I while in the other part (F_O) the fronts are generated by the refracted rays coming from O.

2.5.3. Strong asymmetry

The arrangement is called strongly asymmetric, if the caustic C is entirely outside the obstacle. In this case the wave portrait is more complicated, especially in the region between the O and C where front parts generated by O and I may rotate in opposite direction. As we saw in the moderate case the common tangents are crucial, and for the strongly asymmetric arrangement there are two pairs of them; the inner common tangents also exist. Theoretical and experimental study of this complex situation should be the subject of further research.

3. Experiments

3.1. Materials

3.1.1. Chemicals

All of the chemicals used here were of reagent grade and used without further purification: 4,7-diphenil-1,10-phenantroline (bathophenantroline) triethanolamine, glacial acetic acid (Reanal), ammonium peroxidisulphate (Merck), acrylamide, N,N'-methylenebis(acrylamide), malonic acid, sodium bromate, sodium bromide, sulfuric acid and $(NH_4)2SO_4$ (Fluka).

3.1.2. Other materials

Fumed silica (Wacker HDK T-30), polysulphone membrane filter discs (Gelman Sciences HT-450) diameter: 47 mm, pore size: 0.45 μm.

3.1.3. Solutions applied in the experiments

Four different solutions were prepared for the experiments. First a stock BZ solution was prepared under the hood in the following way. Malonic acid (4.16 g, 0.04 mol), $NaBrO_3$ (6.04 g, 0.04 mol) and NaBr (3.14 g, 0.05 mol) were dissolved in 110 ml water in a 500 ml stoppered flask. Then 20 ml of 5 M sulfuric acid was added and the stopper was closed to prevent any escape of bromine. It was opened again only when the yellow color indicating the presence of bromine had disappeared. For the experiments two "fast" solutions, F1 and F2, and two "slow" solutions, S1 and S2, were prepared starting from the stock applying the following recipes. F1: 2 ml water was added to 15 ml stock. F2: 3 ml water + 14 ml stock. S1: 8 g $(NH_4)2SO_4$ was dissolved in 30 ml water and was mixed with 65 ml stock solution. S2: 20 ml S1+1.4 g $(NH_4)2SO_4$. The solutions were slowly ageing thus wave velocities measured with the same solutions decreased by about 4–5 % each day. This was due to a gradual decomposition of brominated products and to a slow reaction between malonic acid and bromate in the acidic medium. As the duration of each experiment was always less than 1 h, this effect could be neglected during one experiment.

3.2. Methods and preparations for the experiments

3.2.1. Preparation of catalytic membranes

A dry polysulphone membrane disc was placed in a Petri dish containing a solution of 250 mg bathophenantroline dissolved in 5 ml glacial acetic acid for 3 min. Next, it was immersed in an aqueous solution (0.1 M sulfuric acid, 0.01 M ammonium sulphate) for another 3 min. The membranes were washed with and stored in a 0.1 M aqueous ammonium hydrogen sulphate solution appropriate sized membrane rings were cut from these catalytic membrane discs before the experiments.

3.2.2. Preparation of gel slabs

Five ml fumed silica was mixed with 10 ml of solution 1 (16% acrylamide, 0.8% N,N'-methylenebis(acrylamide) and 0.8% triethanolamine in water) and air bubbles (caused by the air originally adsorbed on the fumed silica) were removed by vacuum. (Fumed silica is an inert filling material, and it was applied only to improve the mechanical properties of the gel.) Then 3–4 drops of solution 2 (20% ammonium peroxidisulphate in water prepared freshly) was added and the mixture was poured on a glass plate. To produce a gel slab a second glass plate was placed on the top. The two plates were separated by 3 mm spacers. After 10 min the glass plates were removed and the gel slab was washed and stored in water. The freshly produced gel slab was 3 mm thick but in the water it swelled to reach a final thickness of about 4 mm, the size of which remained the same and within experimental error even in the reagent solutions. The swollen gel slab was cut to proper sized rings and disks fitting the reactor.

3.2.3. Establishing slow and fast regions in membrane reactors – a novel reactor design

The aim of the present work is to compare the predictions of the geometrical wave theory with the experimental observations made in quasi-two-dimensional asymmetric geometrical arrangements. To this end an appropriate membrane reactor should be constructed where the experimental conditions approximate the theoretical assumptions well. The simplest theory assumes piecewise homogeneous regions (an inner "slow" and an outer "fast" region) surrounding a circular obstacle.

Applying our membrane technique [6] the central obstacle can be easily realized by making a hole in the membrane [7]. However, realization of piecewise homogeneous regions in the membrane is a more delicate task. This is because a parameter "jump" should be established at the borderline of the slow and fast wave conducting zones. For chemical waves the possible control parameters are mostly concentrations. Consequently, an uneven concentration distribution should be generated and maintained in the reactor. In our previous experiments this was achieved by a controlled accumulation of certain reaction products in an annular zone around the obstacle [9] which slowed down the wave propagation there. Accumulation of the products was achieved by a special barrier membrane placed below the catalytic one on the top of the gel slab [8]. While this technique worked well creating slow regions above the barrier membrane, steady-state profiles of the control parameters were far from a well-defined jump [9]. Moreover, reaching this steady state required several hours. Thus, it seemed reasonable to develop a better technique for the present more delicate experiments.

The basis of the new technique is a reactor design where the thin (0.1 mm) membrane forms a bridge over a 0.3 mm gap between two thicker (about 4 mm thick) gel layers (see the insert of Fig. 10). The membrane is in close contact with both gel slabs which, on the other hand, are isolated from each other by a 0.3 mm wide PVC barrier. The concentrations (here especially the acidity) are different in the two slabs. Thus, a concentration ramp is created in the catalytic membrane where it bridges over the gap between the two gels. In the case of a narrow gap the resulting steep concentration ramp is a good approximation of a concentration jump. As diffusion from one gel to the other is possible only via the thin membrane bridge, the concentration difference can be maintained without substantial changes for several hours. The two gel slabs form reservoirs for the reagents and they are big enough for experiments running for days. After the experiments the gels were stored in the appropriate BZ solutions again.

Advantages of the present technique compared to the previous one are the following: (i) concentration changes can be located at narrow regions approximating better the jump assumption,

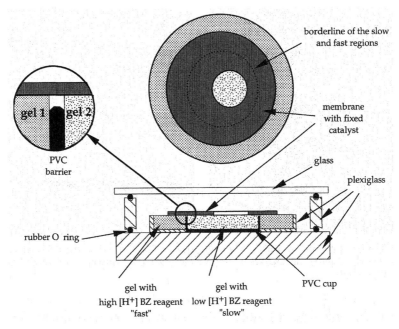

Fig. 10. Cross-sectional view of the reactor used in the experiments. The reactor was mounted on a brass slab thermostated to 25°C (not shown in the figure).

(ii) quasi-steady-state concentration profiles can be reached within few minutes and most importantly

(iii) wave velocities can be controlled in a continuous fashion and independently in both regions simply by changing the reagent concentrations inside the two gel slabs.

The new reactor is depicted in Fig. 10. Before conducting the experiments, the inner gel disk was soaked in a less acidic ("slow" S1 or S2) and the outer gel ring in a more acidic ("fast" F1 or F2) solution for 2 h. Then the gel disk was placed into the central PVC cup of the reactor and the gel ring was placed around it. Finally, a catalytic membrane ring was placed on the top of the gel slabs. In a symmetric arrangement the circular hole in the membrane and the perimeter of the reactor cup formed concentric circles. Otherwise the arrangement was asymmetric.

3.2.4. Creation of chemical pinwheels on a membrane ring

After the membrane ring is placed into the reactor it takes few minutes while enough BZ reagents diffuse into it from the gel slabs to make the membrane excitable. Then a wavefront was initiated with a silver wire. A part of this front was eliminated by touching it with an iron paper clip. The result of these perturbations should be a wavefront with a single free end which forms the core of an evolving spiral. This rotating spiral sends wavefronts both in clockwise and counterclockwise directions on the ring. Chemical pinwheels can be generated by deleting clockwise or counterclockwise waves

Fig. 11. Chemical waves in a moderately asymmetric reactor. Minimal loop: obstacle. Picture (a) and (b) show two consecutive states. The time elapsed between the first and second snapshot: 70 s. See Table 1 and text for further details.

again with the paper clip. To avoid drying out of the membrane and the underlying gels the top of the reactor was opened only temporally for the perturbations.

When all these manipulations were successfully completed the glass top of the reactor was closed and experimental observations started after about 20–30 min. Asymptotic state of the rotating pinwheels could be maintained for several hours without any measurable changes.

3.3. Results and discussion

Our aim here was to realize experimentally all the three theoretical cases of moderate asymmetry given in Figs. 7–9 and to confirm the validity of the qualitative identifications by calculations. The results of such experiments are shown in Figs. 11–13.

In Fig. 11 the minimal loop is the obstacle. In reality, in the close neighborhood of the obstacle there are only wavefronts which are its involutes. In this respect, the situation is similar to the symmetric and the weakly asymmetric case. In these cases, however, the whole inner slow region is ruled by wavefronts which are involutes of the obstacle and there are no wavefronts which would be reverse involutes of the interface. This is because there is only one leading point rotating on the perimeter of the obstacle. In Fig. 11 the situation is qualitatively different: the sharp break points are signs of short wavefront sections which are reverse involutes of the interface. This situation is shown theoretically in Fig. 7. In this figure a second leading point appears together with its enclave which penetrates into the inner slow region. (We remark here that the inner slow regions in the experimental figures can be seen weakly as darker areas. This is due to the less acidic chemistry in these slow regions.) The second leading

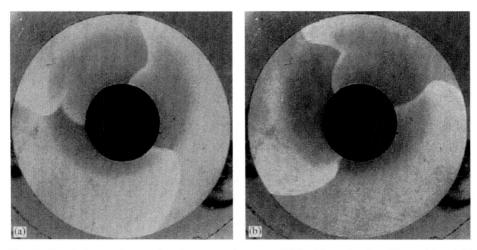

Fig. 12. Chemical waves in a moderately asymmetric reactor. Minimal loop: interface. Picture (a) and (b) show two consecutive states. The time elapsed between the first and second snapshot: 131 s. See Table 1 and text for further details.

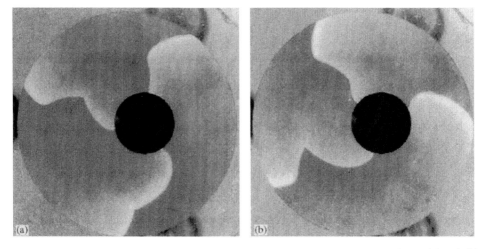

Fig. 13. Chemical waves in a moderately asymmetric reactor. Minimal loop: mixed. Picture (a) and (b) show two consecutive states. The time elapsed between the first and second snapshot: 187 s. See Table 1 and text for further details.

point, unlike the first one, has no closed orbit, however, it disappears and reappears periodically.

In Fig. 12 the minimal loop is the interface of the slow and fast regions. In the outer fast region all wavefronts are involutes of the interface. In the slow inner region the wavefronts are mainly reverse involutes of the interface but a sharp break point indicates the presence of an enclave where the fronts are involutes of the obstacle. Fig. 8 gives a theoretical description of this case. Now, it is the outer leading point which has a closed orbit (this is the minimal loop in this case) and the inner leading point appears only periodically.

Table 1
Experimental conditions and some measured and calculated data for Figs. 11–13. Explanation of the symbols: Inner sol., Outer sol.: Inner and outer solutions. (The composition of the "slow" (S1 and S2) and the fast (F1 and F2) solutions are given in the material section.) Age: The time elapsed since the preparation of the solutions. V_S, V_F: Wave velocities measured in the slow, fast regions respectively. a: Asymmetry, the distance between the center of two circles (the obstacle and the interface). R_0, R_I, R_C: Radius of the obstacle, the interface and the caustic circle, respectively. τ_0, τ_I, τ_M: Virtual rotation times calculated for the obstacle, the interface and the mixed loop, respectively

	Fig. 11	Fig. 12	Fig. 13
Inner sol.	S1	S2	S2
Outer sol.	F1	F1	F2
Age [days]	2	7	14
V_S [mm/min]	2.50 ± 0.01	2.05 ± 0.02	1.28 ± 0.02
V_F [mm/min]	5.19 ± 0.01	4.50 ± 0.01	3.06 ± 0.01
a [mm]	3.40 ± 0.05	3.05 ± 0.05	4.55 ± 0.05
R_O [mm]	4.00 ± 0.05	5.00 ± 0.05	4.00 ± 0.05
R_i [mm]	9.75 ± 0.05	9.75 ± 0.05	9.75 ± 0.05
K [mm]	4.68	4.44	4.07
τ_0 [min]	10.0	15.3	19.6
τ_i [min]	11.8	13.6	20.0
τ_m [min]	11.0	15.3	19.3

Finally in Fig. 13 the minimal loop is a mixed one. This can be realized from the fact that neither the inner nor the outer leading point has a closed orbit. In Fig. 13a the wavefront at the bottom of the figure has only one leading point which is on the obstacle while the other one is missing. The opposite case can be observed for the wavefront on the right-hand side of Fig. 13b where the only leading point is on the interface. Thus both leading points should disappear temporally. Such a situation is also shown in the theoretical Fig. 9.

The above qualitative considerations are supported by calculations, the results of which are given in Table 1. The same table contains data used in the calculations and some other information on the three experiments depicted in Figs. 11–13. First, the data show that the condition of moderate asymmetry

$$|R_C - R_O| < a < R_C + R_O$$

is satisfied for all the three cases. Second, the results verify quantitatively that the minimal loops are the ones which were assumed previously. For example in Fig. 11, from the three possible loops, it is the one around the obstacle for which the time is minimal; consequently, this is the minimal loop. While such calculations are simple for the obstacle and for the interface they are more cumbersome for the mixed loop. Thus the main line of this calculation is given in the appendix.

Regarding the qualitative similarities between the theoretical, Figs. 7–9, and the experimental ones, Figs. 11–13, and the results of the calculations displayed in Table 1. we can conclude that there is a good agreement between the theory and the experiments.

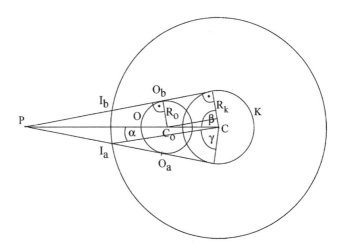

Fig. 14. Calculation of the time period belonging to the mixed loop.

Acknowledgements

This work was partially supported by OTKA (T-030110, F-022228) and FKFP (0287/1997) grants, and CARNET-II (ICOP - DISS - 2168 - 96) project.

Appendix

To calculate the period time of the mixed loop let us see Fig. 14. First, we assume $R_C \geqslant R_O$, as shown in the figure. Let us introduce the notations $a = C_O C$, $e = I_b O_b$. The mixed loop consists of the arc $O_a O_b$ of the obstacle boundary, the tangential segment $O_b I_b$, the arc $I_b I_a$ of the interface and the tangential segment $I_a O_a$. Using $R_C/R = v_1/v_2$ one obtains for the period time of the mixed loop:

$$\tau_M = \frac{2[(\pi - \beta)R_O + e + \alpha R_C]}{v_1}. \tag{A.1}$$

The angles α and β and the length e can be expressed by the geometrical data (R, R_O, R_C, a) of the arrangement as follows:

$$\cos \beta = \frac{R_C - R_O}{a}, \quad \alpha = \beta - \gamma, \quad \cos \gamma = \frac{R_C}{R},$$

$$e = \sqrt{R^2 - R_C^2} - \sqrt{a^2 - (R_C - R_O)^2}.$$

Substituting these expressions into (A.1) we get

$$\tau_M = \frac{2}{v_1} \left[(R_C - R_O)\cos^{-1}\frac{R_C - R_O}{a} + \sqrt{R^2 - R_C^2} - \sqrt{a^2 - (R_C - R_O)^2} \right.$$

$$\left. + \pi R_O - R_C \cos^{-1}\frac{R_C}{R} \right]. \tag{A.2}$$

In the case $R_O > R_C$ the calculation is similar, we get for the period time:

$$\tau_M = \frac{2}{v_1}\left[(R_C - R_O)\cos^{-1}\frac{R_O - R_C}{a} + \sqrt{R^2 - R_C^2} + \sqrt{a^2 - (R_O - R_C)^2}\right.$$
$$\left. + \pi R_O - R_C\cos^{-1}\frac{R_C}{R}\right]. \tag{A.3}$$

Differentiating τ_M with respect to a it is easy to see that in the case $R_C \geqslant R_O$, the period time τ_M decreases as the asymmetry a increases. If a is small, then $\tau_O < \tau_M$; if a is large (close to $R_I - R_O$), then $\tau_M < \tau_O$. Hence there exists a unique value a^* of a for which $\tau_O = \tau_M$; below this value the minimal loop is O and above this value the minimal loop is mixed. In the case $R_C < R_O$ τ_M is an increasing function of a.

References

[1] R.J. Field, M. Burger (Eds.), Oscillations and Travelling Waves in Chemical Systems, Wiley, New York, 1985.

[2] S.K. Scott, Oscillations, Waves, and Chaos in Chemical Kinetics, Oxford Univ. Press, Oxford, 1994.

[3] R. Kapral, K. Showalter (Eds.), Chemical Waves and Patterns, Kluwer, Dordrecht, Netherlands, 1995.

[4] J.D. Murray, Mathematical Biology, Springer, Berlin, 1989.

[5] A.T. Winfree, When Time Breaks Down, Princeton University press, Princeton, NJ, 1987.

[6] A. Lázár, Z. Nosziczius, Zs. Nagy-Ungvárai, H.D. Försterling, Physica D 84 (1995) 112.

[7] A. Lázár, Z. Nosziczius, H. Farkas, H.D. Försterling, Chaos 5 (1995) 443.

[8] A. Lázár, H.D. Försterling, Z. Nosziczius, A. Volford, J. Chem. Soc. Faraday Trans. 92 (1996) 2903.

[9] A. Lázár, H.D. Försterling, H. Farkas, P.L. Simon, A. Volford, Z. Nosziczius, Chaos 7 (1997) 731.

[10] A. Volford, Z. Nosziczius, V. Krinsky, Ch. Dupont, A. Lázár, H.-D. Försterling, J. Phys. Chem. 102 (1998) 8355.

[11] N. Wiener, A. Rosenblueth, Arch. Inst. Card. De Mexico 16 (1946) 205.

[12] D.P. Zipes, J. Jalife, Cardiac Electrophysiology, W.B. Saunders Co. Philadelphia, 1990, pp. 543–546.

[13] W. Schoels, N. El-Sherif, in: M. Shenasa, M. Borggrefe, G. Breithardt (Eds.), Cardiac Mapping, Futura Publishing, Mount Kisco, NY, 1993, pp. 281–290.

[14] Z. Nosziczius, W. Horsthemke, W.D. McCormick, H.L. Swinney, W.Y. Tam, Nature 329 (1987) 619.

[15] V.I. Arnold, Singularities of Caustics and Wave Fronts, Kluwer Academic Publishers, Dordrecht, 1990.

[16] Volford, P.L. Simon, H. Farkas, Banach Center Publications, 1999, in press.

[17] M. Born, E. Wolf, Principles of Optics, 6th Edition, Pergamon Press, Oxford, 1980.

[18] P.L. Simon, H. Farkas, J. Math. Chem. 19 (1996) 301.

[19] S. Janeczko, I. Stewart, in: M. Roberts, I. Stewart (Eds.), Singularity Theory and its Applications, Springer, Berlin, 1990, pp. 220–255.

[20] D. Siersma, Banach Center Publications, 1999, in press.

[21] S. Sieniutycz, H. Farkas, Chem. Eng. Sci. 52 (1997) 2927.

[22] P.L. Simon, A retarded differential equation model of wave propagation in a thin ring, SIAM J. Appl. Math., to be published.

[23] J.P. Keener, J.J. Tyson, Physica D 21 (1986) 307.

[24] A.S. Mikhailov, Foudations of synergetics I, Distributed Active Systems, Springer, Berlin, 1990.

[25] P.K. Brazhnik, J.J. Tyson, Phys. Rev. E 54 (1996) 1958.

ELSEVIER

Physica A 274 (1999) 50–59

www.elsevier.com/locate/physa

Formation of Liesegang patterns

Zoltán Rácz

Institute for Theoretical Physics, Eötvös University, Pázmány sétány 1/a, 1117 Budapest, Hungary

Abstract

It has been recently shown that precipitation bands characteristic of Liesegang patterns emerge from spinodal decomposition of reaction products in the wake of moving reaction fronts. This mechanism explains the geometric sequence of band positions $x_n \sim Q(1 + p)^n$ and, furthermore, it yields a spacing coefficient p that is in agreement with the experimentally observed Matalon–Packter law. Here I examine the assumptions underlying this theory and discuss the choice of input parameters that leads to experimentally observable patterns. I also show that the so-called *width law* relating the position and the width of the bands $w_n \sim x_n$ follows naturally from this theory. © 1999 Elsevier Science B.V. All rights reserved.

1. Introduction

Formation of precipitation patterns in the wake of moving reaction fronts (known as the Liesegang phenomenon) has been studied for more than a century [1–3]. The motivation for these studies has been diverse, coming from the importance of related practical problems such as crystal growth in gels, as well as from the fascination with a complex pattern that has eluded a clean-cut explanation (e.g. agate rocks are believed to display Liesegang patterns). From a theoretical point of view, the main factor in the popularity was the belief that much can be learned about the details of precipitation processes (nucleation, growth, coagulation, etc.) by investigating the instabilities underlying this phenomenon. Currently, the Liesegang phenomenon is mainly studied as a nontrivial example of pattern formation in the wake of a moving front [4–7] and there are speculations about the possibility of creating complex mesoscopic structures using this rather inexpensive process.

Liesegang patterns are easy to produce (Fig. 1 shows a particular experiment that we shall have in mind in the following discussion). The main ingredients are two chemicals A and B yielding a reaction product $A+B \to C$ that forms a nonsoluble precipitate $C \to D$ under appropriate conditions [$A = \text{NaOH}$, $B = \text{MgCl}_2$ and $D = \text{Mg(OH)}_2$ in Fig. 1].

E-mail address: racz@poe.elte.hu (Z. Rácz)

0378-4371/99/$ - see front matter © 1999 Elsevier Science B.V. All rights reserved.
PII: S 0378-4371(99)00432-X

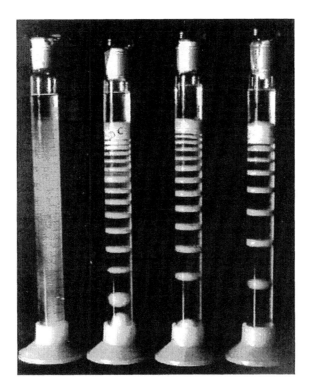

Fig. 1. Liesegang patterns obtained with reagents $A = \text{NaOH}$ and $B = \text{MgCl}_2$ in polyvinylalcohol gel. The white precipitate is $D = \text{Mg(OH)}_2$. The height of the columns is 30 cm and it takes about a 1–2 weeks for the patterns to form. The columns show different patterns due to the difference in the initial concentrations of the outer electrolyte NaOH. The experiments were carried out by M. Zrínyi (Technical University of Budapest).

The reagents are separated initially with one of them (B, inner electrolyte) dissolved in a gel and placed in a test tube. Then at time $t = 0$ an aqueous solution of the other reagent (A, outer electrolyte) is poured over the gel. The initial concentration a_0 of A is chosen to be much larger than that of B (typically $a_0/b_0 \approx 10^2$), thus A diffuses into the gel and a reaction front moves down the tube. Behind the front, a series of stationary precipitation zones (Liesegang bands) appear at positions x_n (x_n is measured from the interface between the gel and the aqueous solution; $n = 1, 2, \ldots, 10$–20, typically). A band appears in a rather short time-interval thus the time of the appearance t_n of the nth band is also a well defined, experimentally measurable quantity. Finally, the widths of the bands w_n can also be determined in order to characterize the pattern in more detail.

The experimentally measured quantities (x_n, t_n and w_n) in *regular* Liesegang patterns satisfy the following *time-*, *spacing-*, and *width* laws.

Time law [8]:

$$x_n \sim \sqrt{t_n} \,. \tag{1}$$

This law is satisfied in all the experiments where it was measured and it appears to be a direct consequence of the diffusive dynamics of the reagents.

Spacing law [9]: The positions of the bands form a geometric series to a good approximation

$$x_n \sim Q(1 + p)^n,$$ (2)

where $p > 0$ is the spacing coefficient while Q is the amplitude of the spacing law. The quantitative experimental observations concern mainly this law. More detailed works go past the confirmation of the existence of the geometric series and study the dependence of the spacing coefficient on a_0 and b_0. The results can be summarized in a relatively simple expression usually referred to as the *Matalon–Packter law* [10,11]:

$$p = F(b_0) + G(b_0)\frac{b_0}{a_0},$$ (3)

where F and G are decreasing functions of their argument b_0.
Width law [12]:

$$w_n \sim x_n.$$ (4)

This is the least established law since there are problems with both the definition and the measurement (fluctuations) of the width. Recent, good quality data [13] does support, however, the validity of (4).

It should be clear that (1)–(4) summarize only those properties of Liesegang patterns that are common in a large number of experimental observations. There is a wealth of additional data on various details such as e.g. the secondary structures or the irregular band spacing [2]. These features, however, appear to be peculiarities of given systems. It is hard to characterize them and their reproducibility is often problematic as well. In view of this, it is not surprising that the theoretical explanations of Liesegang phenomena have been mainly concerned with the derivation of (1)–(4).

The theoretical approaches to quasiperiodic precipitation have a long history and the two main lines of thoughts are called as pre- and post-nucleation theories (for a brief overview see Ref. [7]). They all share the assumption that the precipitate appears as the system goes through some nucleation or coagulation thresholds. The differences are in the details of treating the intermediate steps "...C..." in the chain of reactions $A + B \rightarrow \ldots C \ldots \rightarrow D$ producing the precipitate D. In general, all the theories can explain the emergence of distinct bands but only the pre-nucleation theories can account [4–6,14–16] for the time- and spacing laws of normal patterns. These theories are rather complicated, however, and have been developed only recently [1] to a level that the dependence of p on the initial concentrations a_0 and b_0 can be investigated quantitatively, and connection can be made to the Matalon–Packter law [10,11].

Unfortunately, there are several problems with the theories mentioned above. First, they employ a large number of parameters and some of these parameters are hard to grasp theoretically and impossible to control experimentally (an example is the lower threshold in the density of C's below which aggregation $C + D \rightarrow 2D$ ceases [5,6]). Second, some of the mechanisms invoked in the explanations are too detailed and tailored to a given system in contrast to the generality of the resulting pattern in diverse systems. A real drawback of the too detailed description is that quantitative deductions are difficult to make even with the present computer power [4]. A final problem we should mention is the absence of an unambiguous derivation of the width law in any of the theories.

In order to avoid the above problems, we have recently developed a simple model of band formation [17] based on the assumption that the main ingredients of a macroscopic description should be the presence of a moving reaction front and the phase separation that takes place behind the front. This theory contains a minimal number of parameters, it accounts for the spacing law, and it is simple enough that the existence of the Matalon–Packter law can be established numerically. The apparent success warrants a closer look at the model and, in this lecture, I will describe in detail how one arrives at such a model and what are the underlying assumptions of the theory. Then I would like to discuss the choice of input parameters that yield experimentally observable patterns and, finally, I will show that the derivation of width law is straightforward in this theory.

2. The model

Let us begin building the model by taking a look at Fig. 1. It shows alternating high- and low-density regions of the chemical $Mg(OH)_2$ and the systems appear to be in quasi-steady states (actually, there are experiments that suggest that the pattern does not change over a 30-year period [2]). We shall take this picture as an evidence that phase separation [18] underlies the formation of bands and, furthermore, that the phase separation takes place at a very low effective temperature (no coarsening is observed).

The phase separation, of course, must be preceded by the production of C's. This is the least understood part of the process and it is particular to each system. What is clear is that due to the condition $a_0 \gg b_0$, a reaction front ($A + B \rightarrow something$) moves down the tube diffusively (note that this is the point where the role of the gel is important since it prevents convective motion). The result of the reaction may be rather complex (intermediate products, sol formation, etc.) and one of our main assumptions is that all these are irrelevant details on a macroscopic level. Accordingly, the production of C will be assumed to be describable by the simplest reaction scheme $A + B \rightarrow C$.

Once $A + B \rightarrow C$ is assumed, the properties of the front and the production of C's are known [19]. Namely, the front moves diffusively with its position given by $x_f = \sqrt{2D_f t}$, the production of C's is restricted to a slowly widening narrow interval $[w_f(t) = w_0 t^{1/6}]$ around x_f, and the rate of production $S(x,t)$ of C's can be approximated by a Gaussian (the actual form is not a Gaussian, see Ref. [20] for details about a nonmoving front)

$$S(x,t) = \frac{S_0}{t^{2/3}} \exp\left[-\frac{[x - x_f(t)]^2}{2w_f^2(t)} \right] . \tag{5}$$

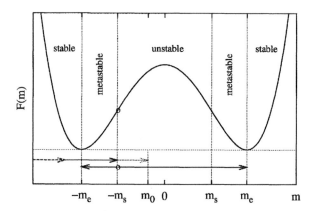

Fig. 2. The homogeneous part of the free energy as a function of $m = c - (c_l + c_h)/2$. The phase separation is an activated process in the metastable regimes while it goes by spinodal decomposition in the (linearly) unstable regime. The spinodal point (○) separates these regimes. The arrows are meant to illustrate how the density in the front increases towards m_0 and how the phase separation to the steady states ($\pm m_e$) takes place when the density reaches the spinodal value $-m_s$.

The parameter D_f can be expressed through a_0, b_0, and the diffusion coefficients of the reagents (D_a, D_b) while S_0 and w_0 depend also on the rate constant, k, of the reaction $A + B \to C$.

An important property of the front is that it leaves behind a constant density c_0 of C's [7] and c_0 depends only on a_0, b_0, D_a and D_b. This is important because the relevant parameters in the phase separation are D_f and c_0 (where and how much of the C's are produced[1]) and thus the least available parameter (k) does not play a significant role in the pattern formation.

Having a description of the production of C's, we must now turn to the dynamics of their phase separation. Since the emerging pattern is macroscopic, we shall assume that, on a coarse-grained level, the phase separation can be described by the simplest 'hydrodynamical' equation that respects the conservation of C's. This is the Cahn–Hilliard equation [21,22] or, in other contexts, it is the equation for model B in critical dynamics [23]. This equation, however, requires the knowledge of the free-energy density $(\mathscr{F})$ of the system. For a homogeneous system, $\mathscr{F}$ must have two minima corresponding to the low- (c_l) and high-density (c_h) states being in equilibrium (Fig. 1). The simplest form of $\mathscr{F}$ having this property and containing a minimal number of parameters is the Landau–Ginzburg free energy (Fig. 2)

$$\mathscr{F} = -\tfrac{1}{2}\varepsilon m^2 + \tfrac{1}{4}\gamma m^4 + \tfrac{1}{2}\sigma(\nabla m)^2 \,, \tag{6}$$

where $m = c - (c_l + c_h)/2$ is the density, c, of the C's measured from the average of the two steady-state values (we are following the notation in Ref. [17] where the 'magnetic

---

[1] For usual parameter values, the width is much smaller than all the other length-scales and its actual value does not significantly affect the onset of phase separation.

language' has its origin in a connection to Ising lattice gases). The parameters ε, γ, and σ are system dependent with $\varepsilon > 0$ ensuring that the system is in the phase-separating regime, $\sigma > 0$ provides stability against short-wavelength fluctuations, and requiring $\sqrt{\varepsilon/\gamma} = (c_h - c_l)/2$ fixes the minima of $\mathscr{F}$ at $\pm m_e$ corresponding to c_l and c_h. Note that the $m \rightarrow -m$ symmetry is usually not present in a real system and $\mathscr{F}$ could contain e.g. an m^3 term. The presence or absence of the $m \rightarrow -m$ symmetry, however, is not relevant for the discussion that follows.

Using (6) and including the source term, the Cahn–Hilliard equation takes the form

$$\partial_t m = -\lambda\Delta(\varepsilon m - \gamma m^3 + \sigma\Delta m) + S, \qquad (7)$$

where λ is a kinetic coefficient. The above equation should contain two noise terms. One of them should be the thermal noise while the other should originate in the chemical reaction that creates the source term. Both of these noise terms are omitted here. The reason for neglecting the thermal noise is the low effective temperature of the phase separation as discussed in connection with Fig. 1. The noise in S, on the other hand, is dropped since the $A + B \rightarrow C$-type reaction fronts have been shown to be mean-field like above dimension two [24].

The absence of noise means that the phase separation can occur only through spinodal decomposition [18].[2] Thus the assumption behind omitting the noises is that the characteristic time of nucleation is much larger than the time needed by the front to increase the density of C's beyond the spinodal value ($-m_s$ in Fig. 2) where the system is unstable against linear perturbations. Since there are examples where the bands appear to be formed by nucleation and growth [2], the spinodal decomposition scenario is clearly not universally applicable, and one should explore the effects of including noise (this becomes, however, an order of magnitude harder problem).

Eq. (7) together with the form of the source (5) now defines our model [17] that produces regular Liesegang patterns (Fig. 3) satisfying the spacing law (2) and, furthermore, the spacing coefficient is in agreement with the Matalon–Packter law (3). Fig. 3 shows a rather general picture that is instructive in understanding the pattern formation. The last band acts as a sink for neighboring particles above $-m_e(c_l)$ density. Thus, the C's produced in the front end up increasing the width of the last band. This continues until the front moves far enough so that the density in it reaches the spinodal value. Then the spinodal instability sets in and a new band appears. Remarkably, the above picture is rather similar to the phenomenological 'nucleation and growth' scenario [7] with the density at the spinodal point playing the role of threshold density for nucleation. It is thus not entirely surprising that both of these theories do equally well in producing the spacing- and the Matalon–Packter law.

One should note that the actual form of $\mathscr{F}$ does not play an important role in the picture developed above. The crucial feature is the existence of a spinodal density

---

[2] The spinodal decomposition takes place in the regime where the homogeneous solutions of Eq. (7) are linearly unstable.

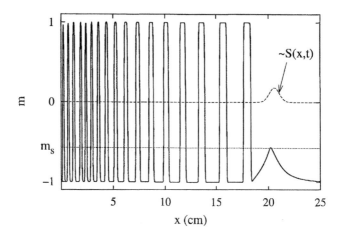

Fig. 3. Liesegang pattern obtained for front parameters $D_f = 21.72$, $w_0 = 4.54$, and $S_0 = 0.181$ with length, time, and m (concentration) measured in units of $\sqrt{\sigma/\varepsilon} = 2 \times 10^{-4}$ m, $\sigma/(\lambda\varepsilon^2) = 40$s, and $\sqrt{\varepsilon/\gamma}$, respectively. The dashed line denotes the rate of production of C's (S), measured in units of $\lambda\varepsilon^{5/2}/(\gamma^{1/2}\sigma)$ and magnified by a factor 2×10^5. The dotted line represents the density at spinodal point, $m_s = -1/\sqrt{3}$.

above which phase separation occurs. This is the meaning of our previous remark about the irrelevance of the m^3 term in the free energy (of course, one should also realize that explanations of details in experiments may require the inclusion of such terms).

3. Choice of parameters

Fig. 3 shows the results of numerical solution of Eq. (7) with the same parameter values as in Fig. 3 of Ref. [17] but stopped at an earlier time so that the visual similarity to the experiments (number of bands in Fig. 1) would be greater. In this section, we shall examine whether the parameters used for obtaining this resemblance have any relevance to real Liesegang phenomena.

The experimental patterns have a total length of about $\ell_{\exp} \approx 0.2$ m and the time of producing such a pattern is about 1–2 weeks (we shall take $\tau_{\exp} \approx 10^6$ s). Since our model has a length-scale $\ell_{th} = \sqrt{\sigma/\varepsilon}$ and a time-scale $\tau_{th} = \sigma/(\lambda\varepsilon^2)$, they can be chosen so $[\sqrt{\sigma/\varepsilon} = 2 \times 10^{-4}$ m and $\sigma/(\lambda\varepsilon^2) = 40$ s$]$ that $\ell_{\exp} \approx \ell_{th}$ and $\tau_{\exp} \approx \tau_{th}$. Once we have chosen ℓ_{th} and τ_{th} we can start to calculate other quantities and see if they have reasonable values.

It is clear from Fig. 3 that the widths of the bands are in agreement with the experiments, they are of the order of a few mm at the beginning and approach ~ 1 cm at the end. The width of the front is also of the order of 1 cm after 10^6 s. Unfortunately, there is no information on the reaction zone in this system. In a study of a different system [25], it was found that $w_f(t=2$ h$) \approx 3$ mm. Extrapolating this result to $t=10^6$ s

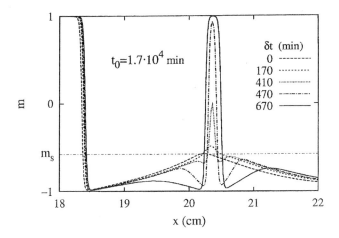

Fig. 4. Details of the time-evolution of the formation of the last band in Fig. 3. The time t_0 is the moment when the concentration at the front reached the spinodal value and δt is measured from t_0.

one finds $w_f \approx 1$ cm [note that the exponent of the increase of $w_f(t) \sim t^{1/6}$ is small] in agreement with the observed value.

Next, we calculate the diffusion coefficient of the front, $D_f = 21.72 \cdot \ell_{\text{th}}^2/\tau_{\text{th}} \approx 2 \times 10^{-8}$ m²/s. This value appears to be an order of magnitude larger than the usual ionic diffusion coefficients ($D \approx 10^{-9}$ m²/s). One should remember, however, that Fig. 3 is the result for initial conditions $a_0/b_0 = 10^2$ [17] and, for this ratio of a_0/b_0, the diffusion coefficient of the front D_f is about 10 times larger than D_a ($D_f/D_a \approx 10$ see Fig. 4 in Ref. [7]). Thus D_f also comes out to be the right order of magnitude.

We do not have information on the amplitude (S_0) of the source but, once the concentrations (a_0, b_0) are given and D_f and w_f are known then S_0 is fixed by the conservation law for the C's. Thus, the correct order of magnitude for D_f and w_f should ensure that S_0 is also of the right order of magnitude.

Finally, we shall calculate the time it takes for a band to form. It is well known that the bands appear rather quickly. From the visual observation of the beginning of the band formation it takes about $\tau_{\text{ini}} = 30$–60 min for the band to be clearly seen and then it takes much longer to increase its width to the final value. In order to calculate τ_{ini} let us consider the formation of the last band in Fig. 3 (see Fig. 4). The lower limit of the density that can be visually noticed is, of course, not well defined. We shall assume that this density corresponds to $m = 0$, i.e., it is the half-way density from c_l to c_h. This means that we see the beginnings of the band at $\delta t = 410$ min and the density reaches well above 90% of its final value by $\delta t = 470$ min. Consequently, we obtain again an estimate ($\tau_{\text{ini}} \approx 60$ min) for an observed quantity that agrees with the experiments. As a result of the above estimates, we feel that the parameters in our model can indeed be chosen so that they are relevant to real Liesegang experiments.

4. Width law

The width law is problematic from an experimental point of view since the fluctuations in the widths appear to be large. Part of the difficulties are undoubtedly due to the fact that the boundaries of the bands are not sharply defined and high-resolution digitizing methods are needed in a precise analysis. The most thorough experiment to date has been carried out recently [13] with the result $w_n \sim x_n^\alpha$ where $\alpha \approx 0.9$–1.0.

As to the theories, they also have their share of difficulties since, on a microscopic level, the growth of the width involves precipitation processes in the presence of large concentration gradients, while a macroscopic treatment must elaborate on the dynamics of the interfaces between two phases. Accordingly, there are only a few works to report on. Dee [4] used reaction–diffusion equations supplemented by terms coming from nucleation and growth processes and obtained $w_n \sim x_n$ from a rather limited (6 bands) numerical result. Chopard et al. [5,6] employed cellular automata simulations of a phenomenological version of the microscopic processes and found $w_n \sim x_n^\alpha$ with $\alpha \approx 0.5$–0.6. Finally, Droz et al. [13] combined scaling considerations with the conservation law for the number of C particles to obtain α in terms of the scaling properties of the density of precipitates in the bands. Assuming constant density they found $\alpha = 1$. Our derivation below parallels this last work in that the same conservation law is one of the main ingredients in it.

In our theory, the derivation of the width law is straightforward. One combines the facts that (i) the reaction front leaves behind a constant density (c_0) of C's, (ii) the C's segregate into low (c_l) and high (c_h) density bands, (iii) the number of C's is conserved in the segregation process; and writes down the equation expressing the conservation of C's

$$(x_{n+1} - x_n)c_0 = (x_{n+1} - x_n - w_n)c_l + w_n c_h . \tag{8}$$

Using now the spacing law (2) that has been established for this model one finds

$$w_n = \frac{p(c_0 - c_l)}{c_h - c_l} x_n = \zeta x_n . \tag{9}$$

We have thus derived the width law and obtained the coefficient of proportionality, ζ, as well. The importance of ζ lies in the fact that measuring it provides a way of assessing c_0 that is not easily measured otherwise.

5. Final remarks

In summary, we have seen that the spinodal decomposition scenario for the formation of Liesegang patterns performs well whenever quantitative comparison with experiments is possible. It remains to be seen if the applicability of this model extends beyond the *regular* patterns. One should certainly try to use this theory to explain the *exotic* patterns (e.g. inverse patterns, helixes) that are experimentally reproducible and lack even qualitative understanding.

Acknowledgements

I thank M. Droz, M. Zrínyi, T. Antal, P. Hantz, J. Magnin, and T. Unger for useful discussions. This work has been supported by the Hungarian Academy of Sciences (Grant No. OTKA T 029792).

References

[1] R.E. Liesegang, Naturwiss. Wochenschr. 11 (1896) 353.
[2] H.K. Henisch, Periodic Precipitation, Pergamon Press, Oxford, 1991.
[3] K.H. Stern, A Bibliography of Liesegang Rings, 2nd Edition, U.S. Government Printing Office, Washington, 1967.
[4] G.T. Dee, Phys. Rev. Lett. 57 (1986) 275.
[5] B. Chopard, P. Luthi, M. Droz, Phys. Rev. Lett. 72 (1994) 1384.
[6] B. Chopard, P. Luthi, M. Droz, J. Stat. Phys. 76 (1994) 661.
[7] T. Antal, M. Droz, J. Magnin, Z. Rácz, M. Zrínyi, J. Chem. Phys. 109 (1998) 9479.
[8] H.W. Morse, G.W. Pierce, Diffusion and Supersaturation in Gelatine, Proceedings of the American Academy of Arts and Sciences, Vol. 38, 1903, pp. 625–647.
[9] K. Jablczynski, Bull. Soc. Chim. France 33 (1923) 1592.
[10] R. Matalon, A. Packter, J. Colloid Sci. 10 (1955) 46.
[11] A. Packter, Kolloid Zeitschrift 142 (1955) 109.
[12] S.C. Müller, S. Kai, J. Ross, J. Phys. Chem. 86 (1982) 4078.
[13] M. Droz, J. Magnin, M. Zrínyi, J. Chem. Phys. 110 (1999) 9618.
[14] C. Wagner, J. Colloid Sci. 5 (1950) 85.
[15] S. Prager, J. Chem. Phys. 25 (1956) 279.
[16] Ya.B. Zeldovich, G.I. Barrenblatt, R.L. Salganik, Sov. Phys. Dokl. 6 (1962) 869.
[17] T. Antal, M. Droz, J. Magnin, Z. Rácz, Phys. Rev. Lett. 83 (1999) 2880.
[18] J.D. Gunton, M. San Miguel, P.S. Sahni, in: C. Domb, J.L. Lebowitz, The Dynamics of First order Transitions, in Phase Transition and Critical Phenomena, Vol. 8, Academic Press, New York, 1983.
[19] L. Gálfi, Z. Rácz, Phys. Rev. A 38 (1988) 3151.
[20] H. Larralde, M. Araujo, S. Havlin, H.E. Stanley, Phys. Rev. A 46 (1992) 855.
[21] J.W. Cahn, J.E. Hilliard, J. Chem. Phys. 28 (1958) 258.
[22] J.W. Cahn, Acta Metall. 9 (1961) 795.
[23] P.C. Hohenberg, B.I. Halperin, Rev. Mod. Phys. 49 (1977) 435.
[24] S. Cornell, M. Droz, Phys. Rev. Lett. 70 (1993) 3824.
[25] Y.-E. Lee Koo, R. Kopelman, J. Stat. Phys. 65 (1991) 893.

ELSEVIER

Physica A 274 (1999) 60–66

www.elsevier.com/locate/physa

Applications of statistical physics to the oil industry: predicting oil recovery using percolation theory

P.R. King[a,*], S.V. Buldyrev[b], N.V. Dokholyan[b], S. Havlin[c], Y. Lee[b], G. Paul[b], H.E. Stanley[b]

[a]*BP Amoco Exploration, Sunbury-on-Thames, Middx., TW16 7LN, UK & Department of Engineering, Cambridge University, Cambridge, UK*
[b]*Center for Polymer Studies, Boston University, Boston, MA 02215, USA*
[c]*Minerva Center & Department of Physics, Bar-Ilan University, Ramat Gan, Israel*

Abstract

In this paper we apply scaling laws from percolation theory to the problem of estimating the time for a fluid injected into an oil field (for the purposes of recovering the oil) to breakthrough into a production well. The main contribution is to show that percolation theory, when applied to a realistic model, can be used to obtain the same results as calculated in a more conventional way but significantly more quickly. Specifically, we found that a previously proposed scaling form for the breakthrough time distribution when applied to a real oil field is in good agreement with more time consuming simulation results. Consequently these methods can be used in practical engineering circumstances to aid decision making for real field problems. © 1999 Elsevier Science B.V. All rights reserved.

1. Introduction

There are many practical ways in which statistical physics is being used by the oil industry. To list just a few, real space renormalisation ideas are used to derive large-scale flow parameters given a detailed geological description of an oil reservoir [1,2], concepts similar to Potts models are used to represent detailed geological structure in reservoirs [3–5], understanding pattern formation in granular materials helps understand geological processes [6], non-linear dynamics of drill strings is being used to reduce drilling costs [7], simulated annealing is used to optimise business decisions adding hundreds of millions of dollars to the value of projects [8]. And so the list goes

---

* Corresponding author.
E-mail address: youngki@bu.edu (Y. Lee)

0378-4371/99/$ - see front matter © 1999 Elsevier Science B.V. All rights reserved.
PII: S 0378-4371(99)00327-1

on. There are too many applications to go into any detail in this presentation. Instead we shall concentrate on the problem of how to predict the uncertainty in recovery given statistical information of the underlying disorder in an oil reservoir.

Oil reservoirs are extremely complex, containing geological heterogeneities on all length scales from centimetres to kilometres. These heterogeneities have a significant impact on hydrocarbon recovery by perturbing the displacement front, leading to a reduction in recovery by around 25% (a further 25% being lost by trapping at the pore scale by interfacial tension).

The most common method of oil recovery is by displacement. Either water or a miscible gas is injected into wells, to push the oil to production wells. Ultimately, the injected fluid breaks through at the production wells reducing the amount of oil that can be produced. These fluids must be separated from the produced oil and suitably disposed of, which can be very costly. For economic purposes it is important to know when the injected fluid will break through and what the rate of decline in oil production will be. In this paper we concentrate on the first of these problems.

The reservoir rock is highly heterogeneous because of the sedimentary processes that deposited it. In many cases it is possible to distinguish between "good" rock (high permeability) and "poor" rock (low or zero permeability). For all practical purposes, the flow takes place just in the good rock. For example, the good rock may be ancient river channels producing sandbodies of length tens of kilometres, width tens to hundreds of metres and thickness up to tens of metres. It is the interconnectivity of these channels that controls the flow. The spatial distribution of the sandbodies is also governed by the geological process but can frequently be considered to be statistically independent or of a short-range correlation. Hence, the problem of the connectivity of the sandbodies is precisely a continuum percolation problem. The place of the occupancy probability p of percolation theory is taken by the volume fraction of good sand (known as the net-to-gross ratio in the oil industry literature).

We have very little direct knowledge about the distribution of rock properties in the reservoir. Direct measurements are limited to samples that represent around 10^{-13} of the total reservoir volume. Away from the wells, rock properties must be inferred from statistical models, based on knowledge of the general geological environment and measurements made from modern day examples or from surface outcrops of a similar type. Hence, there is a great deal of uncertainty in the prediction of the rock properties. This leads to a large uncertainty in our prediction of time to breakthrough of the injected fluid. We need to assess this uncertainty for economic risk evaluation.

The conventional approach is to build detailed numerical models of the reservoir and then perform flow simulations to predict breakthrough. This is repeated with many realisations of the subsurface models to build up an estimate of the uncertainty. Unfortunately this process is very computer intensive with each flow simulation taking typically many hours of CPU time. In order to get good statistics one must repeat the calculation many hundreds or thousands of times. Additional calculations are necessary in order to estimate the uncertainty associated with an alternative pattern of injection and production wells, which is required to determine the optimal well pattern. The purpose of this study is to use the methods of percolation theory to make the estimation of uncertainty much more efficient. Previously percolation theory has been used to estimate static connectivity [9], and we now apply similar concepts to the dynamic displacement problem.

2. Flow model

To simplify the model, we shall assume that the permeability is either zero (shale) or one (sand). The sandbodies are cuboidal (and in the first instance isotropic in shape). They are distributed independently and randomly in space to a volume fraction of p. Further we shall assume that the displacing fluid has the same viscosity and density as the displaced fluid. This has the advantage that as the injected fluid displaces the oil the pressure field is unchanged. This pressure field is determined by the solution of the single-phase flow equations. That is, the local flow rate is given by Darcy's law $v = -K\nabla P$, which coupled with the conservation condition for an incompressible fluid ($\nabla \cdot v = 0$) gives the equation for pressure as $\nabla \cdot (K\nabla P) = 0$. The injected flow then just follows the streamlines (normals to the isobars) of this flow. The permeability, K, is either zero or one as assumed above. The boundary conditions are fixed pressure of $+1$ at the injection well and 0 at the production well. In this work we shall only consider a single well pair separated by a Euclidean distance r. The breakthrough time then corresponds to the first passage time for transport between the injector and the producer.

For a given geometry of the reservoir we can then sample for different locations of the wells (or equivalently for the same well locations for different models of the reservoir with the same underlying statistics) and plot the distribution of breakthrough times. This is the conditional probability for the breakthrough time, t_{br}, given that the reservoir size (measured in dimensionless units of sandbody length) is L, the distance between wells is r, and the net-to-gross ratio is p, i.e., $P(t_{br}|r, L, p)$. In previous studies [10–14] we have demonstrated by extensive simulations that this distribution obeys the following scaling Ansatz:

$$P(t_{br}|r, L, p) \sim \frac{1}{r^{d_t}} \left(\frac{t_{br}}{r^{d_t}}\right)^{-g_t} f_1\left(\frac{t_{br}}{r^{d_t}}\right) f_2\left(\frac{t_{br}}{L^{d_t}}\right) f_3\left(\frac{t_{br}}{|p - p_c|^{-vd_t}}\right), \quad (1)$$

where

$$f_1(x) = \exp(-ax^{-\phi}), \quad (2)$$

$$f_2(x) = \exp(-bx^{\psi}), \quad (3)$$

$$f_3(x) = \exp(-cx^{\theta}). \quad (4)$$

Hence d_t is the exponent characterizing how $\langle t_{br} \rangle$ scales with r, and v is the correlation length exponent. In this paper we will not discuss the background to this scaling relationship, but concentrate on how well it succeeds in predicting the breakthrough time for a realistic permeability field.

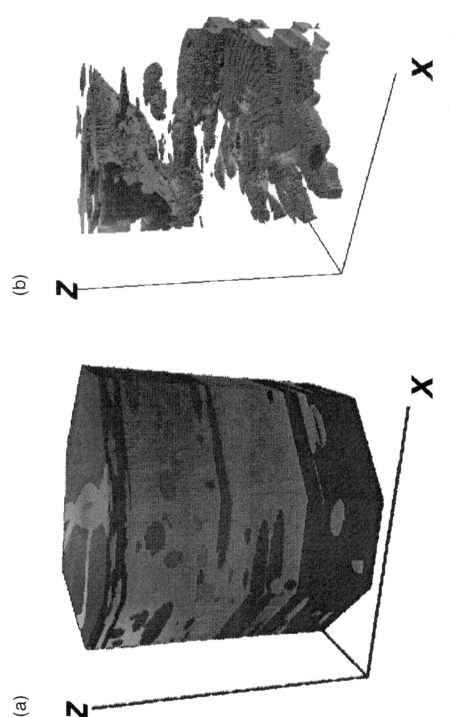

Fig. 1. (a) Permeability map for North Sea field. Note the vertical exaggeration; the true dimensions are 700 m × 700 m × 170 m vertically. Colors are chosen from a rainbow scale, from red (high permeability of ≈900 mD) to blue (low permeability of ≈100 mD) (b) Subset of Fig. 1a. showing only the region with permeability larger than a cutoff of 273 mD.

3. Application to a real oil field

We take as an example a model of a North Sea oil field. The permeability map is shown in Fig. 1a. This is a turbidite reservoir. That is, the sands were deposited by submarine turbidity currents ("avalanches") in a deep sea environment. The resulting sands consist of channels (from the avalanche scour) and lobe-shaped sandbodies from the resulting avalanche plumes. The permeability is bi-modal (Fig. 2). We apply a cutoff in permeability at 273 mD to separate the sand into good and bad. The resulting sandbodies are shown in Fig. 1b. This permeability cutoff was chosen as it corresponds to the threshold value for the system at which the incipient infinite cluster appears. Thus for this cutoff value the sandbodies are just connected. We then carried out simulations of the flow to determine the distribution of breakthrough times for well pairs at different locations and a variety of separations (Fig. 3a, solid lines).

From the scaling result, Eq. (1), and using the fact that

$$d_t = 1.33 , \tag{5}$$

we would expect to get data collapse if we plot $r^{1.33} P(t_{br}|r, L, p)$ against $t_{br}/r^{1.33}$ [11]. Fig. 3b shows this reasonably well but there is a lot of noise which somewhat obscures it. However, we can determine what we would expect the breakthrough time distribution to be from the scaling relationship. We plot this in Fig. 3a, using dashed lines. The agreement with the Monte Carlo predictions is certainly good enough for engineering purposes. The main point is that the scaling predictions took a fraction of a second of CPU time compared with the hours required for the conventional Monte Carlo approach, making this a practical tool to be used for engineering and management decisions.

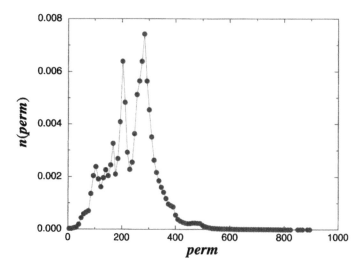

Fig. 2. Permeability distribution for North Sea Field. The approximate bimodal form is often found.

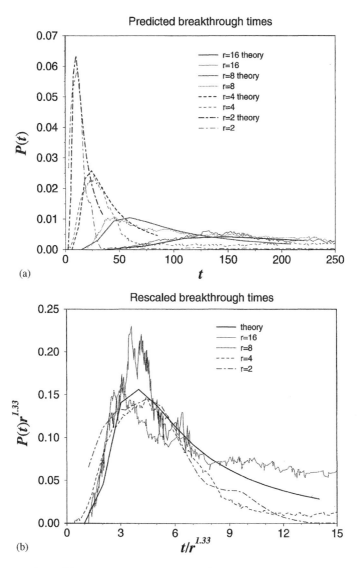

Fig. 3. (a) Distributions of breakthrough times for four different values of the well spacing r. Lines are determined by explicit Monte Carlo calculations, and from the theoretical scaling law of Eq. (1). (b) Rescaled distributions of breakthrough time for the same data, with the exponent value discussed in Section 3.

4. Conclusions

We have applied results obtained earlier for the scaling law [10,11] for breakthrough time distributions for oil field recovery to realistic data from a North Sea field. We have shown that agreement between the theory and the conventional Monte Carlo approach is accurate for engineering purposes and therefore makes it a practical tool for decision making.

Acknowledgements

The authors thank J. Andrade for many helpful discussions, and BP Amoco for financial support and permission to publish this paper.

References

[1] P.R. King, Transport Porous Media 4 (1989) 37.
[2] P.R. King, A.H. Muggeridge, W.G. Price, Transport Porous Media 12 (1993) 237.
[3] B.D. Ripley, Stochastic Simulation, Wiley, New York, 1987.
[4] A.T. Walden, P. Guttorp (Eds.), Statistics in the environmental and earth sciences, Edward Arnold, Paris, 1992.
[5] H. Tjelmeland, Statistics no. 5 report, 1996.
[6] H. Makse, S. Havlin, P.R. King, H.E. Stanley, Nature 386 (1997) 379.
[7] A.R. Champneys, Physica D 62 (1993) 347.
[8] T.J. Harding, N.J. Radcliffe, P.R. King, SPE J. 3 (1998) 99.
[9] P.R. King, in: A.T. Buller, E. Berg, O. Hjelmel, J. Kleppe, O. Torsaeter, J.O. Aasen (Eds.), North Sea Oil and Gas Reservoirs III, Graham and Trotman, London, 1990.
[10] N.V. Dokholyan, Y. Lee, S.V. Buldyrev, S. Havlin, P.R. King, H.E. Stanley, J. Stat. Phys. 93 (1998) 603.
[11] Y. Lee, N.V. Dokholyan, Y. Lee, S.V. Buldyrev, S. Havlin, P.R. King, H.E. Stanley, Phys. Rev. E 60 (1999) 3425–3428.
[12] N.V. Dokholyan, S.V. Buldyrev, S. Havlin, P.R. King, Y. Lee, H.E. Stanley, Physica A 266 (1999) 53.
[13] P.R. King, J. Andrade, N.V. Dokholyan, S.V. Buldyrev, S. Havlin, Y. Lee, H.E. Stanley, Physica A 266 (1999) 107.
[14] N.V. Dokholyan, Y. Lee, S.V. Buldyrev, S. Havlin, P.R. King, H.E. Stanley, preprint, 1999.

ELSEVIER

Physica A 274 (1999) 67–84

www.elsevier.com/locate/physa

Applications of statistical mechanics to non-brownian random motion

Ryszard Kutner*, Krzysztof Wysocki

Institute of Experimental Physics, Warsaw University, Hoża 69, Pl-00681 Warsaw, Poland

Abstract

We analysed discrete and continuous Weierstrass–Mandelbrot representations of the Lévy flights occasionally interrupted by spatial localizations. We chose the *discrete* representation to easily detect by Monte Carlo simulation which stochastic quantity could be a candidate for describing the real processes. We found that the particle propagator is able to reveal surprisingly close, stable long-range algebraic tail. Unfortunately, long flights present in the system make, in practice, the particle mean-square displacement an irregular step-like function; such a behavior was expected since it is an experimental reminiscence of divergence of the mean-square displacement, predicted by the theory. We developed the *continuous* representation in the context of random motion of a particle in an amorphous environment; we established a correspondence between the stochastic quantities of both representations in which the latter quantities contain some material constants. The material constants appear due to the thermal average of the space-dependent stretch exponent which defines the probability of the particle passing a given distance. This averaging was performed for intermediate or even high temperatures, as well as for low or even intermediate internal friction regimes where long but *not* extremely long flights are readily able to construct a significant part of the Lévy distribution. This supplies a kind of self-cut-off of the length of flights. By way of *example*, we considered a possibility of observing the Lévy flights of hydrogen in amorphous low-concentration, high-temperature $Pd_{85}Si_{15}H_{7.5}$ phase; this conclusion is based on the results of a real experiment (Driesen et al., in: Janot et al. (Eds.), Atomic Transport and Defects in Metals by Neutron Scattering, Proceedings in Physics, Vol. 10, Springer, Berlin, 1986, p. 126; Richter et al., Phys. Rev. Lett. 57 (1986) 731; Driesen, Doctoral Thesis, Antwerpen University, 1987), performed by detecting the incoherent quasielastic scattering of thermal neutrons. We emphasize that the observed HWHM $\sim k^{\beta}$, where exponent β is distinctly smaller than 2, could be caused by these long flights of hydrogen. © 1999 Elsevier Science B.V. All rights reserved.

---

* Corresponding author. Fax: +48-22-6287252.
E-mail address: ryszard.kutner@fuw.edu.pl (R. Kutner)

0378-4371/99/$ - see front matter © 1999 Elsevier Science B.V. All rights reserved.
PII: S 0378-4371(99)00313-1

1. Introduction

In the last two decades a great effort was made to find physical and chemical [1–14] geophysical [15] (and refs. therein), ecological [16] (and refs. therein), biophysical, biological [17–19], and medical [20–23] (and refs. therein), economical or financial [4,24–27] (and refs. therein) phenomena undertaking scaling and particularly the Lévy processes. Simultaneously, an effort was made for deeper understanding of the mathematical and physical foundations of these processes. By our paper we would like to contribute more to both the trends as there is no systematic statistical mechanics theory which includes rare events of the Lévy type. [1]

In principle, we try to answer two related questions: (i) which stochastic quantity describing the Lévy flights has a chance to describe real processes and (ii) why is it so? We suppose that to answer these questions we supply arguments why, for example, in *amorphous* materials we can expect Lévy flights. It should be emphasized that our work was stimulated by the result of a real experiment on hydrogen diffusion in amorphous low-concentration, high-temperature $Pd_{85}Si_{15}H_{7.5}$ [31–33] observed by quasielastic, incoherent scattering of thermal neutrons.

Our paper consists of two parts. The first, general and a purely stochastic one (Sections 2 and 3), where we point out, based on the Weierstrass–Mandelbrot representation of the Lévy flights [2,4,7,11,34,35], that there exists a possibility in experimental realization range of parameters where the *propagator* for the Lévy flights exhibits quite closely the well-defined algebraic, long-range tail as distinguished from the mean-square displacement which is an improper quantity here.

In the second part (Section 4) we connect these parameters to stochastic and thermodynamic quantities that describe diffusion in some amorphous materials which explains, for example, the unusual experimental result on diffusion of hydrogen in amorphous $Pd_{85}Si_{15}H_{7.5}$ mentioned above.

2. Waiting-time distributions

We apply in this paper a separable and thermalized version of the continuous-time random walk (CTRW) formalism [36] (and refs. therein). The basic quantity used by the CTRW is the waiting-time distribution (WTD), $\psi(r,t)$, which is the probability density of transition by a single displacement, r, exactly at time t after the former one. Since space and time variables are treated here as stochastic ones the WTD obeys the following normalization condition:

$$\int dr \int_0^\infty dt\, \psi(r,t) = 1 \,. \tag{1}$$

---

[1] The promising Tsallis thermostatistics for non-extensive physical systems, based on the Renyi–Tsallis entropy, [28–30] (and refs. therein) seems to be still at a beginning stage.

For three-dimensional amorphous materials it seems natural to assume the space-time factorization

$$\psi(r, t) = p(r)\phi(t),\tag{2}$$

since random (sometimes long) flights over a disordered energetic landscape cannot depend on the depth of a single trap. The single-jump distribution, $p(r)$, is specified and discussed in the next paragraph, while the (properly normalized) waiting-time distribution in a single trap $\phi(t)$ assumes the Poisson form

$$\phi(t) = \frac{1}{\tau}\exp\left(-\frac{t}{\tau}\right),\tag{3}$$

where τ is treated as the mean residence time of a particle in the average trap (in Section 4.1 we consider the condition under which the Poisson WTD given by (3) could be valid for amorphous material) and $\phi(t)$ is automatically thermalized (cf. [36] and refs. therein). Since the experimental sample used in [31–33] is in thermal equilibrium, thermalization is necessary as the time origin can be chosen arbitrarily.

2.1. Weierstrass–Mandelbrot random flights

Following Montroll and Shlesinger [4] we consider several useful random walk quantities.

2.1.1. Jump distributions

As the first quantity we consider a spherically symmetric jump distribution, $p(r)$, in the form

$$p(r) = \frac{1}{4\pi r^2} p_0(r),\tag{4}$$

which is a generalization of the Pearson–Rayleigh [4] random walk (we used the notation $r = |r|$), where $p_0(r)$ is the transition probability to find the particle, after its single jump, at distance r (independently of the direction of vector r); from Eq. (4) follows the normalization

$$\int dr\, p(r) = \int_0^\infty dr\, p_0(r) = 1.\tag{5}$$

To generate a trajectory possessing a stochastic self-similar structure we choose a spherical part of the jump probability, $p_0(r)$, in the form of the discrete Weierstrass–Mandelbrot function

$$p_0(r) = \left(1 - \frac{1}{N}\right) \sum_{j=0}^{\infty} \frac{1}{(Nb)^j} f\left(\frac{r}{b^j}\right),\tag{6}$$

where normalization obeyed is

$$\int_0^\infty dr\, \frac{1}{b^j} f\left(\frac{r}{b^j}\right) = 1,\tag{7}$$

the f-function is specified and discussed elsewhere (cf. Section 4.2 and Appendix A) but as a simple example we assume in Section 3.1 that $f(x) = \delta(x - 1)$.

Fourier transforming of the jump distribution yields a one-step random walk structure factor

$$\tilde{p}(k) = \int_0^\infty dr \frac{\sin(kr)}{kr} p_0(r) = \left(1 - \frac{1}{N}\right) \sum_{j=0}^\infty \frac{1}{N^j} \int_0^\infty dr \frac{\sin(b^j kr)}{b^j kr} f(r) = g(k),$$

(8)

where Eq. (6) (again with notation $k = |k|$) was used. Moreover, we assume that the integral

$$\int_0^\infty dr \frac{\sin(qr)}{qr} f(r) < \infty$$

(9)

for any real $0 < q < \infty$, which seems to be a reasonable assumption saying, in fact, that on each level j we have to deal separately with usual random motions. The structure factor, $g(k)$, satisfies the inhomogeneous scaling relation

$$g(bk) = Ng(k) - (N - 1) \int_0^\infty dr \frac{\sin(kr)}{kr} f(r),$$

(10)

which directly follows from Eq. (8). Hence, one can expect that $g(k)$ includes a singular term; only this term is explicitly shown below together with the usual, regular ones.

Using expression (A.2) obtained in Appendix A, we can immediately write

$$g(k) \approx 1 - \frac{1 - 1/N}{\ln N} \beta \Gamma(-1 - \beta) \cos\left(\frac{\pi}{2}\beta\right) \langle r^\beta \rangle k^\beta - \frac{1}{6} \frac{1 - 1/N}{1 - b^2/N} \langle r^2 \rangle k^2,$$

(11)

where $\langle r^\beta \rangle = \int_0^\infty dr\, r^\beta f(r)$ (see also definition (A.4) in Appendix A; here, as usual, we simply put $Q(\ln k) = 0$ in Eq. (A.2) besides restricting, the regular part in Eq. (A.2) to two terms only, since the case $k \ll 1$ is considered); as seen, the non-analytical part of $\tilde{p}(k)$ is proportional to k^β and plays a role only for $\beta < 2$. For such a range of β one obtains from Eq. (11), after tedious calculations by using the inverse Fourier transform (cf. again Appendix A), the asymptotic result in a real space:

$$p(r) \approx \frac{1}{4\pi} \frac{1 - 1/N}{\ln N} \beta \frac{\langle r^\beta \rangle}{r^{d+\beta}}$$

(12)

(here Euclidean dimension $d = 3$); hence and by considering (4) one has simply

$$p_0(r) \approx \frac{1 - 1/N}{\ln N} \beta \frac{\langle r^\beta \rangle}{r^{1+\beta}},$$

(13)

which is already independent of the Euclidean dimension.

2.1.2. Propagators

Analogously, by combining for $\beta < 2$ Eqs. (8), (11) and (C.5) one obtains, following the derivations in Appendix B, the asymptotic form of the propagator

$$P(R, t) \approx \frac{1}{4\pi} \frac{1 - 1/N}{\ln N} \beta \frac{\langle r^\beta \rangle}{R^{d+\beta}} \frac{t}{\tau}$$

(14)

and the related form

$$P(R, t) \approx \frac{1 - 1/N}{\ln N} \beta \frac{\langle r^\beta \rangle}{R^{1+\beta}} \frac{t}{\tau},$$

(15)

i.e. the probability density of finding a particle at distance R from the origin at time t (defined at the same initial condition as the propagator). Here $R(=|R|) \rightarrow \infty$ and simultaneously $t \rightarrow \infty$ in such a way that the combined variable $t/R^{1+\beta} \rightarrow 0$. It is important to know how close the long-range tail is visible in the $P(R, t)$. In Section 3 we study this problem by numerical means and in Section 4.2 analytically.

2.1.3. Dynamic structure function

From relations (8), (11) and (C.6), the dynamic structure function or incoherent quasielastic neutron scattering law (called so, for example, in the case of thermal neutrons scattered by hydrogen dissolved in a solid), $S_{\text{inc}}(k, \omega)$, is found for $\beta < 2$ to be the Lorentzian-shaped line

$$S_{\text{inc}}(k, \omega) = \frac{1}{\pi} \frac{D'k^{\beta}}{\omega^2 + [D'k^{\beta}]^2} , \tag{16}$$

where the "fractional diffusion coefficient"

$$D' = \frac{1 - 1/N}{\ln N} \beta \Gamma(-1 - \beta) \cos\left(\frac{\pi}{2}\beta\right) \frac{\langle r^{\beta} \rangle}{\tau} \tag{17}$$

and half-width at half-maximum (HWHM) of the above dynamic structure function is given by

$$HWHM = D'k^{\beta} . \tag{18}$$

This result is considered in Section 5 in the context of the continuous Weierstrass–Mandelbrot representation of the Lévy flights for amorphous materials.

3. Physical window – a numerical recipe and results

Since result (14) is an asymptotic one, we want to find for which values of R and t the long-range tail begins to dominate. We answer this question mainly numerically by performing Monte Carlo simulation according to the algorithm presented below.

3.1. The algorithm

Our algorithm is essentially based on the discrete Weierstrass–Mandelbrot representation (6) of the Lévy flights given in Section 2.1, where we assumed for simplicity that $f(x) = \delta(x - 1)$. Since this algorithm was already considered in our earlier paper [35], we present here only its most important stages.

The basic step of our algorithm is to choose index j with probability $(1 - 1/N)/N^j$. Hence, one can simulate a simple probabilistic game of random casting a two-side coin. As a single success we define the situation where the coin falls on its abverse side with an a priori probability $1/N$, i.e. when the random number $rn < 1/N$. As a single defeat, occurring of course with probability $1 - 1/N$, we define the opposite situation when the coin falls on its reverse side (then $1/N \leqslant rn < 1$). We always cast

Table 1
Correspondence between discrete and continuous representations of the Lévy flights

Discrete representation	Continuous representation
j	$\frac{E}{\Delta}$
N	$\exp(\frac{\Delta}{k_B T})$
Space unit	a
b	$\exp(\zeta\Delta)$
$f(\frac{r}{b^j})$	$\exp(\frac{r}{a\exp(\zeta E)})$

our coin j-times till the first defeat occurs; then the number j of successive successes is just the index we are seeking. Thus, the concrete distance b^j of a flight is calculated; since the direction of the flight is random we draw it in a standard way and then we calculate the current position $R(t)$ of the particle after t draws of index j together with the random orientation of a flight or after time t counted in Monte Carlo steps (MCS).

3.2. Results

The above algorithm allows us to numerically illustrate (for example, in two dimensions) the difference between the Lévy flights and Brownian motions (Fig. 1). Besides, we calculated two quantities: (i) the mean-square displacement of the particle $\langle R(t)^2\rangle$ vs. time t (Fig. 2), and (ii) the statistics $P(R,t)$ vs. distance R for fixed time (Fig. 3), (both, for example, in one dimension).

In Fig. 1 we present by way of example the result of Monte Carlo simulations of Brownian motions and of the Levy flights for (i) $1 < \beta < 2$ and for (ii) $\beta < 1$; note that for both cases theory predicts a divergence of the mean-square displacement. However, for case (i) the mean absolute value of displacement $0 < \langle|R(t)|\rangle < \infty$ in contrast to case (ii) where this quantity also diverges. Therefore, a drastic difference between these three types of random motions is clearly visible.

In Fig. 2 we present the particle mean-square displacement $\langle R(t)^2\rangle$ vs. time t counted in MCS (Monte Carlo steps) for three different statistical ensembles including $M = 10^5, 10^6$ and 10^7 individual realizations of the random flights up to time $t = 6 \times 10^2$ MCS. Here we used, for example, the following basic parameters: $N = 1.158$ and $b = 1.10$ which gives $\beta = (\ln N/\ln b) = 1.54$; it is this value of exponent β that was obtained in the real experiment on hydrogen diffusion in the amorphous, low-concentration and high-temperature $Pd_{85}Si_{15}H_{7.5}$ phase [31–33]. Our choice of parameter N slightly higher than 1 came from the suggestion given in [31–33] that for such high-temperature phase also inequality (27) in Section 4.1 (particularly, $\Delta \ll k_B T$) should be obeyed; this means, according to the correspondence between discrete and continuous representations of the Lévy flights shown in Table 1 (Section 4.2), that $N \equiv \exp(\Delta/k_B T) = 1 + \varepsilon$, where $\varepsilon \ll 1$.

As seen, the mean-square displacement is an irregular step-like function where the increase of the magnitude of the statistical ensemble leads to an increase of the height

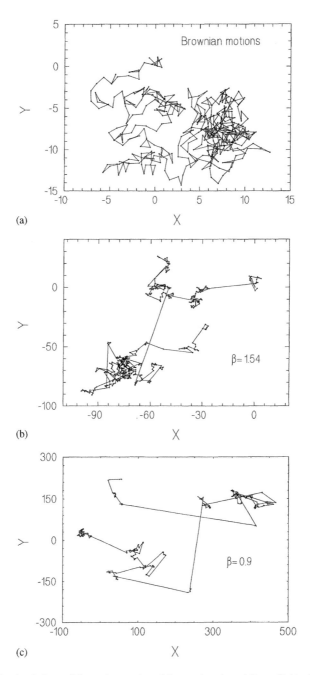

Fig. 1. Monte Carlo simulations of Brownian motions (picture above), and Levy flights for: (i) $1 < \beta < 2$ (picture in the middle) and (ii) $\beta < 1$ (picture below). In all cases flights are denoted by straight intervals and turning points by dots.

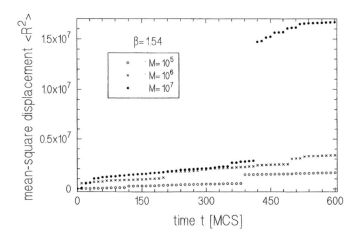

Fig. 2. Mean-square displacement $\langle R(t)^2 \rangle$ vs. time t for three different statistical ensembles including $M = 10^5$, 10^6 and 10^7 individual realizations of random flights up to 6×10^2 MCS (Monte Carlo steps). The following basic parameters: $N = 1.158$ and $b = 1.10$ were used when $\beta = 1.54$.

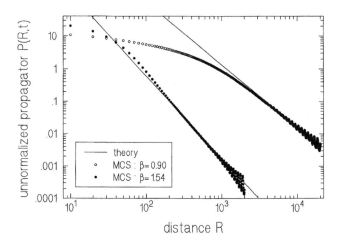

Fig. 3. Comparison of the Monte Carlo simulations (circles) of the statistics $P(R,t)$ with the theoretical predictions (straight lines) for $\beta = 1.54$ (at $N = 1.158$ and $b = 1.10$), $t = 10^2$ MCS, $M = 10^8$ realizations, and for $\beta = 0.90$ (at $N = 1.090$ and $b = 1.10$), $t = 4 \times 10^2$ MCS, $M = 10^6$ realizations.

of these steps. If we were to increase the time the effect would be even enhanced since the longer the time the higher the fault of steps appear. This means that there is no chance to obtain the mean-square displacement as a smooth function of time. Such a behavior was expected since it is an experimental reminiscence of divergence of the mean-square displacement, predicted by theory.

In Fig. 3 we compare the results of our MC simulation of statistics $P(R,t)$ (collected in a histogramic form with discreteness $\Delta R = 10$ much smaller than R in the interesting, intermediate range) with the theoretical prediction given by Eq. (15), for example, for

$\beta=0.90$ (at $N=1.090$ and $b=1.10$), $t=4\times10^2$ MCS, $M=10^6$ realizations; and $\beta=1.54$ (at $N=1.158$ and $b=1.10$), $t=10^2$ MCS, $M=10^8$ realizations. For the intermediate range of R the algebraic, long-range tail is well defined in a numerical experiment, and the agreement between both independent approaches is sufficiently good.

Although in simulations we always deal with finite j (here, the maximal value of j, i.e. $j_{\max}=178$) the influence of long flights on the mean-square displacement of the particle $\langle R(t)^2\rangle$ is well visible in the form of steps. Also, the long-range tail can be easily observed in statistics $P(R,t)$ which is the advantage of our approach, since, from the analytical point of view (within discrete representation) to obtain a singular term in (A.2) by using the Mellin transformation the summation in (8) must be performed up to infinity.

4. Bridge between stochasticity and reality

In this section we consider distributions $\phi(t)$ and $p(r)$ in form which are motivated by a physical picture imposed by random motions in an amorphous material. However, we study the situation where only $p(r)$ (and not $\phi(t)$) becomes a broad distribution; this seems to be a proper approach to describe hydrogen long flights in amorphous $Pd_{85}Si_{15}H_{7.5}$.

4.1. Stretch exponent and Poisson distribution

We find a constraint that makes the mean residence time, τ, of a particle in a trap a finite quantity.

We take the random site energy model with the distribution of site energies in the exponential form [37] (and refs. therein) which is (together with the Gaussian distribution) comprehensively applied to amorphous or glassy systems [38,39],

$$P(E) \approx \frac{1}{\bar{E}} \exp\left(-\frac{E}{\bar{E}}\right) , \tag{19}$$

where it is assumed that the average depth of a trap, $\bar{E}$, determining the width of the distribution, is much smaller than the maximal (energetic) depth of a trap, $E_{\max}^{tp}$, i.e.

$$\bar{E} \ll E_{\max}^{tp} . \tag{20}$$

Besides, we assume as usual the Poisson waiting-time distribution within a given trap of depth E:

$$\phi_E(t) = \frac{1}{\tau(E)} \exp\left(-\frac{t}{\tau(E)}\right). \tag{21}$$

As usual, in this context [36] we treat the process as being thermally activated; so the Hopf–Arrhenius law is valid:

$$\tau(E) = \tau_0 \exp\left(\frac{E}{k_B T}\right) . \tag{22}$$

From the above three relations we find that

$$\tau = \langle t \rangle = \int_0^{E_{\max}^{tp}} dE\, P(E) \int_0^\infty dt\, t\phi_E(t) = \int_0^{E_{\max}^{tp}} dE\, P(E)\tau(E) \approx \frac{\tau_0}{1 - \bar{E}/k_B T}\,, \tag{23}$$

the above result was obtained under two additional assumptions that: (i) Δ, the average spacing between consequent energy levels, obeys inequality

$$\Delta \ll \bar{E}\,, \tag{24}$$

which defines the required quasicontinuous energy spectrum, and (ii)

$$k_B T \ll E_{\max}^{tp}\,. \tag{25}$$

The mean residence time, τ, is a positive finite quantity if and only if the crucial inequality

$$\bar{E} < k_B T \tag{26}$$

is obeyed; this means that the Poisson form of $\phi(t)$ given by (3) is then valid. If inequality (26) is violated (in the sense that $\bar{E} > k_B T$), i.e. that we are passing to the low-temperature regime, then τ is infinite and $\phi(t)$ has no more the Poisson form but exhibits a long-time tail [38].

As seen, inequalities (24) together with (26) and (25) define an intermediate-temperature regime

$$\Delta \ll \bar{E} < k_B T \ll E_{\max}^{tp} \tag{27}$$

which limits our considerations. Relation (27) also says that the traps are too shallow (in average) to force a broad waiting-time distribution $\phi(t)$. Deep traps are (rather) singular and are not able to change this situation. Similar considerations regarding the jump distribution, $p(r)$, give additional constraints consistent with those introduced above.

4.2. Thermal average of the space-dependent stretch exponent

Based on the continuous-space representation, we calculate the spherical part of the jump probability $p_0(r)$; this demonstrates a calculation complementary to that presented in Section 2.1 and provides a possibility of interpreting a real physical situations.

Since typically a translatory motion of the particle is possible when its energy $E > \bar{E}$, one can write the following average:

$$p_0(r) = \int_{\bar{E}}^{E_{\max}} dE\, B(E)w_E(r)\,, \tag{28}$$

with the associated constraint easy to obey:

$$\bar{E} \ll E_{\max}\,, \tag{29}$$

where energy $E_{\max}$ is the maximal one the particle can achieve (in general $E_{\max} \neq E_{\max}^{tp}$), $B(E) = A\exp(-E/k_B T)/k_B T$ is the Boltzmann distribution (normalized within the energy range $\bar{E} < E \leqslant E^{\max}$ which defines a translatory motion) with $A = 1/(\exp(-(\bar{E}/k_B T) - \exp(-E^{\max}/k_B T))$, and function $w_E(r) = \lambda(E) \exp(-\lambda(E)r)$ is the probability density that the particle passes exactly the distance r without being meanwhile caught in a trap; here $\lambda(E)$ is the (spectral) probability density per unit distance that the particle with energy E is caught by a random trap. Again, by a stochastic analysis where a Poisson process is assumed in the energy space (since this process is somehow based on the Boltzmann distribution and the particle scattering by a disorder potential), one finds that

$$\lambda(E) = \exp(-\zeta E)/a \, , \tag{30}$$

where $\zeta\, dE$ is the probability that the particle with energy in the range $(E, E + dE)$ is caught in a trap. Eq. (30) was directly obtained from the relation for change of the probability density, $d\lambda(E) = -\lambda(E)\zeta\, dE$, in the energy space, where ζ is an already slowly varying function of E. One can treat coefficient ζ as a monotonic function of the particle mobility or $1/\zeta$ as a monotonic function of an internal friction in the system.

We suggest furthermore that coefficient a somehow measures the distance between the nearest traps. Hence, one can rewrite (28) as follows:

$$p_0(r) = \int_{\bar{E}}^{E_{\max}} dE \frac{A}{k_B T} \exp\left(-\frac{E}{k_B T}\right) \frac{1}{a} \exp(-\zeta E) \exp\left(-\exp(-\zeta E)\frac{r}{a}\right) , \tag{31}$$

which is a thermal average of the stretched exponent. To recognize the asymptotic (in space) behavior of $p_0(r)$ we change the variable in (31), i.e. $y = \exp(-\zeta E)r/a$, which makes it possible to transform (31) into

$$p_0(r) = A\beta \frac{a^\beta}{r^{1+\beta}} \int_{r/r_2}^{r/r_1} dy \, y^\beta \exp(-y) , \tag{32}$$

where we found that exponent $\beta = 1/k_B T\zeta$, here $r_1 = a \exp(\zeta\bar{E})$ and $r_2 = a \exp(\zeta E_{\max})$. The scaling form of the integral ($\sim F(r/a)$) again emphasizes a self-similar nature of this type of random flights. [2]

To perform the final step of these calculations, we assume that the value of variable y at which the integrand in (32) assumes maximum, i.e. when $y = y_{\max} = \beta$, obeys inequality

$$\frac{r}{r_2} \ll y_{\max} \ll \frac{r}{r_1} , \tag{33}$$

or equivalently

$$\beta r_1 \ll r \ll \beta r_2 ; \tag{34}$$

---

[2] More precisely, the integral in (32) can be represented in two equivalent scaling forms: $F(r/r_1; r_1/r_2)$ or $F(r/r_2; r_2/r_1)$, which emphasizes that we have to deal with lower and upper characteristic limits r_1 and r_2 (cf. inequality (34) for details).

additionally we use the obvious thermodynamic relation

$$k_B T \ll E_{max} . \tag{35}$$

Constraint (34) can be easily fulfilled for the assumed temperature range since it is sufficient to keep E_{max} and r large enough (cf. also inequalities (35) and (37)). It is (34) that defines the range of r in which the Lévy flights can be realized.

For the range of r defined by Eq. (34) expression (32) takes finally the approximate form

$$p_0(r) \approx A\Gamma(1 + \beta)\beta \frac{a^\beta}{r^{1+\beta}} . \tag{36}$$

One deals here with a long-range tail in the distribution $p_0(r)$ when $\beta < 2$, i.e. when

$$\frac{1}{2\zeta} < k_B T , \tag{37}$$

which is a thermodynamic inequality that says that we study a low or at most an intermediate friction regime and therefore long flights of a particle in the system are permitted.

Besides, we assume for self-consistency of our scheme that

$$\Delta \ll \frac{1}{\zeta} . \tag{38}$$

The quantity βr_1 present in (32), is an "ultraviolet" short-distance cut-off; for distance r which is not much greater than this cut-off we have the chance to see the Gauss distribution (the Brownian motion). The next quantity βr_2 defines an "infrared" cut-off which, e.g. protects long flights from the influence of the finite size of the sample. Inequality (34) opens such a range of distance r where the integral in (32) can be already replaced with sufficient approximation by $\Gamma(1+\beta)$ (nevertheless, the lower and upper limits of integral in (32) are larger than zero and smaller than infinity, respectively). Thus for the continuous representation of the Lévy flights the non-analytical term can be seen even before the integration is extended to zero and infinity. This means that, in practice, *long but not extremely long flights readily construct the main part of the Lévy distribution*; this is a kind of practical self-cut-off.

Finally, one can define by direct comparison of Eq. (6) with (31) the correspondence between discrete and continuous representations of the Lévy flights.

The above correspondence leads, of course, to the correspondence between normalization factors, since instead of $1 - 1/N$ in the discrete representation one has factor A in the continuous one. Besides, one has a formal correspondence between other multiplicative factors $\langle r^\beta \rangle / \ln N$ and $a^\beta \Gamma(1 + \beta)$ where the difference between them comes from essentially different methods of extracting the long-range tail (indirect one in the case of the discrete representation and direct method used within the continuous representation).

The indirect analytical method requires summation up to infinity in expression (6) and the related ones since only infinite geometric series (used at intermediate steps (cf. Appendix A)) is able to produce a pole responsible for the singularity. However, our numerical results, which are obtained directly from (6) (cf. Section 3), show that already at a finite index j, whose maximal value is given here by $j_{max} = 178$, the long-range tail is well seen (see Fig. 3). Such a dichotomous situation for discrete representation calls for a more refined mathematical extraction of the singular term.

5. Concluding remarks

The key relation which makes possible the application of the Lévy flights to an amorphous, thermalized system is, in fact, (30) which follows from the assumption that we have to deal with the Poisson process in the energy space since this process is based, in principle, on the Boltzmann distribution. In fact, the Poisson process controls here each physical stage which leads to the thermal average of the space-dependent stretched exponent (cf. (31)) and then to the algebraic long-range tail. This is an analog of the corresponding time-dependent stretched exponent developed for localized motion (cf. [39] and extension of the derivation in Section 4.1 for infinite τ). The analog is possible, since we have to deal with capture induced by the scattering of the particle by a disorder potential landscape, i.e. we have to deal not only with the thermally activated escape but also with a thermally activated capture.

The discrete representation (6) of the Lévy flights is, as proved in Section 3, particularly helpful in Monte Carlo simulations, since in this way one can easily recognize the algebraic, long-range tail for a finite value of index j. This tail is seen for the intermediate range of distance R (cf. Fig. 3) which is confirmed by considerations within the continuous representation of the Lévy flights (cf. inequality (34)).

We can conclude that this work provides an answer to the question why $\beta < 2$ was found in a real experiment on hydrogen diffusion in amorphous low-concentration, high-temperature $Pd_{85}Si_{15}H_{7.5}$ [31–33] observed by quasielastic, incoherent scattering of thermal neutrons. By this experimental method anomalous diffusion of hydrogen in amorphous and in the metallic glass $Ni_{24}Zr_{76}H_8$ was also found [40,41] to be caused, on contrary, by the broad (instead of the Poisson) distribution of $\phi(t)$, however, at a regular jump distribution $p(r)$. It is an open question whether a window of physical parameters exists which makes it possible to observe an anomalous waiting together with the Lévy flights in amorphous or glassy materials. Prior to projecting a real experiment answering to this question a mechanical model, possessing the desired statistical mechanics features, is needed and seems possible [42].

Acknowledgements

The authors are indebted to Rolf Hempelmann from the Universität Saarbrücken and Izabela Sosnowska from the Warsaw University for stimulating discussions. This work is partially supported by the Polish KBN Grant No.2 P03B 059 12 and UM-766/98.

Appendix A

In this section we derive $\tilde{p}(\boldsymbol{k})$ in explicit form as a sum of non-analytic and regular terms by extending the derivation given in [4]. We begin our considerations with the one-step random walk structure factor $g(k)$ given by Eq. (8). To perform the summation over j in (8) we replace $\sin(b^j kr)$ by its Mellin transform so that we can write (after interchanging the sum and the integrals)

$$g(k) = \left(1 - \frac{1}{N}\right) \int_0^\infty dr\, f(r) \frac{1}{2\pi i} \int_{c-i\infty}^{c+i\infty} ds\, k^{-(s+1)} r^{-(s+1)}$$

$$\times \frac{1}{1 - (b^{-(s+1)}/N)} \frac{\pi/2}{\cos(\pi/2/s)\Gamma(1-s)}, \tag{A.1}$$

where $-1 < \Re s < 1$. The integrand has the first-order poles at negative odd integers $s = -(2l+1)$, $l = 0, 1, 2, \ldots$, taken from the denominator $\cos((\pi/2)s)$, and from the denominator $1 - b^{-(s+1)}/N$ at $s = -1 - \beta(1 - 2\pi in/\ln N)$, $n = 0, \pm 1, \pm 2, \ldots$. Translating the contour of integration to $\Re s = -\infty$ and taking into account the residues at the poles enclosed on a real non-positive axis, we obtain for $k < 1$ the formula

$$g(k) = 1 - k^\beta \frac{1 - \frac{1}{N}}{\ln N} \beta \Gamma(-1 - \beta) \cos\left(\frac{\pi}{2}\beta\right) \langle r^\beta \rangle - k^\beta Q(\ln k)$$

$$+ \sum_{l=1}^\infty \frac{(-1)^l}{(2l+1)!} \frac{1 - 1/N}{1 - b^{2l}/N} \langle r^{2l} \rangle k^{2l}, \tag{A.2}$$

where auxiliary quantity

$$Q(\ln k) = \frac{1 - 1/N}{\ln N} \beta \sum_{n=-\infty}^{\infty\prime} \left\langle \exp\left(\beta\left(1 - 2\pi i \frac{n}{\ln N}\right) \ln r\right)\right\rangle \exp\left(-2\pi i \beta \frac{n}{\ln N} \ln k\right)$$

$$\times \Gamma\left(-1 - \beta\left(1 - 2\pi i \frac{n}{\ln N}\right)\right) \cos\left(\frac{\pi}{2}\beta\left(1 - 2\pi i \frac{n}{\ln N}\right)\right), \tag{A.3}$$

where $\sum_{n=-\infty}^{\infty\prime}(\ldots)$ means summation over all integers except $n = 0$. When $\beta \leqslant 1/2$, series (A.3) has to be summed by using a convergence factor. The borderline case $\beta = 2$ is easily analyzed by similar means (cf. [35] and refs. therein). In the derivation of Eqs. (A.2) and (A.3) we assumed that for any complex finite z the absolute value of the moment is finite

$$|\langle r^z \rangle| \left(= \left|\int_0^\infty dr\, r^z f(r)\right|\right) < \infty. \tag{A.4}$$

This is only the reason why function f is called here the regular one.

Appendix B

By using result (11) for $\beta < 2$ we derive asymptotic forms of $p(r)$ and also $p_0(r)$. This derivation concerns the three-dimensional case but it is essentially a straightforward generalization of the one-dimensional calculations presented in [1].

B.1. Derivation for $0 < \beta < 1$

For large r one can write the Fourier transform of (11) in the useful form

$$p(\mathbf{r}) = -\frac{1}{2\pi^2}\frac{1}{r^2}\int_0^\infty dk\, k\,\exp(-D''k^\beta)\frac{d\cos(kr)}{dk}\,, \tag{B.1}$$

where from (11) one has $D'' = [(1 - 1/N)/\ln N]\beta\Gamma(-1-\beta)\cos((\pi/2)\beta)\langle r^\beta\rangle$ (one has from the above and Eq. (17) that $D' = D''/\tau$). This derivation concerns the case where $0 < \beta < 1$; the opposite case is considered in the second part of this appendix.

After performing the above integration by parts (twice) one obtains two terms:

$$p(\mathbf{r}) \approx \frac{1}{2\pi^2}\frac{\beta(1+\beta)D''}{r^{d+\beta}}\int_0^\infty dy\, y^{\beta-1}\sin(y)\exp(-D''(y/r)^\beta)$$

$$+\frac{1}{2\pi^2}\frac{(\beta D'')^2}{r^{d+2\beta}}\int_0^\infty dy\, y^{2\beta-1}\sin(y)\exp(-D''(y/r)^\beta)\,, \tag{B.2}$$

which should be compared. As seen, the second term, which plays the role of a correction to scaling, vanishes quicker with r than the first one by the factor of the order $\sim 1/r^\beta$. Hence, one obtains finally the useful formula

$$p(\mathbf{r}) \approx \frac{1}{4\pi}\frac{1-1/N}{\ln N}\beta\frac{\langle r^\beta\rangle}{r^{d+\beta}}\,, \tag{B.3}$$

which is further used in the derivation of related expressions.

B.2. Derivation for $1 < \beta < 2$

In this case the derivation is slightly more complicated than in the previous one. The starting point is given by the formula

$$p(\mathbf{r}) = \frac{1}{2\pi^2}\frac{1}{r^d}\int_0^\infty dy\,\cos(y)\exp(-D''(y/r)^\beta)$$

$$-\frac{1}{2\pi^2}D''\beta\frac{1}{r^{d+\beta}}\int_0^\infty dy\, y^\beta\cos(y)\exp(-D''(y/r)^\beta)\,, \tag{B.4}$$

which consists of two terms obtained from (B.1) after performing the derivation over k (which is a part of the integrand) and next changing the integration variable. The next step in calculations is based on the following identity:

$$\int_0^\infty dy\, y^\alpha \cos(y)\exp(-D''\xi y^\beta)$$

$$= \int_0^\infty dy\, y^\alpha \cos(y)\exp(-D''\xi y)$$

$$\times\left\{1 - \xi D''(y^\beta - y) + \frac{\xi^2}{2}D''^2(y^\beta - y)^2 + \cdots\right\}dy\,, \tag{B.5}$$

where $\alpha = 0$ or β; finally, only linear terms in $\xi(=1/r^\beta)$ are taken into account in (B.4) (since zero-order terms canceled). After calculations based on the relation

$$\int_0^\infty dy\, y^\nu \cos(y)\exp(-cy) = \frac{\Gamma(\nu+1)}{(c^2+1)^{(\nu+1)/2}}\cos\left((\nu+1)\arctan\left(\frac{1}{c}\right)\right)\,, \tag{B.6}$$

and (B.5), the first term in (B.4), is divided into two parts: (i) the one which already decays asymptotically as $\sim 1/r^{d+\beta}$, and (ii) the second one which supplies asymptotically only the correction to scaling. Hence, one reaches (up to the leading term only) the same final result for $p(r)$ as already given by expression (B.3).

Appendix C

In the frame of the CTRW formalism an auxiliary sojourn probability distribution (SPD), $\Phi(t)$, is introduced such that the particle remains within a trap without any transition till the elapse of time t:

$$\Phi(t) = \int_t^\infty dt'\, \phi(t') = \exp\left(-\frac{t}{\tau}\right) . \tag{C.1}$$

Both WTD, $p(r)\phi(t)$, and SPD, $\Phi(t)$, quantities are helpful to construct the central quantity, i.e. the propagator $P(R,t)$ defined as the conditional probability density to find the particle at site R at time t when the initial condition is specified. Now we can write the following convolution:

$$P(R,t) = \delta(R)\Phi(t) + \int_0^t dt_1\, \phi(t_1)p(R)\Phi(t-t_1)$$

$$+ \int dr_1 \int_0^t dt_2 \int_0^{t_2} dt_1\, \phi(t_1)p(r_1)\phi(t_2-t_1)p(R-r_1)\Phi(t-t_2) + \cdots .$$

$$\tag{C.2}$$

The first term in the above convolution denotes the probability density to reach location R in time t without any flight (after an initial transition). The second term describes the probability density that a particle reaches by a single flight (from the initial location at the origin) site R at time $t_1 (< t)$ and next stays there at least until time t elapses. The third term is richer than the second one since it describes besides the initial localization of the particle at $R=0$ till the elapse of an arbitrary time $t_1 (< t_2)$, the single flight event by displacement r_1 and its waiting until time $t_2 (< t)$ elapses, when the displacement $R - r_1$ occurs the particle occupies site R for time t. Analogously, the higher integral convolutions describe multiple waitings and flights. According to the Fourier–Laplace transform (where $R \to k$, $t \to u$) the above (infinite) sum of convolutions transforms into a infinite geometric series so that we easily obtain in the closed form:

$$\tilde{P}(k,u) = \frac{\tilde{\Phi}(u)}{1 - \tilde{p}(k)\tilde{\phi}(u)} , \tag{C.3}$$

where $\tilde{\Phi}(u)$ $(=1/(u + 1/\tau))$ is the Laplace transform of $\Phi(t)$ given by (C.1). Since $\phi(t)$ has the Poisson form (3), Eq. (C.3) can be easily rewritten in the form

$$\tilde{P}(k,u) = \frac{1}{u + [1 - \tilde{p}(k)]/\tau} \tag{C.4}$$

or in time domain

$$\tilde{P}(\boldsymbol{k},t) = \exp(-[1 - \tilde{p}(\boldsymbol{k})]t/\tau) \, .$$ (C.5)

Hence, dynamic structure function takes the form

$$S_{\text{inc}}(\boldsymbol{k},\omega) = \frac{1}{\pi} \Re \, \tilde{P}(\boldsymbol{k}, u = \mathrm{i}\omega) = \frac{1}{\pi} \frac{[1 - \tilde{p}(\boldsymbol{k})]/\tau}{\omega^2 + \{[1 - \tilde{p}(\boldsymbol{k})]/\tau\}^2} \, ,$$ (C.6)

which must be further elaborated.

References

[1] E.W. Montroll, B.J. West, in: E.W. Montroll, J.L. Lebowitz (Eds.), Fluctuation Phenomena, SSM Vol. VII, North-Holland, Amsterdam, 1979, p. 61.

[2] B.B. Mandelbrot, The Fractal Geometry of Nature, Freeman, San Francisco, 1982.

[3] G.H. Weiss, R.J. Rubin, in: I. Prigogine, S.A. Rice (Eds.), Advances in Chemical Physics, Vol. LII, J. Wiley, New York, 1983, p. 363.

[4] E.W. Montroll, M.F. Shlesinger, in: J.L. Lebowitz, E.W. Montroll (Eds.), Nonequilibrium Phenomena II. From Stochastics to Hydrodynamics, SSM Vol. XI, North-Holland, Amsterdam, 1984, p. 1.

[5] A. Blumen, J. Klafter, G. Zumofen, in: I. Zschokke (Eds.), Optical Spectroscopy of Glasses, Reidel, Dordrecht, 1986.

[6] S. Havlin, D. Ben-Avraham, Adv. Phys. 36 (1987) 695.

[7] T. Vicsek, Fractal Growth Phenomena, World Scientific, Singapore, 1989.

[8] J.-P. Bouchaud, A. Georges, Phys. Rep. 195 (1990) 127.

[9] M.B. Isichenko, Rev. Mod. Phys. 64 (1992) 962.

[10] T.H. Solomon, E.R. Weeks, H. Swinney, Phys. Rev. Lett. 71 (1993) 3975.

[11] J. Klafter, G. Zumofen, M.F. Shlesinger, in: M.F. Shlesinger, G.M. Zaslavsky, U. Frisch (Eds.), Lévy Flights and Related Topics in Physics, Lecture Notes in Physics, Vol. 450, Springer, Berlin 1995, p. 196.

[12] E. Barkai, J. Klafter, in: S. Benkadda, G.M. Zaslavsky (Eds.), Chaos, Kinetics and Nonlinear Dynamics in Fluids and Plasmas, Springer, Berlin, 1997.

[13] R. Metzler, E. Barkai, J. Klafter, Phys. Rev. Lett. 82 (1999) 3563.

[14] A.E. Hansen, E. Schröder, P. Alstrom, J.S. Andersen, M.T. Levinsen, Phys. Rev. Lett. 79 (1997) 1845.

[15] A. Tsinober, in: M.F. Shlesinger, G.M. Zaslavsky, U. Frisch (Eds.), Lévy Flights and Related Topics in Physics, Lecture Notes in Physics, Vol. 450, Springer, Berlin, 1995, p. 3.

[16] T.H. Keitt, H.E. Stanley, Nature 393 (1998) 257.

[17] S.V. Buldyrev, A.L. Goldberger, S. Havlin, C.-K. Peng, M. Simons, H.E. Stanley, Phys. Rev. E 47 (1993) 4514.

[18] R.N. Mantegna, S.V. Buldyrev, A.L. Goldberger, S. Havlin, C.-K. Peng, M. Simons, H.E. Stanley, Phys. Rev. Lett. 73 (1994) 3169.

[19] S.V. Buldyrev, A.L. Goldberger, S. Havlin, R.N. Mantegna, M.E. Matsa, C.-K. Peng, M. Simons, H.E. Stanley, Phys. Rev. E 51 (1995) 5084.

[20] C.-K. Peng, J. Mietus, J. Hausdorff, S. Havlin, H.E. Stanley, A.L. Goldberger, Phys. Rev. Lett. 70 (1993) 1343.

[21] P.Ch. Ivanov, M.G. Rosenblum, C.-K. Peng, J.E. Mietus, S. Havlin, H.E. Stanley, A.L. Goldberger, Physica A 249 (1998) 587.

[22] P.Ch. Ivanov, L.A. Amaral, A.L. Goldberger, H.E. Stanley, Europhys. Lett. 43 (1998) 363.

[23] T.F. Nonnenmacher, G.A. Losa, E.R. Weibl (Eds.), Fractals in Biology and Medicine, Birkhäuser, Basel, 1993.

[24] M.H.R. Stanley, L.A.N. Amaral, S.V. Buldyrev, S. Havlin, H. Leschhorn, P. Maass, M.A. Salinger, G.E. Stamley, Nature 379 (1996) 804.

[25] B.B. Mandelbrot, Fractals and Scaling in Finance: Discontinuity, Concentration, Risk, Springer, Berlin, 1997.

[26] B.B. Mandelbrot, Multifractals and $1/f$ Noise: Wild Self-Affinity in Physics, Springer, Berlin, 1999.

[27] R.N. Mantegna, H.E. Stanley, An Introduction to Econophysics. Correlations and Complexity in Finance, Cambridge, MA, 1999, preprint.
[28] C. Tsallis, A.M.C. de Souza, R. Maynard, in: M.F. Shlesinger, G.M. Zaslavsky, U. Frisch (Eds.), Lévy Flights and Related Topics in Physics, Lecture Notes in Physics, Vol. 450, Springer, Berlin 1995, p. 269.
[29] C. Tsallis, S.V.F. Levy, A.M.C. de Souza, R. Maynard, Phys. Rev. Lett. 75 (1995) 3589. Erratum: Phys. Rev. Lett. 77 (1996) 5442.
[30] C. Tsallis, A.M.C. de Souza, Phys. Lett. A 235 (1997) 444.
[31] G. Driesen, R. Hempelmann, D. Richter, in: C. Janot, W. Petry, D. Richter, T. Springer (Eds.), Atomic Transport and Defects in Metals by Neutron Scattering, Proceedings in Physics, Vol. 10, Springer, Berlin, 1986, p. 126.
[32] D. Richter, G. Driesen, R. Hempelmann, I.S. Anderson, Phys. Rev. Lett. 57 (1986) 731.
[33] G. Driesen, Hydrogen-diffusion mechanism in amorphous $Pd_{1-y}Si_yH_x$. A neutron-scattering study, Doctoral Thesis, Antwerpen University, 1987.
[34] R. Kutner, Physica A 264 (1999) 84.
[35] R. Kutner, M. Regulski, Physica A 264 (1999) 107.
[36] J. Haus, K.W. Kehr, Phys. Rep. 150 (1987) 263.
[37] T. Wichmann, K.G. Wang, K.W. Kehr, J. Phys. A 27 (1994) L263.
[38] C. Monthus, J.-P. Bouchaud, J. Phys. A 29 (1996) 3847.
[39] J.-P. Bouchaud, in: R. Kutner, A. Pękalski, K. Sznajd-Weron (Eds.), Anomalous Diffusion. From Basics to Applications, Lecture Notes in Physics, Vol. 519, Springer, Berlin, 1999, p. 140.
[40] J.B. Suck, H. Rudin, H.U. Künzi, A. Heidemann, in: S. Steeb, H. Warlimont (Eds.), Rapidly Quenched Metals, Elsevier, Amsterdam, 1985, p. 1545.
[41] W. Schirmacher, M. Prem, J.-B. Suck, A. Heidemann, Europhys. Lett. 13 (1990) 523.
[42] S.V. Buldyrev, Private communication.

ELSEVIER

Physica A 274 (1999) 85–90

www.elsevier.com/locate/physa

Population dynamics and Burgers' equation

David R. Nelson

Lyman Laboratory of Physics, Harvard University, Cambridge, MA 02138, USA

Abstract

A brief review of recent results [D.R. Nelson, N.M. Shnerb, Phys. Rev. E 58 (1998) 1383] connecting population dynamics in the presence of disorder and convection with a noisy Burgers' equation is given. In the limit of high convection velocity, linearized population growth in dimensions is described by a $(d-1)$-dimensional Burgers' equation with random noise. The Burgers mapping leads to universal results for the spreading and transverse profile of populations drifting downstream through a disordered medium with spatially varying growth rates. © 1999 Published by Elsevier Science B.V. All rights reserved.

Bacterial growth in a petri dish has long been the basic experiment of microbiology. Although much has been learned by this technique, most bacteria of course do not live in petri dishes, but rather in inhomogeneous environments characterized by, e.g., spatially varying growth rates. Often as in the soil after a rain storm, bacterial diffusion and growth are accompanied by convective drift in an aqueous medium through the disorder.

A delocalization transition in inhomogeneous biological systems has recently been proposed, focusing on a single species continuous growth model, in which the population disperses via diffusion and convection [1]. The Fisher equation [2] for the population number density $c(x, t)$ generalized to account for convection and a quenched-random distribution of growth rates reads [1]

$$\frac{\partial c(r,t)}{\partial t} + (v \cdot \nabla)c(r,t) = D\nabla^2 c(r,t) + ac(r,t) + U(r)c(r,t) - bc^2(r,t), \qquad (1)$$

where D is the diffusion constant, v is a spatially homogeneous time-independent convection velocity, and b is a phenomenological parameter responsible for limiting the local concentration to some maximum value. The parameter a is the average growth

E-mail address: nelson@cmts.harvard.edu (D.R. Nelson)

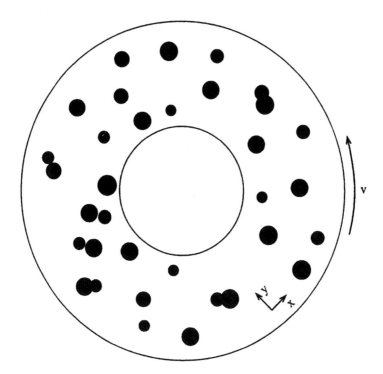

Fig. 1. Schematic of a disordered biological substrate subject to diffusion and convection. The dark spots represent environmental fluctuations, as exemplified by an inhomogeneous pattern of light projected onto a growing population of bacteria. A convective flow with velocity v around the annulus could be simulated by moving the pattern of light.

rate in the region, and $U(r)$ describes quenched-random fluctuations in the growth rate about this average value. It is convenient, but not essential, to imagine that the growth and flow take place with periodic boundary conditions in the streamwise direction, as indicated in Fig. 1.

If $v = 0$ and the population is small, we can linearize and investigate its growth or decay via

$$\frac{\partial c(r,t)}{\partial t} = D\nabla^2 c(r,t) + [a + U(r)]c(r,t).$$ (2)

Eq. (2) is just the single-particle Schrödinger equation (in imaginary time) familiar from nonrelativistic quantum mechanics. (For a discussion of the spread of an initially localized population based on this mapping, see Ref. [3], Section 5.6.1, and references therein.) The random growth rates at each position r are assumed to be independent. $U(r)$ is thus a zero mean random variable with variance

$$\langle U(r)U(r')\rangle = 2\Delta\delta(r - r').$$ (3)

In one and two dimensions, it is expected that all eigenfunctions $\phi_n(r)$ of the operator on the right-hand side of (2) are localized. The initial stages of growth can be described by expanding in these localized states,

$$c(r,t) \sum_n c_n \phi_n(r) e^{\Lambda_n t} \tag{4}$$

with the $\{c_n\}$ determined by the initial conditions. Because the linearized growth operator is Hermitian, all eigenvalues $\{\Lambda_n\}$ are real.

When v is nonzero, the growth operator is non-Hermitian and the problem becomes equivalent to the quantum mechanics of a particle in a constant imaginary vector potential [4,5]. The localized states discussed above typically delocalize two at a time with increasing v: Delocalization is accompanied by escape of two eigenvalues from the real axis into the complex plane, where they from a complex conjugate pair [6]. The most well-localized states (corresponding to the most rapidly increasing growth modes) survive the longest, but for large enough v it is expected that *all* states will eventually delocalize [1,4,5]. At intermediate values of v, there is a mobility edge separating localized states from delocalized ones in one dimension. In high dimensions, the mobility edge separates purely localized states from a mixture of localized and delocalized eigenfunctions.

Interesting issues arise in the large v limit for dimension 2 or greater. For large convection velocities and delocalized eigenfunctions, one might have expected the disorder to average itself out in some sense. The eigenfunctions would then be approximately plane waves of wave vector k, and with an eigenvalue spectrum appropriate to diffusion with drift. For a drift velocity in the x-direction, the spectrum near the top of the band (i.e., for the most rapidly growing modes) would take the form

$$\Lambda(k) \approx a - iv_R k_x - D_R k_x^2 - D'_R k_\perp^2 + \mathcal{O}(k^4), \tag{5}$$

where v_R and D_R and D'_R are an effective drift velocity and diffusion constant renormalized by the disorder, and $k_\perp$ is the wave-vector component perpendicular to $\hat{x}$. This is indeed what happens in one dimension [1] where the perpendicular degrees of freedom are absent. In higher dimensions, however, the disorder produces singular corrections to the growth profile perpendicular to the flow, and Eq. (5) is incorrect.

The anomalous disorder profile follows from mapping the long time large distance dynamics onto the physics of a noisy Burgers' equation [7]. To see how this mapping comes about, we rewrite Eq. (1) (with $b = 0$) for a velocity along the x-direction,

$$\partial_t c + v \partial_x c = D \partial_x^2 c + D \nabla_\perp^2 c + [a + U(x, r_\perp)]c, \tag{6}$$

where $r_\perp$ represents all spatial coordinates perpendicular to x. Specifically, we explore the long-time dynamics generated by this equation with a delta function initial condition,

$$\lim_{t \to 0} c(x, r_\perp, t) = \delta(x)\delta^{d-1}(r_\perp), \tag{7}$$

corresponding to a point inoculation of population at the origin. We look for a delocalized solution valid for long times in the limit of large v. The overall exponential time dependence generated by a, the constant part of the growth rate, and the diffusion with drift we expect in the (x, t) variables may be incorporated via the substitution.

$$c(x,r_\perp,t) = \frac{e^{at}}{\sqrt{4\pi Dt}} e^{-(x-vt)^2/4Dt} W(x,r_\perp), \tag{8}$$

where the function $W(x,r_\perp)$ is to be determined. Note that $c(x,r_\perp,t)$ becomes proportional to $\delta(x)$ as $t \to 0$, so the initial condition (7) requires

$$\lim_{x\to 0} W(x,r_\perp) = \delta^{d-1}(r_\perp). \tag{9}$$

Upon inserting Eq. (8) in (6), we find

$$v\partial_x W(x,r_\perp) + \frac{(x-vt)}{t}\partial_x W(x,r_\perp)$$
$$= D\partial_x^2 W(x,r_\perp) + D\nabla_\perp^2 W(x,r_\perp) + U(x,r_\perp)W(x,r_\perp). \tag{10}$$

According to the ansatz (8), $c(x,r_\perp,t)$ is only appreciable for

$$|x - vt| \lesssim 2\sqrt{Dt} \tag{11}$$

so the second term on the left-hand side of Eq. (10) is smaller than the first by a factor of order $\sqrt{D/tv^2}$, and can be neglected in the limit of long times. In the remaining equation for $W(x,r_\perp)$, which has no explicit time dependence, we expect that the term $\partial_x^2 W(x,r_\perp)$ can be neglected for large x and t compared to the single x derivative, which appears on the left-hand side. The result is an imaginary time Schrödinger equation, where x plays the role of "time", namely

$$v\partial_x W(x,r_\perp) \underset{t,x\to\infty}{\approx} D\nabla_\perp^2 W(x,r_\perp) + U(x,r_\perp)W(x,r_\perp). \tag{12}$$

Note that the random "potential" $U(x,r_\perp)$ depends both on the "time" x and on the additional $(d-1)$ spatial directions.

Eq. (12) is a "Schrödinger equation" in the time-variable x, but with a space- and time-dependent random potential. The further substitution (similar to the WKB method of quantum mechanics),

$$W(x,r_\perp) = e^{-\Phi(x,r_\perp)}, \tag{13}$$

leads to

$$\partial_x \Phi + \frac{D}{v}(\nabla_\perp \Phi)^2 = \frac{D}{v}\nabla_\perp^2 \Phi + \frac{1}{v}U(x,r_\perp). \tag{14}$$

Eq. (14) is just the $(d-1)$-dimensional Burgers model of turbulence with random forcing studied by Forster et al. [7], where the turbulent velocity field is given by a "velocity potential", $\Phi(r,t)$, i.e., it is proportional to $\nabla\Phi(x,r_\perp)$.

Consider the application of this mapping to two-dimensional species populations with strong convection. Assume for simplicity $a > 0$, so that the population grows on average as it convects and diffuses downstrream. For fixed x, the solution $W(x,y)$ of the resulting $(1+1)$-dimensional Burgers' equation describes the distribution in y of a growing species population that has traveled through random distribution of growth rates for a time of order $t = x/v$. The results of extensive studies of the Burgers problem [3,7] may be interpreted as follows: For any fixed y value, imagine tracing the genealogy of, say, all bacteria that have reached a particular position (x,y). As

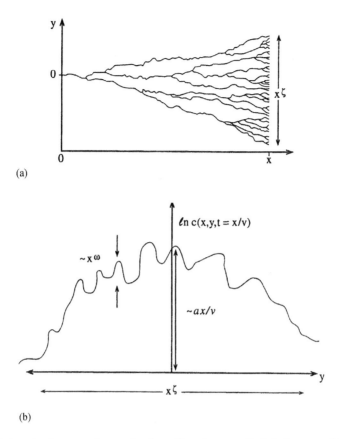

Fig. 2. (a) Trajectories for a growing species that will produce a particularly large population at the point (x, y) at time of order $t = x/v$. The population associated with a given point is spread over a region with typical streamwise size $\sqrt{Dx/v}$. For large x, bacteria that have traveled along such a path of favorable growth rates will dominate the population at (x, y). Amplification along such exceptional paths leads to an anomalous transverse spreading with $y \sim x^\zeta$, $\zeta > \frac{1}{2}$. (b) Schematic of the logarithm of the population $c(x, y, t = x/v)$ discussed above as a function of y. Typical fluctuations in population size away from simple exponential growth are of order $\exp[c'x^\omega]$, and the population from an initial source of point inoculation has spread out a distance of order x^ζ.

$x \to \infty$, the overwhelming majority of bacteria near the point (x, y) will have evolved along a spatially convoluted optimal path of especially favorable growth rates. The fraction of bacteria whose ancestors came along this route is favored over all other routes in $W(x, y)$ by an exponential factor $\sim \exp[c'x^\omega]$, where c' is a constant. The universal exponent ω (which describes the fluctuations in the ground state energy in the analogous problem in the statistical mechanics of flux lines) is known to be $\omega = \frac{1}{3}$ exactly [3].

Any particular path $y(x)$ of optimal evolution itself wanders with typical transverse fluctuations which behave like

$$y(x) \sim x^\zeta, \tag{15}$$

where $\zeta = (\omega + 1)/2 = 2/3$ is a universal critical exponent, independent of the exact probability distribution of the randomness and other details [7]. Because $\frac{2}{3} > \frac{1}{2}$, this optimal path will be well defined even in the presence of diffusion. A schematic of a set of optimal paths is shown in Fig. 2(a). The population distribution $c(x, y, t = x/v)$ is sketched in Fig. 2(b). The exponent ζ also controls the overall transverse spread in y of the spatially varying population for fixed x. This "superdiffusive" spreading ($\zeta > \frac{1}{2}$) arises because "outlying" trajectories that would be rare in conventional diffusion lead to strong amplification if they pass through regions of particularly favored growth. The exponent ω determines the *size* of the fluctuations in $c(x, y, t = x/v)$, which ride on top of an overall exponential growth, $c(x, y, t = x/v) \sim e^{ax/v}$.

Similar behavior is expected for the $(2 + 1)$-dimensional forced Burgers' equation, which results for convecting populations with randomness in $d = 3$, with the universal exponents $\zeta \approx 0.59 \approx 3/5$ and $\omega = 2\zeta - 1 \approx \frac{1}{5}$ [3]. The growing population again becomes streaked out in a streamwise direction, but with a nontrivial wandering transverse to the stream.

See Ref. [1] for a more extensive account of the ideas sketched here, which were developed in collaboration with Nadav Shnerb. For a detailed discussion of how convection forces eigenfunctions to delocalize near an "oasis" of positive growth rates surrounded by a desert, see Ref. [8]. For a more extensive review of convective localization theory and the spectra of non-Hermitian matrices in the context of population biology, see Ref. [9].

It is pleasure to thank Karin Dahmen and Nadav Shnerb for collaborations and discussion on the subject matter contained in this brief review. This work was supported by the National Science Foundation, primarily by the Harvard Materials Research Science and Engineering Center through Grant No. DMR98-09363, and through Grant No. DMR97-14725.

References

[1] D.R. Nelson, N.M. Shnerb, Phys. Rev. E 58 (1998) 1383.
[2] J.D. Murray, Mathematical Biology, Springer, New York, 1993 (Chapter 11).
[3] T. Halpin-Healy, Y. Zhang, Phys. Rep. 254 (1995) 215.
[4] N. Hatano, D.R. Nelson, Phys. Rev. B 56 (1997) 8651.
[5] N. Hatano, D.R. Nelson, Phys. Rev. B 58 (1998) 8384.
[6] N. Shnerb, D.R. Nelson, Phys. Rev. Lett. 80 (1998) 5172.
[7] D. Forster, D.R. Nelson, M. Stephen, Phys. Rev. A 16 (1977) 732.
[8] K. Dahmen, D.R. Nelson, N. Shnerb, Life and death near a windy oasis, J. Math. Biol. con-mat/9807394, in press.
[9] K. Dahmen, D.R. Nelson, N. Shnerb, Population dynamics and non-Hermitian localization, cond-mat/9903276, The Statistical Mechanics of Biocomplexity, The Proceedings of the XV Sitges Conference, June 8–12, 1998, in press.

ELSEVIER

Physica A 274 (1999) 91–98

www.elsevier.com/locate/physa

Statistical physics model of an evolving population

K. Sznajd-Weron, A. Pękalski [*]

Institute of Theoretical Physics, University of Wrocław, pl. Maxa Borna 9, 50-204 Wrocław, Poland

Abstract

There are many possible approaches by a theoretical physicist to problems of biological evolution. Some focus on physically interesting features, like the self-organized criticality (P. Bak, K. Sneppen, Phys. Rev. Lett 71 (1993); N. Vadewalle, M. Ausloos, Physica D 90 (1996) 262). Others put on more effort taking into account factors considered by biologists to be important in determining one or another aspect of biological evolution (D. Derrida, P.G. Higgs, J. Phys. A 24 (1991) L985; I. Mróz, A. Pękalski, K. Sznajd-Weron, Phys. Rev. Lett. 76 (1996) 3025; A. Pękalski, Physica A 265 (1999) 255). The intrinsic complexity of the problem enforces nevertheless drastic simplifications. Certain consolation may come from the fact that the mathematical models used by biologists themselves are quite often even more "coarse grained". © 1999 Elsevier Science B.V. All rights reserved.

In many approaches by physicists, the entities which form the investigated systems are the individuals, characterized either by their genotypes [1–6] or by phenotypes [7,8]. The factors influencing the fate of the populations of such individuals may be natural selection [1–6] or mutations [9,10].

It is however possible to consider a system of many populations (a *metapopulation*). In general each population (a *deme*) should be described by many parameters referring to particular features of the individuals forming the population. For the sake of clarity we have decided here to characterize a deme by the average value of a chosen quantitative (continuous) character, such as body size, color or rate of metabolism.

In sexual populations, the alleles (possible forms of genes) of all individuals form a common genetic pool from which, via sexual reproduction, arbitrary combinations of the alleles are possible. This is the principal source of the life diversity in sexually reproducing populations. Migrations of individuals among the populations bring new

---

[*] Corresponding author. Fax: +48-71-21-44-54.

E-mail address: jpe@ii.uni.wroc.pl (A. Pękalski)

0378-4371/99/$ - see front matter © 1999 Elsevier Science B.V. All rights reserved.
PII: S 0378-4371(99)00316-7

genes into each of the populations, thus changing particular genetic pools. The process is known as *gene flow*. The higher is the gene flow, the more similar are the demes, hence more homogeneous is the metapopulation.

As the result of the gene flow, the allele frequency in the demes changes with time causing changes also in deme's traits.

In the following we shall investigate the evolution, i.e., the dynamics of such a metapopulation. In particular we shall determine

(i) the fate of two, initially separated populations, living in a single, unchanging, environment,

(ii) the fate of two groups of populations, each living in a different habitat, e.g. two mice colonies — one located in the field and the second in the barn and we show how the contacts between them influence the genetic similarity of the two groups,

(iii) the effect of climatic changes on the metapopulation.

Our system is a group of N local populations, or demes, forming one large metapopulation. A deme, located at the site i of the square lattice, is characterized by a number $a_i \in (0,1)$, representing the average value of the investigated character in the population. The distribution of the values a_i in the metapopulation is Gaussian.

As we have mentioned earlier there is always some gene flow between neigboring populations, which can be described as a chemical diffusion of the character [11]. Moreover, the demes interact via competition or predator–prey relationship. These two types of interactions depend on the difference between demes' characters. Finally the characters may be modified by random environmental changes. Therefore we propose a simple nonlinear evolution equation [4–6]

$$a_i(t+1) = a_i(t) + J_1 \sum_{s=1}^{z} [a_i(t) - a_{i+s}(t)]$$
$$+ J_2(t) \sum_{s} \sum_{r>s} [a_i(t) - a_{i+s}(t)][a_i(t) - a_{i+r}(t)] + h_i(t), \qquad (1)$$

where t is the (discrete) time, J_1 and J_2 are dimensionless constants (parameters of the model), describing interactions between the populations.

The last term in Eqs. (2) and (3), $h_i(t)$, describes the temporal changes of the habitat. We assume, as expected in paleoclimatology [12], that the changes are given by a normal distribution with zero mean. The variance, $\sigma(t)$, of the changes, is chosen from the long-term (150 000 yr) data for the temperature deduced from the Antarctic ice [13]. Hence,

$$h_i(t) \approx N(0, \sigma(t)). \qquad (2)$$

From the temperature $T(t)$ data (Fig. 1a) we have constructed the data corresponding to the step-by-step changes of the temperature, $\Delta T(t)$, see Fig. 1b. Then

$$\sigma(t) = s\Delta T(t), \qquad (3)$$

where s is a scaling factor, regulating the pace of the process. In the simulations we took $s = 0.2$.

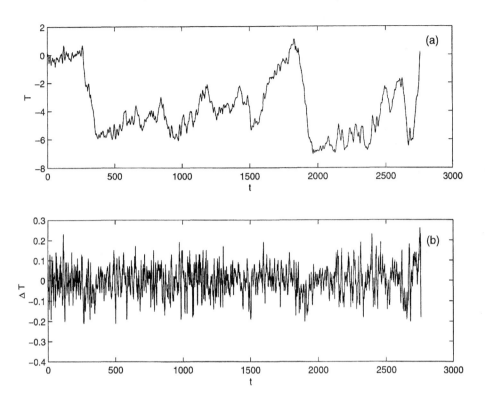

Fig. 1. (a) Temperature T as a function of time t (both in arbitrary units) from the data collected in the Antarctic ice [13] and covering 150 000 yr. (b) Change of the temperature, ΔT calculated from the data of Fig. 1a.

We use the standard Monte Carlo simulations. Our algorithm is as follows:

(i) From a pool of N demes one is picked up at random,

(ii) Its new value of a_i is calculated from Eq. (3). If it lies within the $(0,1)$ range, it is accepted, otherwise the deme is removed from the system and the site (representing here an ecological niche) is repopulated with an arbitrary value of a_i.

(iii) After repeating the above steps N times (the number of demes), one Monte Carlo step (MCS) has been completed.

The quantities of interest, which have been recorded in our simulations, are average value, $A(t)$, of the considered character

$$A(t) = \frac{1}{N} \sum_{i=1}^{N} a_i(t), \qquad (4)$$

the reduced number, $D(t)$, of extinct populations at time t

$$D(t) = \frac{1}{N} \sum_{i=1}^{N} d_i(t), \qquad (5)$$

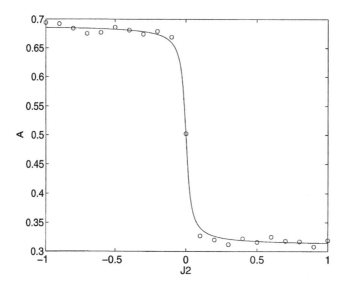

Fig. 2. Mean value of character A as a function of J_2 for $J_1 = 0$.

where $d_i(t)$ is zero if at time t the ith population is still living, and it is equal to one if the population died out at time t, $S(t)$-variance of $A(t)$

$$S(t) = \frac{1}{N} \sum_{i=1}^{N} [A - a_i(t)]^2 \,. \tag{6}$$

To describe the role played by both interaction terms, let us remark that a negative J_1 parameter favors similar values of a_i for neighboring populations, hence leads to the uniformity of the metapopulation. This clearly corresponds to the effect of the gene flow. Positive values of J_1 are biologically uninteresting, since such interactions lead to random values of a_i. As for the second interaction term, let us consider a metapopulation where $J_1 = 0$ and $J_2 \in [-1, 1]$. As seen in Fig. 2, for negative values of J_2 the dynamic rule leads to a metapopulation characterized, after some 1000 MCS, by $A \sim 0.7$, while for positive J_2, the asymptotic value of A is about 0.3. Only when $J_2 = 0$ we have $A = 0.5$. We may therefore regard the non-linear term in Eq. (3), characterized by the J_2 parameter, as representing the Darwinian selection, which may favor either an average value of the character (*stabilizing selection*, $J_2 = 0$), or out of average values (*directional selection*, $J_2 \neq 0$).

First model

Let us first consider the case of a single, constant environment, i.e. $\sigma(t) = 0$ and J_2 having the same value everywhere on the lattice. This simple model will allow us to set the basis for the next modifications. Simulations using the dynamic rule (Eq. (3)) produced the results shown in Figs. 3 and 4. They show the existence of three different types of behavior of the metapopulation.

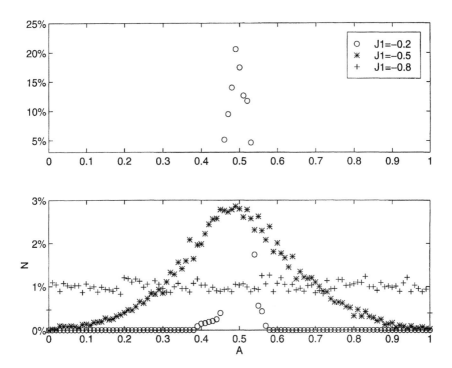

Fig. 3. Distribution, $P(A)$, of the quantitative character A.

In the first one the asymptotic value of the variance of the character, $S(t \to \infty)$ equals zero. This means that the metapopulation reached a static, homogeneous state. Nearly all populations have the same value of the character and there is just one genetic pool. In Fig. 3 it corresponds to $J_1 = -0.2$. For longer times the distribution of the character approaches a delta function.

In the second region the asymptotic variance stabilizes at a certain, smaller than the initial one, value. This means that the character has no longer, as before, a single value, but it is distributed inside an interval. In this case the system is dynamically stable. The shape of the distribution is exemplified by the second curve ($J_1 = -0.5$) in Fig. 3.

Finally, there is a third region, where the asymptotic value of the variance is the same as the initial one, indicating that nothing important happened during the simulation and the populations have kept the initial values of the character. In Fig. 3 it corresponds to the flat curve ($J_1 = -0.8$). We put $J_2 = 0$ for each plot.

This kind of behavior is corroborated by the time dependence of the fraction D of the population which becomes extinct during one interaction (Fig. 4).

We have thus shown that, depending on the strength of the gene flow, the fate of the metapopulation may be one of the following:
(i) If the gene flow is strong, a complete union of all populations sharing a common genetic pool. All previous differences vanish.

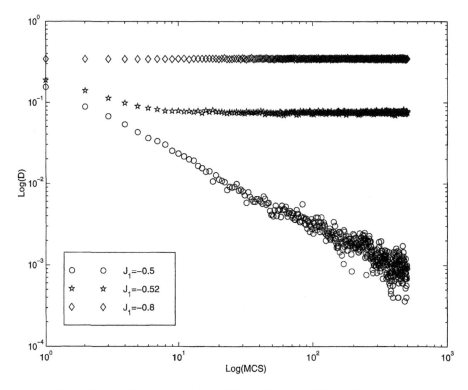

Fig. 4. Number of extinct populations $D(t)$ as a function of time on log–log scale.

(ii) For moderate gene flow there is a partial mixing of the populations, which retain their individual character,

(iii) For weak gene flow there is practically no mixing of the populations, which remain isolated.

The existence of just three such scenarios was also found in nature [14].

Second model

The initial model can be made more complex by considering two groups of populations, each living in a different habitat and due to the natural selection, having different values of the considered character. During the whole evolution process the two environments will remain unchanged, therefore each group will have to adjust to a different habitat. It is clearly seen from the results of our simulations shown in Fig. 5, that if there is a strong exchange of genes between the two groups, even large initial differences in the populations could be eliminated. In another words, for the whole metapopulation the homogenizing effect of the gene flow wins over the differentiating tendency of the natural selection. For weaker gene flow, a hybrid zone may appear. In this region the individuals are neither well adapted to the first nor to the second region. If the flow is very weak, there is no mixing and the populations remain completely different.

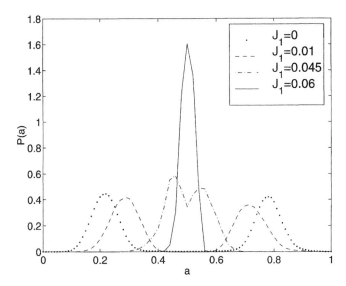

Fig. 5. Distribution, $P(a)$, of the character a. The curves are parametrized by the values of the J_1 parameter. The dash line corresponds to $J_1 = 0$ (complete isolation of the two groups of populations – no gene flow), the dotted line corresponds to $J_1 = 0.01$ (weak gene flow), the dash-dot line – to a stronger gene flow ($J_1 = 0.045$). A hybrid zone at $a = 0.5$ appears but the two groups still retain their individual features. The solid line ($J_1 = 0.06$) represents the case of nearly total unification of the two groups of demes. In both the majority of the individuals have the average (~ 0.5) value of the character.

Third model

Finally let us consider the metapopulation in a changing habitat. Here again, depending on the strength of the gene flow, three asymptotic scenarios are possible. Moreover, for the stabilizing selection, our results (Fig. 6) show that large changes of the character and increased extinctions are not necessarily caused by significant changes in the environment. It means that an increased extinction may not necessarily follow from an abrupt global change of the climate caused, e.g., by a single event, like a meteorite hitting the Earth. Comparison of the available paleoclimatic data [12] to the moments of great extinctions shows indeed that the relation between these two is not straightforward. It is possible in our model that the combined effect of a correlated series of climatic changes and of interactions among the populations (gene flow and natural selection) leads to significant alterations of the metapopulation. We have found out that both — interactions between populations and correlations in the changes of the environments are the necessary ingredients needed to produce complex dependence between changes of the habitat and extinctions.

The situation in which only cataclysms may provoke largely increased rate of extinctions, exists in our model only for non-interacting populations. It should be however stressed that big-scale changes in the environment (cataclysms) always produce significant extinctions.

This lecture is based on the series of papers published recently by us in Physica A [4–6].

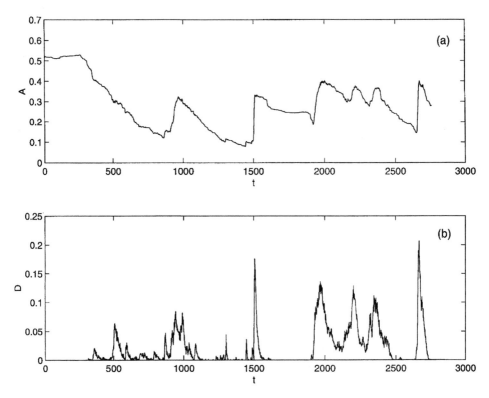

Fig. 6. The effect of the changes of the environment on a) average value of the character $A(t)$, b) extinction rate $D(t)$. Initially the selection was stabilizing, i.e. $J_2 = 0$. The value of σ is taken from Fig. 1b.

References

[1] D. Derrida, P.G. Higgs, J. Phys. A 24 (1991) L985.
[2] I. Mróz, A. Pękalski, K. Sznajd-Weron, Phys. Rev. Lett. 76 (1996) 3025.
[3] A. Pękalski, Physica A 265 (1999) 255.
[4] K. Sznajd-Weron, A. Pękalski, Physica A 252 (1998) 336.
[5] K. Sznajd-Weron, A. Pękalski, Physica A 259 (1998) 457.
[6] K. Sznajd-Weron, A. Pękalski, Physica A 269 (1999) 527.
[7] A.T. Bernardes, in Annual Reviews of Computational Physics, vol. IV, D. Stauffer (Ed.), World Scientific Publ., Singapore, 1996.
[8] K. Malarz, D. Tiggemann, Int. J. Mod. Phys. C 9 (1998) 481.
[9] T.J.P. Penna, J. Stat. Phys. 78 (1995) 1629.
[10] S.M. de Oliveira, P.M.C. de Oliveira and D. Stauffer, Evolution, Money, War and Computers, Teubner, Stuttgart-Leipzig, 1999.
[11] T. Hamilton, Process and Pattern in Evolution, The Macmillan Company, New York, 1967.
[12] R.S. Bradley, Quaternary Paloeclimatology, Chapman and Hall, London, 1994.
[13] J.R. Petit, L. Mounier, J. Jozuel, Y. Korotkievitch, V. Kotlyakov, C. Lovins, Nature 343 (1990) 56.
[14] E. Mayr, Populations, Species, and Evolution, Harvard University Press, Cambridge, Massachusetts, 1970.

ELSEVIER

Physica A 274 (1999) 99–110

www.elsevier.com/locate/physa

Application of statistical physics to heartbeat diagnosis

S. Havlin[a,b,*], L.A.N. Amaral[b], Y. Ashkenazy[a], A.L. Goldberger[c],
P.Ch. Ivanov[b,c], C.-K. Peng[b,c], H.E. Stanley[b]

[a] *Gonda-Goldschmied Center, Department of Physics, Bar-Ilan University, Ramat-Gan 52900, Israel*
[b] *Center for Polymer Studies and Department of Physics, Boston University, Boston, MA 02215, USA*
[c] *Cardiovascular Division, Harvard Medical School, Beth Israel Hospital, Boston, MA 02215, USA*

Abstract

We present several recent studies based on statistical physics concepts that can be used as diagnostic tools for heart failure. We describe the scaling exponent characterizing the long-range correlations in heartbeat time series as well as the multifractal features recently discovered in heartbeat rhythm. It is found that both features, the long-range correlations and the multifractility, are weaker in cases of heart failure. © 1999 Elsevier Science B.V. All rights reserved.

1. The human heartbeat

It is common to describe the normal electrical activity of the heart as "regular sinus rhythm". However, cardiac interbeat intervals fluctuate in a complex, apparently erratic manner in healthy subjects even at rest. Analysis of heart rate variability has focused primarily on short time oscillations associated with breathing (0.15–0.40 Hz) and blood pressure control (~ 0.1 Hz) [1]. Fourier analysis of longer heart-rate sets from healthy individuals typically reveals a $1/f$-like spectrum for frequencies < 0.1 Hz [2–7].

Peng et al. [8–10] studied scale-invariant properties of the human heartbeat time series. The analysis is based on beat-to-beat heart rate fluctuations over very long time intervals (up to 24 h $\approx 10^5$ beats) recorded with an ambulatory monitor.[1] The time

---

* Correspondence address: Gonda-Goldschmied Center, Department of Physics, Bar-Ilan University, Ramat-Gan 52900, Israel. Fax: +972-3-535-3298.

E-mail address: havlin@phys9.ph.biu.ac.il (S. Havlin)

[1] Heart Failure Database (Beth Israel Deaconess Medical Center, Boston, MA). The database includes 18 healthy subjects (13 female and 5 male, with ages between 20 and 50, average 34.3 yr), and 12 congestive heart failure subjects (3 female and 9 male, with ages between 22 and 71, average 60.8 yr) in sinus rhythm.

series obtained by plotting the sequential intervals between beat n and beat $n + 1$, denoted by $B(n)$, typically reveals a complex type of variability. This variability is related to competing neuroautonomic inputs.

To study these dynamics over large time scales, the time series is passed through a digital filter that removes fluctuations of frequencies > 0.005 beat^{-1}. We plot the result, denoted by $B_L(n)$, in Fig. 1. One observes a more complex pattern of fluctuations for a representative healthy adult (Fig. 1a) compared to the "smoother" pattern of interbeat intervals for a subject with severe heart disease (Fig. 1b). These heartbeat time series produce a contour reminiscent of the irregular landscapes that have been widely studied in physical systems.

To quantitatively characterize such a "landscape", Peng et al. [8–10] study a mean fluctuation function $F(n)$, defined as

$$F(n) \equiv \overline{|B(n' + n) - B(n')|}, \tag{1}$$

where the bar denotes an average over all values of n'. Since $F(n)$ measures the average difference between two interbeat intervals separated by a time lag n, $F(n)$ quantifies the magnitude of the fluctuations over different time scales n.

Fig. 1c is a log–log plot of $F(n)$ vs. n for the data in Figs. 1a and b. This plot is approximately linear over a broad physiologically relevant time scale (200–4000 beats) implying that

$$F(n) \sim n^\alpha . \tag{2}$$

It is found that the scaling exponent α is markedly different for the healthy and diseased states: for the healthy heartbeat data, α is close to 0, while α is close to 0.5 for the diseased case. Note that $\alpha = 0.5$ corresponds to a random walk (a Brownian motion), thus the low-frequency heartbeat fluctuations for a diseased state can be interpreted as a stochastic process, in which the interbeat intervals $I(n) \equiv B(n + 1) - B(n)$ are uncorrelated for $n \geqslant 200$.

To investigate these dynamical differences, it is helpful to study further the correlation properties of the time series. Since $I(n)$ is more stationary, one can apply standard spectral analysis techniques [13,14]. The power spectra $S_I(f)$, the square of the Fourier transform amplitudes for $I(n)$, yields

$$S_I(f) \sim \frac{1}{f^\beta} . \tag{3}$$

The exponent β is related to α by $\beta = 2\alpha - 1$ [15]. Furthermore, β can serve as an indicator of the presence and type of correlations:

(i) If $\beta = 0$, there is no correlation in the time series $I(n)$ ("white noise").

(ii) If $0 < \beta < 1$, then $I(n)$ is correlated such that positive values of I are likely to be close (in time) to each other, and the same is true for negative I values.

(iii) If $-1 < \beta < 0$, then $I(n)$ is also correlated; however, the values of I are organized such that positive and negative values are more likely to alternate in time ("anti-correlation").

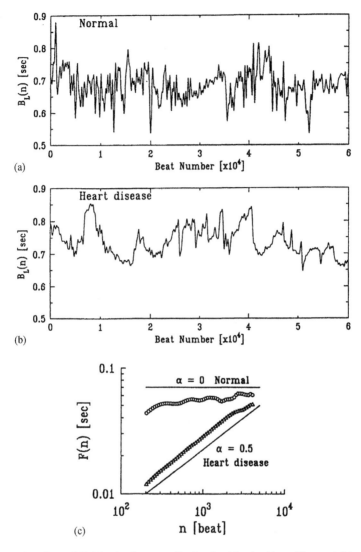

Fig. 1. The interbeat interval $B_L(n)$ after low-pass filtering for (a) a healthy subject and (b) a patient with severe cardiac disease (dilated cardiomyopathy). The healthy heartbeat time series shows more complex fluctuations compared to the diseased heart rate fluctuation pattern that is close to random walk ("brown") noise. (c) Log–log plot of $F(n)$ vs. n. The circles represent $F(n)$ calculated from data in (a) and the triangles from data in (b). The two best-fit lines have slope $\alpha = 0.07$ and $\alpha = 0.49$ (fit from 200 to 4000 beats). The two lines with slopes $\alpha = 0$ and $\alpha = 0.5$ correspond to "$1/f$ noise" and "brown noise", respectively. We observe that $F(n)$ saturates for large n (of the order of 5000 beats), because the heartbeat interval are subjected to physiological constraints that cannot be arbitrarily large or small. After Peng et al. [8–10].

For the diseased data set, we observe a flat spectrum ($\beta \approx 0$) in the low-frequency region confirming that $I(n)$ are not correlated over long time scales (low frequencies). In contrast, for the data set from the healthy subject we obtain $\beta \approx -1$, indicating *nontrivial* long-range correlations in $B(n)$ – these correlations are not the consequence

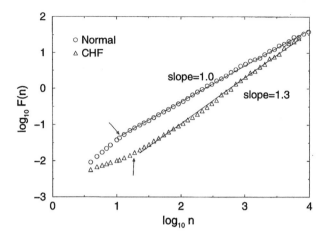

Fig. 2. Plot of $\log F(n)$ vs. $\log n$ for two long interbeat interval time series (~ 24 h). The circles are for a representative healthy subject while the triangles are from a subject with congestive heart failure. Arrows indicate "crossover" points in scaling. Note altered scaling with heart failure, suggesting apparent perturbations of both short and long-range correlation mechanisms. After Peng et al. [8–10].

of summation over random variables or artifacts of non-stationarity. Furthermore, the "anti-correlation" properties of $I(n)$ indicated by the negative β value are consistent with a nonlinear feed-back system that "kicks" the heart rate away from extremes. This tendency, however, does not only operate on a beat-to-beat basis (local effect) but on a wide range of time scales.

A further improvement to the study of the long-range correlation exponent α – detrended fluctuation analysis (DFA) – has been proposed and developed by Peng et al. [11,12]. Fig. 2 compares the DFA analysis of representative 24 h interbeat in-terval time series of a healthy subject ($\bigcirc$) and a patient with congestive heart failure ($\triangle$). Note that for large time scales (asymptotic behavior), the healthy subject shows almost perfect power-law scaling over more than two decades ($20 \leqslant n \leqslant 10\,000$) with $\alpha_{\mathrm{DFA}} = 1$ (i.e., $1/f$ noise) while for the pathologic data set $\alpha_{\mathrm{DFA}} \approx 1.3$ (closer to Brow-nian noise). This result is consistent with our previous finding [8–10] that there is a significant difference in the long-range scaling behavior between healthy and diseased states.

To study the alteration of long-range correlations with pathology, we analyzed cardiac interbeat data from three different groups of subjects: (i) 29 adults (17 male and 12 female) without clinical evidence of heart disease (age range: 20–64 yr, mean 41), (ii) 10 subjects with fatal or near-fatal sudden cardiac death syndrome (age range: 35–82 yr) and (iii) 15 adults with severe heart failure (age range: 22–71 yr; mean 56). Data from each subject contains approximately 24 h of ECG recording encompassing $\sim 10^5$ heartbeats.

For the normal control group, we observed $\alpha_{\mathrm{DFA}} = 1.00 \pm 0.10$ (mean value $\pm$ S.D.). These results indicate that healthy heart rate fluctuations are anticorrelated and exhibit

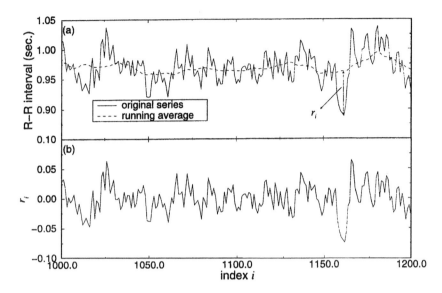

Fig. 3. (a) Segment of RR-interval data vs. beat number (solid curve) and running average based on a local window of 32 heart beats (dashed curve). (b) Detrended curve, i.e., the difference between solid curve and broken curve in top panel. After Askenazy et al. [16].

long-range power-law (fractal) correlation behavior over three decades. Furthermore both pathologic groups show significant deviation of the long-range correlations exponent α_{DFA} from the normal value $\alpha_{DFA} = 1$. For the group of heart failure subjects, we find that $\alpha_{DFA} = 1.24 \pm 0.22$, while for the group of sudden cardiac death syndrome subjects, we find that $\alpha_{DFA} = 1.22 \pm 0.25$.

The different scaling behavior in health and disease must relate to the underlying dynamics of the heartbeat. Applications of this analysis may lead to new diagnostics for patients at high risk of cardiac disease and sudden death.

2. The detrended time series

In this section we present a method of detrending a times series in the following way. First, from the time-series of the raw $B(n) \equiv RR$ data a running average is constructed using an interval-length of 2^m. Next, the running average is subtracted from the original RR-data time series. This procedure is illustrated in Fig. 3a, where the solid curve represents the raw RR-data and the dashed curve represents the running average. The difference between the two curves is denoted by r_i and is shown in Fig. 3b. The resultant time-series r_i is called the detrended time series (DTS) [16] and represents the fluctuations with respect to the local average. This procedure partly removes noise and slow oscillations which should not directly affect the short time scales in the time series [17,18].

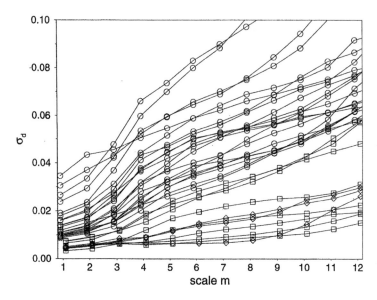

Fig. 4. Standard deviation (in seconds) of the detrended series for a group of 33 subjects vs. the scale factor of the local window used in the detrending. Healthy subjects: Circles, Diabetics: Squares and Heart patients: Rhombohedra. The three topmost diabetics can be regarded clinically as healthy. After Ashkenazy et al. [16].

The standard deviation σ_d of the DTS, using a detrending window of scale m, includes only the behavior of relevant small time scales and may thus be considered a measure of the heart rate variability. To evaluate the discriminating capabilities of σ_d, RR-data for a group of 33 subjects (the same data group as in ref. [19] consisting of 21 healthy subjects, 9 diabetics and 3 heart patients including one heart-transplant patient were examined [16]. Using a time-series consisting of $2^{16} = 65536$ data points, corresponding to 16 hours of measured ECG data, and for the scale values $m = 1$–12 for the detrending window σ_d is calculated. The smallest detrending window is thus 2 and the largest 4096. Results are shown in Fig. 4, which is the analogous of Fig. 4 in Ref. [19] and Fig. 1 in Ref. [20]. In Fig. 4 one notes a clear separation between the group of healthy subjects (circles) on the one hand, and the groups of diabetics (squares) and heart patients (rhombohedra) on the other hand. However, one also notes from this figure that 3 of the diabetics (the three topmost) with as much justification could have been included in the group of healthy subjects thus displacing the separation region for σ_d towards lower values.

In Fig. 4 the largest separation between the healthy subjects and the two other groups is found for the scale $m=8$–11, whereas for an alternative analysis the largest separation was found for the scale $m=4$–6 [19,20]. Indeed, such a separation was earlier identified by Peng et al. [21] (see Fig. 2). Ref. [22] discusses the scale dependence of the different methods. The DFA analysis yields a crossover point for the fractal slope for the scale $m=4$ [21,22]. It should be noted, however, that the crossover point in the DFA analysis

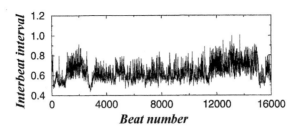

Fig. 5. Consecutive heartbeat intervals measured in seconds are plotted vs. beat number from approximately 3 hrs record of a representative healthy subject. The time series exhibits very irregular and nonstationary behavior.

is not a sharply defined point, rather the change in fractal slope takes place in a gradual way.

3. Multifractality in human heartbeat dynamics

In view of the heterogeneous ("patchy") nature of the heartbeat interval time series (Fig. 5), it has been suggested by Ivanov et al. [23] that a single exponent is not sufficient to characterize the complexity of the cardiac dynamics, and that a multifractal approach may be necessary [24,25]. To test the hypothesis that an infinite number of exponents is required to characterize healthy dynamics [23], a multifractal analysis of heartbeat interval time series has been performed and $D(h)$ has been calculated using wavelet methods [26].

The properties of the wavelet transform make wavelet methods attractive for the analysis of complex nonstationary time series such as one encounters in physiological signals. In particular, wavelets can remove polynomial trends that could lead box-counting techniques to fail to quantify the local scaling of the signal [27]. Additionally, the time–frequency localization properties of the wavelets makes them particularly useful for the task of revealing the underlying hierarchy that governs the temporal distribution of the local Hurst exponents [15,28]. Hence, the wavelet transform enables a reliable multifractal analysis [27,28].

As the analyzing wavelet, we use n-order derivatives of the Gaussian function. Such a wavelet allows us to estimate the singular behavior and the corresponding exponent h at a given location in the time series. The higher the order n of the derivative, the higher the order of the polynomial trends removed and the better the detection of the temporal structure of the local scaling exponents in the signal.

We extract the local value of h through the modulus of the maxima values of the wavelet transform at each point in the time series. We then estimate the scaling of the partition function $Z_q(a)$, which is defined as the sum of the qth powers of the local maxima of the modulus of the wavelet transform coefficients at scale a [28]. For small scales, we expect

$$Z_q(a) \sim a^{\tau(q)} . \tag{4}$$

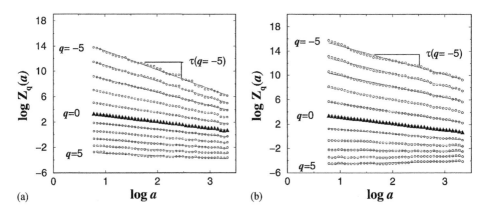

Fig. 6. Heartbeat time series contain densely packed, *non-isolated* singularities which unavoidably affect each other in the time–frequency decomposition. Therefore, rather than evaluating the distribution of the inherently unstable local singularity exponents, we estimate the scaling of an appropriately chosen global measure: the q moments of the probability distribution of the maxima of the wavelet transform $Z_q(a)$ (as analyzing wavelet we use the 3rd derivative of the Gaussian function). Here we show the scaling of the partition function $Z_q(a)$ with scale a obtained from daytime records consisting of $\approx$25,000 beats for (a) a healthy subject and (b) a subject with congestive heart failure. We calculate $\tau(q)$ for moments $q = -5, 4, \ldots, 0, \ldots, 5$ and scales $a = 2 \times 1.15^i$, $i = 0, \ldots, 41$. We display the calculated values for $Z_q(a)$ for scales $a > 8$. The top curve corresponds to $q = -5$, the middle curve (shown heavy) to $q = 0$ and the bottom curve to $q = 5$. The exponents $\tau(q)$ are obtained from the slope of the curves in the region $16 < a < 700$, thus eliminating the influence of any residual small scale random noise due to ECG signal pre-processing as well as extreme, large scale fluctuations of the signal. After Ivanov et al. [23].

For certain values of q, the exponents $\tau(q)$ are related to other familiar exponents. In particular, $\tau(2)$ is related to the scaling exponent of the Fourier power spectra, $S(f) \sim 1/f^{-\beta}$, as $\beta = 2 + \tau(2)$. For positive $q, Z_q(a)$ reflects the scaling of the large fluctuations and strong singularities, while for negative $q, Z_q(a)$ reflects the scaling of the small fluctuations and weak singularities. Thus, the scaling exponents $\tau(q)$ can reveal different aspects of cardiac dynamics.

We define the fractal dimensions $D(h)$ through a Legendre transform of $\tau(q)$ [15],

$$D(h) = qh(q) - \tau(q), \qquad h(q) \equiv \frac{d\tau(q)}{dq} \ . \tag{5}$$

Monofractals display a linear $\tau(q)$ spectrum, $\tau(q) = qH - 1$, where H is the global Hurst exponent. For multifractal signals $\tau(q)$ is a nonlinear function: $\tau(q) = qh(q) - 1$, where $h(q)$ is not constant.

We analyze both diurnal (12:00 to 18:00) and nocturnal (0:00 to 6:00) heartbeat time series records of 18 healthy subjects, and the diurnal records of 12 patients with congestive heart failure. For all subjects, we find that for a broad range of positive and negative q the partition function $Z_q(a)$ scales as a power law (Figs. 6a and b). In Fig. 7, we show $Z_q(a)$ for $q = -2$ and 2 for all 18 healthy subjects in our database. This figure shows that there is good power-law scaling – that is, the data points fall on a straight line in a log–log plot – for all subjects and across nearly two orders of magnitude in a. Also, it is clear that the data have nearly the same slope for all

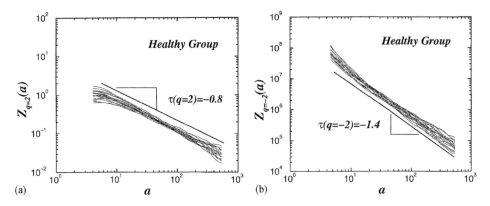

Fig. 7. Scaling of $Z_q(a)$ for the entire healthy group for $q = 2$ and $q = -2$. Courtesy of P.Ch. Ivanov.

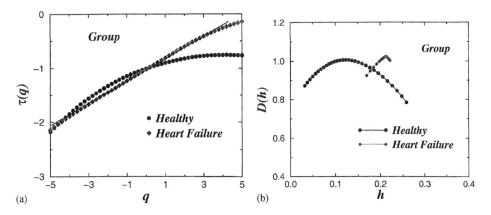

Fig. 8. (a) Multifractal spectrum $\tau(q)$ of the group averages for daytime and night time records for 18 healthy subjects and for 12 patients with congestive heart failure. The results show multifractal behavior for the healthy group and distinct change in this behavior for the heart failure group. (b) Fractal dimensions $D(h)$ obtained through a Legendre transform from the group averaged $\tau(q)$ spectra of (a). The shape of $D(h)$ for the individual records and for the group average is broad, indicating multifractal behavior. On the other hand, $D(h)$ for the heart failure group is very narrow, indicating monofractality. The different form of $D(h)$ for the heart failure group may reflect perturbation of the cardiac neuroautonomic control mechanisms associated with this pathology. After Ivanov et al. [23].

subjects. The exponent of the DFA is $\alpha_{DFA} = [\tau(r) + 3]/2$ [11,12]. This result yields $\alpha_{DFA} \simeq 1.1$ for healthy records and a similar analysis yields $\alpha_{DFA} \simeq 1.3$ for congestive heart failure in agreement with [21].

For all healthy subjects, we find that $\tau(q)$ is a nonlinear function (Figs. 8a), which indicates that the heart rate of healthy humans is a multifractal signal. Fig. 8b shows that for healthy subjects, $D(h)$ has nonzero values for a broad range of local Hurst exponents h. The multifractality of healthy heartbeat dynamics cannot be explained by activity, as we analyze data from subjects during nocturnal hours. Furthermore, this multifractal behavior cannot be attributed to sleep stage transition, as we find

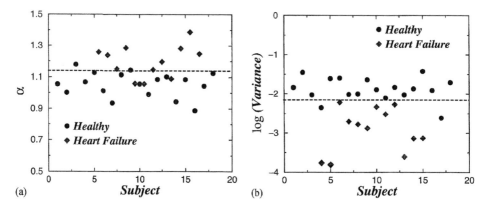

Fig. 9. (a) Correlation exponent and (b) variance of wavelet coefficients. Courtesy of P.Ch. Ivanov and L.A.N. Amaral.

multifractal features during daytime hours as well. The range of scaling exponents– $0 < h < 0.4$ – with nonzero fractal dimension $D(h)$, suggests that the fluctuations in the healthy hearbeat dynamics exhibit strongly anti-correlated behavior (as opposed to $h = \frac{1}{2}$, which corresponds to uncorrelated behavior).

In contrast, subjects with a pathological condition – congestive heart failure – show a loss of multifractality (Figs. 8a and b). We find for the heart failure subjects that $D(h)$ is supported only over a narrow range of exponents h, which indicates weaker multifractality or even monofractality. Moreover, even when the same exponent h is present for both healthy and heart failure subjects, the fractal dimension $D(h)$ associated with this particular exponent has a smaller value for the heart failure subjects (Fig. 8).

Our results show that, for healthy subjects, local Hurst exponents in the range $0.1 < h < 0.2$ are associated with fractal dimensions close to one. This means that the subsets characterized by these local exponents are statistically dominant. On the other hand, for the heart failure subjects, we find that the statistically dominant exponents are confined to a narrow range of local Hurst exponents: $h \approx 0.3$. These results suggest that for heart failure the fluctuations are less anti-correlated than for healthy dynamics, since the dominant scaling exponents h are closer to $\frac{1}{2}$. Our findings support previous reports on long-range anti-correlations of heart beat intervals [21], and can be used as an alternative diagnostic tool.

We compare our method with other widely used methods of heart rate time series analysis. As an example, we show in Fig. 9 two well-established methods. The first is based on the measurement of long-range correlations on the fluctuations in heartbeat intervals [8–10]. These correlations have been quantified with both the power spectrum and the detrended fluctuation analysis. Fig. 9(a) shows the values of the correlation exponent α measured through the detrended fluctuation analysis. The dashed line represents an in-sample threshold for discrimination of the healthy and heart failure groups.

A second method based on Ref. [21] and recently applied in Refs. [16,20] measures the variance of the coefficients of the wavelet transform of the heartbeat signal at a

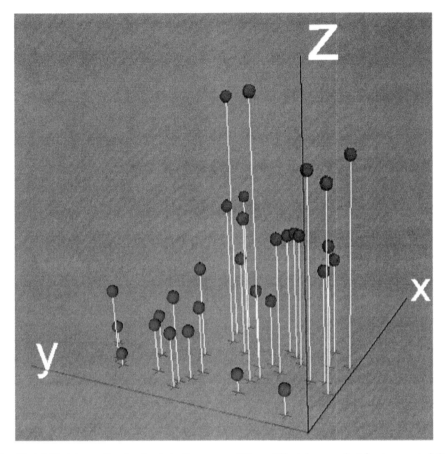

Fig. 10. A 3-D plot revealing the discriminating power of the multifractal approach. After Ivanov et al. [23].

wavelet scale of $a = 32$, shown in Fig. 9(b). The dashed line represents an in-sample threshold for discrimination of the healthy and heart failure groups. These two methods do not result in a fully consistent assignment, e.g., two heart failure subjects are assigned to the diseased group by the first method and to the healthy group by the second method.

Results of the multifractal method [23] are shown in Fig. 10. Each subject's dataset is characterized by three quantities. The first quantity (z-axis) is the degree of multi-fractality which is the difference between the maximum and minimum values of local Hurst exponent h for each individual [Fig. 8(b)]. Note that the degree of multifractality takes value zero for a monofractal.

The second quantity (y-axis) is the exponent value τ ($q = 3$) obtained from the scaling of the third moment $Z_3(a)$. The third quantity (x-axis) is the standard deviation of the interbeat intervals. The healthy subjects are represented with blue balls and the heart failure subjects in red. We see from Fig. 10 that the multifractal approach has the potential to robustly discriminate the healthy subjects from heart failure subjects.

Acknowledgement

Supported by the NIH/National Center for Research Resources (grant P41 13622).

References

[1] W.F. Doolittle, in: E. Stone, R. Schwartz (Eds.), Intervening Sequences in Evolution and Development, Oxford University Press, New York, 1990, p. 42.
[2] R.I. Kitney, O. Rompelman, The Study of Heart-Rate Variability, Oxford University Press, London, 1980.
[3] S. Akselrod, D. Gordon, F.A. Ubel, D.C. Shannon, A.C. Barger, R.J. Cohen, Science 213 (1981) 220.
[4] M. Kobayashi, T. Musha, IEEE Trans. Biomed. Eng. 29 (1982) 456.
[5] A.L. Goldberger, D.R. Rigney, J. Mietus, E.M. Antman, S. Greenwald, Experientia 44 (1988) 983.
[6] J.P. Saul, P. Albrecht, D. Berger, R.J. Cohen, Computers in Cardiology, IEEE Computers Society Press, Washington DC, 1987, pp. 419–422.
[7] D.T. Kaplan, M. Talajic, Chaos 1 (1991) 251.
[8] C.-K. Peng, J.E. Mietus, J.M. Hausdorff, S. Havlin, H.E. Stanley, A.L. Goldberger, Phys. Rev. Lett. 70 (1993) 1343.
[9] C.-K. Peng, Ph.D. Thesis, Boston University, 1993.
[10] C.-K. Peng, S.V. Buldyrev, J.M. Hausdorff, S. Havlin, J.E. Mietus, M. Simons, H.E. Stanley, A.L. Goldberger, in: G.A. Losa, T.F. Nonnenmacher, E.R. Weibel (Eds.), Fractals in Biology and Medicine, Birkhauser Verlag, Boston, 1994.
[11] C.-K. Peng, S.V. Buldyrev, S. Havlin, M. Simons, H.E. Stanley, A.L. Goldberger, Phys. Rev. E 49 (1994) 1685.
[12] C.-K. Peng et al., J. Electrocardiol. 28 (1995) 59.
[13] A. Bunde, S. Havlin (Eds.), Fractals and Disordered Systems, Springer, Berlin, 1991.
[14] A. Bunde, S. Havlin (Eds.), Fractals in Science, Springer, Berlin, 1994.
[15] J. Feder, Fractals, Plenum Press, New York, 1988.
[16] Y. Ashkenazy, M. Lewkowicz, J. Levitan, S. Havlin, K. Saermark, H. Moelgaard, P.E. Bloch Thomsen, Fractals 7 (1999) 85.
[17] H. Moelgaard, 24-hour heart rate variability: methodology and clinical aspects, Doctoral Thesis, University of Aarhus, 1995.
[18] H. Moelgaard, P.D. Christensen, H. Hermansen et al., Diabetologia 37 (1994) 788.
[19] Y. Ashkenazy, M. Lewkowicz, J. Levitan, H. Moelgaard, P.E. Bloch Thomsen, K. Saermark, Fractals 6 (1998) 197.
[20] S. Thurner, M.C. Feurstein, M.C. Teich, Phys. Rev. Lett. 80 (1998) 1544.
[21] C.K. Peng, S. Havlin, H.E. Stanley, A.L. Goldberger, Chaos 5 (1995) 82.
[22] L.A.N. Amaral, A.L. Goldberger, P.Ch. Ivanov, H.E. Stanley, Phys. Rev. Lett. 81 (1988) 2388.
[23] P.Ch. Ivanov, L.A.N. Amaral, A.L. Goldberger, S. Havlin, M.G. Rosenblum, Z. Struzik, H.E. Stanley, Nature 399 (1999) 461.
[24] T. Vicsek, Fractal Growth Phenomena, 2nd Edition, World Scientific, Singapore, 1993.
[25] H. Takayasu, Fractals in the Physical Sciences, Manchester University Press, Manchester UK, 1997.
[26] I. Daubechies, Ten Lectures on Wavelets, S.I.A.M., Philadelphia, 1992.
[27] J.F. Muzy, E. Bacry, A. Arneodo, Phys. Rev. Lett. 67 (1991) 3515.
[28] J.F. Muzy, E. Bacry, A. Arneodo, Int. J. Bifurc. Chaos 4 (1994) 245.

ELSEVIER

Physica A 274 (1999) 111–119

www.elsevier.com/locate/physa

'Sausage-string' deformations of blood vessels at high blood pressures

P. Alstrøm[a,*], R. Mikkelsen[a], F. Gustafsson[b], N.-H. Holstein-Rathlou[b]

[a]*CATS, The Niels Bohr Institute, DK-2100 Copenhagen, Denmark*
[b]*Department of Medical Physiology, The Panum Institute, DK-2200 Copenhagen, Denmark*

Abstract

A new instability is proposed to explain the 'sausage-string' patterns of alternating constrictions and dilatations formed in blood vessels at high blood pressure conditions. Our theory provides predictions for the conditions under which the cylindrical geometry of a blood vessel becomes unstable. The theory is related to experimental observations in rats, where high blood pressure is induced by intravenous infusion of angiotensin II. © 1999 Elsevier Science B.V. All rights reserved.

1. Introduction

Despite the development of blood pressure lowering drugs, high blood pressure remains a major risk factor for development of stroke and heart disease. The high blood pressure is primarily caused by an elevation of resistance to blood flow due to contraction of the smooth muscle cells surrounding the blood vessels. This contraction may for example be induced by the octapeptide angiotensin II. Patients with a kidney disease or pregnant women suffering from preeclampsia are in danger of suffering from a sudden and extreme increase in blood pressure. In almost every case the elevation in blood pressure is accompanied by considerable elevations of the blood levels of angiotensin II [1]. The vascular damage associated with substantial increase in blood pressure is confined to small arteries and arterioles, and it is preceded by a deformation, where the blood vessels develop alternating constriction and dilatation, giving the vessel a 'sausage-string appearance' (Fig. 1).

In the experimental studies presented here of extreme high blood pressures in rats, high blood pressure is induced by intravenous infusion of angiotensin II [3–6]. When

---

* Corresponding author.
E-mail address: alstrom@cats.nbi.dk (P. Alstrøm)

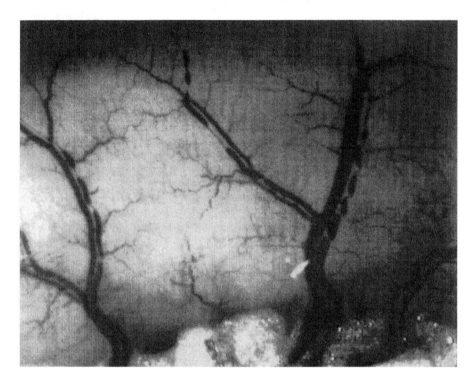

Fig. 1. In vivo micrograph of rat intestinal arterioles showing a typical 'sausage-string' pattern following an acute increase in blood pressure induced by intravenous infusion of angiotensin II. The neighboring vessels not showing constrictions and dilatations are the corresponding venules. From Ref. [2].

the infusion is at a certain level, the vessels may develop the sausage-string pattern (Fig. 1). Despite several decades of experimental research of the phenomenon, the mechanism causing the 'sausage-string' pattern has remained unknown [2]. It has been suggested that the pattern is due to a 'blow out' of the vessel wall caused by the high blood pressure [7], but the sausage-string pattern is observed only in the small arteries and large arterioles (Fig. 2), and here the pressure elevation is small compared to that in the larger arteries [8]. Also, the phenomenon is highly reproducible [5]. If we stop the infusion of angiotensin II, the normal, uniform cylindrical geometry is restored. Again starting the infusion, the sausage-string pattern reappears. Another feature of the pattern is its overall spatial periodicity. We argue that the sausage-string pattern is caused by an instability, and not by a mechanical breakdown.

Here, we use a simple anisotropic, elastic model of the vessel wall. We show that under certain high pressure conditions a new type of instability occurs which leads to a periodic sausage-string pattern of constrictions and dilatations along the vessel. Our theory provides predictions for the conditions under which the cylindrical form of a blood vessel becomes unstable. We show that the appearance of the sausage-string pattern is limited to smaller blood vessels, because the pressure elevation there is small. Also, we find that the instability only occurs for vessels where the wall-to-lumen ratio

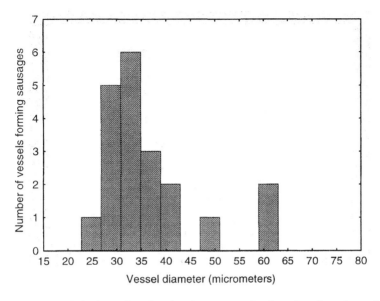

Fig. 2. Size distribution of blood vessels undergoing the sausage-string formation, from observations of gut arterioles (15–80 μm) in rats with high blood pressure.

(vessel-wall thickness divided by the inner radius) is sufficiently small, and therefore not in the small arterioles, where the wall-to-lumen ratios are larger than those found in large blood vessels. The instability borders, between which the sausage-string pattern appears, are determined by two different conditions, the upper border is related to the pressure, the lower border is related to the geometry.

2. Instability theory

Consider an initial small axial symmetric perturbation of the inner radius, $r \rightarrow r + u(z)$ (Fig. 3a). To determine the stability, we must know the evolution equation for $u(z, t)$. We invoke the continuity equation,

$$\partial_t(\pi r^2) = -\partial_z J ,$$ (1)

a change of cross-sectional area at a downstream site z gives rise to a fluid flux $J(z)$. The flux is related to the pressure P,

$$J = -c(r)\partial_z P ,$$ (2)

where $c(r)$ is the fluid conductance. The evolution equation to lowest order in the perturbation follows:

$$\partial_t u = \frac{c(r)}{2\pi r} \partial_z^2 P .$$ (3)

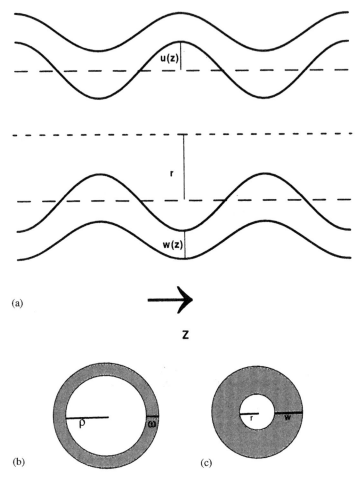

Fig. 3. (a) A schematic picture of a blood vessel of inner radius r undergoing a perturbation $u(z)$. The wall thickness $w(z)$ is larger at smaller radii since the circumference is smaller. (b) Schematic cross-section of a blood vessel in relaxed state. (c) Schematic cross-section of a blood vessel in activated state. The cross-sectional area is assumed fixed, so the wall thickness is larger than in (b).

Assuming that the only stresses of the vessel wall are in the principal angular and vessel directions [10,11], the pressure is given by an integral of a Laplacian form [9–11],

$$P = \int_{r}^{r+w} \left[S \frac{1}{\tilde{r}[1 + (\partial_z \tilde{r})^2]^{1/2}} - S_z \frac{\partial_z^2 \tilde{r}}{[1 + (\partial_z \tilde{r})^2]^{3/2}} \right] d\tilde{r}, \tag{4}$$

where w is the width of the vessel wall, and S and S_z are the stresses in the angular direction and in the vessel direction. The stresses, S and S_z (forces per actual cross-sectional area), are related to experimentally measured idealized stresses, σ (Fig. 4) and σ_z, the forces per relaxed cross-sectional area [14],

$$S = \gamma \gamma_z \sigma, \quad S_z = \gamma \gamma_z \sigma_z. \tag{5}$$

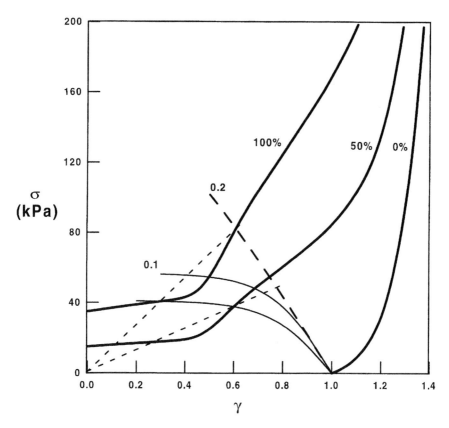

Fig. 4. A schematic plot of typical stress–strain relations for arterioles (adapted from Refs. [12,13]). The three heavy solid curves correspond to a completely relaxed vessel (0%), a vessel where the smooth muscle cells are half maximally activated (50%), and a vessel where the smooth muscle cells are maximally activated (100%). The thin solid lines indicate how the points $(\gamma_r, \sigma(\gamma_r))$ and $(\gamma_w, \sigma(\gamma_w))$ [marked 0.1] move with muscle cell activation for an arteriole with wall-to-lumen ratio $\omega/\rho = 0.1$. The point of instability for the cylindrical form of the blood vessel can be illustrated geometrically by thin dashed lines from $(0,0)$ through $(\gamma_r, \sigma(\gamma_r))$ [see text]. The instability point is where $\sigma(\gamma_w)/\gamma_w$ equals $\sigma(\gamma_r)/\gamma_r$. The thick dashed line [marked 0.2] shows how the point $(\gamma_w, \sigma(\gamma_w))$ move with muscle cell activation for an arteriole with $\omega/\rho = 0.2$, keeping the same curve for $(\gamma_r, \sigma(\gamma_r))$.

Here, γ and γ_z are the normalized lengths in the angular and vessel directions (e.g. γ_z is equal to L/L_0 where L and L_0 is the actual and resting length of a tissue strip in the angular direction). We assume that γ_z is constant, $\gamma_z = \gamma_0$, because the length of a vessel remains almost constant during contraction. The stress σ_z is therefore also replaced by a constant σ_0.

The width w of the vessel wall is related to the inner radius r, assuming that the cross-sectional area of the vessel wall is constant. When the inner radius ρ and wall thickness ω are known for the angularly relaxed state ($\gamma = 1$), we have (Fig. 3b and c)

$$(r + w)^2 - r^2 = (\rho + \omega)^2 - \rho^2 \,. \tag{6}$$

Eq. (6) suggests a useful change of variable, from $\tilde{r}$ to $\tilde{\rho}$, where

$$\tilde{r}^2 - r^2 = \tilde{\rho}^2 - \rho^2 . \tag{7}$$

When $\tilde{r}$ varies between values r and $r+w$, $\tilde{\rho}$ varies between the fixed values ρ and $\rho + \omega$. The normalized length γ at a radius $\tilde{r}$ is the ratio between $\tilde{r}$ and its relaxed value $\tilde{\rho}$.

The integral for the pressure reduces for small perturbations to

$$P = \gamma_0 \int_\rho^{\rho+\omega} [\sigma - \sigma_0 r \partial_z^2 r][\tilde{\rho}^2 - \rho^2 + r^2]^{-1/2} \, d\tilde{\rho} . \tag{8}$$

Also the stress σ (angular direction) depends on the normalized length

$$\gamma = [\tilde{\rho}^2 - \rho^2 + r^2]^{1/2} / \tilde{\rho} . \tag{9}$$

From the above equations, one finds to lowest order in the perturbation $u(z,t)$,

$$P = P_0(r) + I(r)u - I_0(r)\partial_z^2 u , \tag{10}$$

where

$$P_0(r) = \gamma_0 \int_\rho^{\rho+\omega} \sigma[\tilde{\rho}^2 - \rho^2 + r^2]^{-1/2} \, d\tilde{\rho} , \tag{11}$$

$$I_0(r) = \gamma_0 \sigma_0 r \int_\rho^{\rho+\omega} [\tilde{\rho}^2 - \rho^2 + r^2]^{-1/2} \, d\tilde{\rho}$$

$$= \gamma_0 \sigma_0 r \log[1 + (\omega + w)/(\rho + r)] , \tag{12}$$

and $I(r)$ can be expressed in terms of the normalized length γ,

$$I(r) = \frac{\gamma_0 \gamma_r}{\rho(1 - \gamma_r^2)} \left[\frac{\sigma(\gamma_w)}{\gamma_w} - \frac{\sigma(\gamma_r)}{\gamma_r} \right] , \tag{13}$$

where

$$\gamma_r = r/\rho, \quad \gamma_w = (r+w)/(\rho + \omega) , \tag{14}$$

are the normalized inner and outer radius.

Now, the evolution, Eq. (3), for $u(z,t)$ takes the form

$$\partial_t u = -\frac{c(r)}{2\pi r}[-I(r)\partial_z^2 u + I_0(r)\partial_z^4 u] . \tag{15}$$

For a perturbation of the form $u = \sum_k u_k(t)\cos(kz)$, we have $u_k(t) \sim u_k(0)\,e^{\lambda_k t}$ with

$$\lambda_k = \frac{c(r)}{2\pi r} k^2[-I(r) - I_0(r)k^2] . \tag{16}$$

Since the value of $I_0(r)$ is positive, it is the sign of $I(r)$ that determines the stability of the vessel wall. When $I(r)$ becomes negative, the cylindrical geometry becomes unstable to modes with $k^2 < |I|/I_0$.

3. Experimental results

The fastest growing mode, where λ_k is maximal, is at the value $k = [|I|/(2I_0)]^{1/2}$, corresponding to 'sausages' of length

$$\ell = 2\pi[2I_0/|I|]^{1/2} . \tag{17}$$

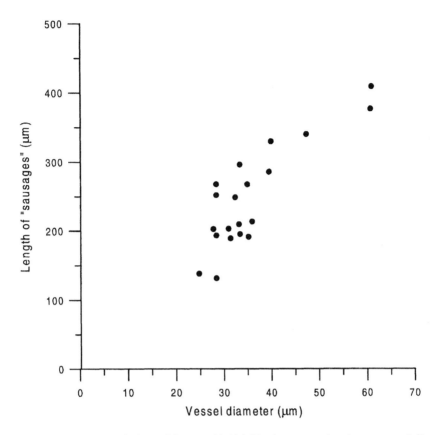

Fig. 5. Length of the 'sausages' observed in rats with high blood pressure, plotted versus vessel diameter before infusion of angiotensin II.

For $\omega/\rho \sim 0.1$, and $\sigma_0 \sim 100$ kPa, we have $|I|\rho \sim I_0/\rho \sim 10$ kPa. Hence, $\ell \sim 10\rho$, thus the length of the 'sausages' will be about 10 times the radius of the relaxed vessel, in good agreement with our experimental observations (Fig. 5) (see also Ref. [5]).

The cylindrical form becomes unstable when σ/γ calculated at the inner radius equals the value of σ/γ at the outer radius [Eq. (13)]. This is illustrated geometrically (Fig. 4), by drawing a line (thin dashed) in the plot of σ versus γ from $(0,0)$ through $(\gamma_r, \sigma(\gamma_r))$. If the point $(\gamma_w, \sigma(\gamma_w))$ lies above this line, $\sigma(\gamma_w)/\gamma_w > \sigma(\gamma_r)/\gamma_r$, and the cylindrical form is stable. If on the other hand the point $(\gamma_w, \sigma(\gamma_w))$ lies below the line, $\sigma(\gamma_w)/\gamma_w < \sigma(\gamma_r)/\gamma_r$, and the cylindrical form is unstable.

The angular stress σ in blood vessels increases exponentially with the normalized length at normal physiological conditions [10–13] (Fig. 4). The value of $I(r)$ is positive and the blood vessel keeps its cylindrical form. When extreme hypertension is induced by infusion of angiotensin II, the operating point for the vessel will move to the less steep part of the $\sigma - \gamma$ curve (Fig. 4), and the value of $I(r)$ becomes negative. The cylindrical form becomes unstable, and the 'sausage-string' pattern appears.

In our experimental studies, we find that the 'sausage-string' pattern following infusion of angiotensin II occurs in arterioles of diameter 30–40 μm (Fig. 2). In accordance herewith, our analysis predicts that large vessels will be stable, because the operating point $(\gamma, \sigma(\gamma))$ lies on the steep and stable part of the $\sigma - \gamma$ curve. For small arterial vessels, the wall-to-lumen ratio ω/ρ is generally large [10,4]. From the expression for $I(r)$, Eq. (13), we find that the instability point decreases with increasing wall-to-lumen ratio ω/ρ. In Fig. 4, the thin solid lines are examples of how the points $(\gamma_r, \sigma(\gamma_r))$ and $(\gamma_w, \sigma(\gamma_w))$ [marked 0.1] may move with muscle cell activation for an arteriole with wall-to-lumen ratio $\omega/\rho = 0.1$. For a given activation of the smooth muscle cells (50%, 100%), a thin dashed line is drawn from $(0,0)$ through $(\gamma_r, \sigma(\gamma_r))$. When the point $(\gamma_w, \sigma(\gamma_w))$ lies above the thin dashed line, the cylindrical form is stable; when it lies below, the cylindrical form is unstable. For $\omega/\rho = 0.1$, we see from Fig. 4 that the cylindrical form is stable at 50% activation, but unstable at 100% activation of the smooth muscle cells surrounding the blood vessel. For $\omega/\rho = 0.2$ ($(\gamma_w, \sigma(\gamma_w))$ illustrated by thick dashed line marked 0.2), the cylindrical form is only barely unstable at 100% activation. The 'sausage-string' instability does not appear in small blood vessels with large wall-to-lumen ratios. The pressure and the contractile potential sets an upper limit, and the wall-to-lumen ratio sets a lower limit, for vessels that undergo the 'sausage-string' instability in response to an acute increase in blood pressure.

4. Conclusion

In conclusion, due to a large increase in the smooth muscular tone of blood vessels, the normal cylindrical vessel geometry may become unstable, giving rise to the appearance of a sausage-string pattern of alternating constrictions and dilatations. The sausage-string pattern is the expression of a novel instability, involving the nonlinear stress–strain characteristics of the vessel wall, and not caused by a mechanical failure of the vessel wall due to a high blood pressure. The developed theory explains some key features which we observe experimentally, particularly why the instability is only observed in small arteries and large arterioles. Our theory demonstrates that the instability predominantly arises in blood vessels with high contractile potential, limited pressure elevation, and small wall-to-lumen ratios. The observed correspondence between the 'sausage' length and the vessel diameter is also consistent with the theory.

Acknowledgements

The present study was supported by grants from the Danish Natural Science Research Council, the Danish Medical Research Council, and the Danish Heart Association.

References

[1] C. Kitiyakara, N.J. Guzman, J. Amer. Soc. Nephrol. 9 (1998) 133.
[2] F. Gustafsson, Blood Pressure 6 (1997) 71.

[3] F.B. Byrom, Lancet 2 (1954) 201.
[4] F.B. Byrom, Prog. Cardiovasc. Dis. 1 (1974) 31.
[5] J. Giese, Acta Pathol. Microbiol. Scand. 62 (1964) 497.
[6] J. Giese, The Pathogenesis of Hypertensive Vascular Disease, Munksgaard, Copenhagen, 1966.
[7] L.J. Beilin, F.S. Goldby, Clin. Sci. Mol. Med. 52 (1977) 111.
[8] G.A. Meininger, K.L. Fehr, M.B. Yates, J.L. Borders, H.J. Granger, Hypertension 8 (1986) 66.
[9] A.E. Green, J.E. Adkins, Large Elastic Deformations, Clarendon Press, Oxford, 1960.
[10] Y.C. Fung, Biomechanics. Mechanical Properties of Living Tissues, 2nd Edition, Springer, New York, 1990.
[11] Y.C. Fung, Biomechanics. Motion, Flow, Stress, and Growth, Springer, New York, 1990.
[12] R.W. Gore, Circ. Res. 34 (1974) 581.
[13] M.J. Davis, R.W. Gore, Amer. J. Physiol. 256 (1989) H630.
[14] R. Feldberg, M. Colding-Jørgensen, N.-H. Holstein-Rathlou, Amer. J. Physiol. 269 (1995) F581.

ELSEVIER

Physica A 274 (1999) 120–131

www.elsevier.com/locate/physa

Fractality, chaos, and reactions in imperfectly mixed open hydrodynamical flows

À. Péntek[a], G. Károlyi[b], I. Scheuring[c], T. Tél[d], Z. Toroczkai[e,*], J. Kadtke[a], C. Grebogi[f]

[a] Marine Physical Laboratory, Scripps Institution of Oceanography, University of California at San Diego, La Jolla, CA 92093-0238, USA
[b] Department of Structural Mechanics, Technical University of Budapest, Műegyetem rkp. 3., H-1521 Budapest, Hungary
[c] Department of Plant Taxonomy and Ecology, Research Group of Ecology and Theoretical Biology, Eötvös University, Ludovika tér 2, H-1083 Budapest, Hungary
[d] Institute for Theoretical Physics, Eötvös University, Puskin u. 5-7, H-1088 Budapest, Hungary
[e] Department of Physics, University of Maryland, College Park, MD 20742-4111, USA
[f] Institute for Plasma Research, University of Maryland, College Park, MD 20742, USA

Abstract

We investigate the dynamics of tracer particles in time-dependent open flows. If the advection is passive the tracer dynamics is shown to be typically transiently chaotic. This implies the appearance of stable fractal patterns, so-called unstable manifolds, traced out by ensembles of particles. Next, the advection of chemically or biologically active tracers is investigated. Since the tracers spend a long time in the vicinity of a fractal curve, the unstable manifold, this fractal structure serves as a catalyst for the active process. The permanent competition between the enhanced activity along the unstable manifold and the escape due to advection results in a steady state of constant production rate. This observation provides a possible solution for the so-called "paradox of plankton", that several competing plankton species are able to coexists in spite of the competitive exclusion predicted by classical studies. We point out that the derivation of the reaction (or population dynamics) equations is analog to that of the macroscopic transport equations based on a microscopic kinetic theory whose support is a fractal subset of the full phase space. © 1999 Elsevier Science B.V. All rights reserved.

1. Introduction

The advection of particles in time-dependent hydrodynamical flows is known to be typically *chaotic* [1–33]. If the particle takes on the velocity of the flow very rapidly,

---

* Corresponding author.
E-mail address: toro@k2.umd.edu (Z. Toroczkai)

i.e., inertial effects are negligible, we call the advection *passive*, and the particle a *passive tracer*. Its equation of motion is then

$$\dot{r} = v(r, t), \tag{1}$$

where v is the known velocity field. In stationary flows when the right-hand side is independent of t, problem (1) is integrable and the particle trajectories coincide with the streamlines of the flow. In time-dependent cases, however, streamlines and particle trajectories are *different*, and the latter ones can be much more complicated.

Here we consider passive advection in *open* flows [18–31], where the time-dependent region of the flow is assumed to be restricted to a *finite* domain, called the *mixing region*. While the tracer trajectories are simple outside the mixing region, they are typically chaotic inside of it.

It is worth emphasizing that a complicated flow field (turbulence) inside the mixing region is *not* required for a complex tracer dynamics. Even simple forms of time dependence, e.g. a periodic repetition of the velocity field in the mixing region is sufficient. The flow model we use here is of this type: in a finite domain the flow is time periodic with a period T, and outside this domain it is stationary.

In the following Section we show that the chaotic advection by open flows is associated with the appearance of stable *fractal* patterns. Then in Section 3 we argue that if the advected tracers are *active* in a chemical or biological sense, i.e. they can either react with neighboring particles or are subjected to a growth dynamics, then the activity mainly takes place along the same fractal set. This implies a new, novel form of surface reaction, which includes the parameters of the chaotic advection. Finally, in the concluding Section 4 we point out some possible applications of these results.

2. Advection of passive tracers in open flows

We consider open flows, where the flow is simple and stationary everywhere but in the mixing region. In the inflow and outflow regions the particle's motion is simple, as they just follow the streamlines. In the mixing region, however, the particle paths can be very complicated and typically chaotic. As this chaotic behavior is restricted to a finite region both in space and time, it is necessarily of transient type. This *transient chaos* [34] is the most ubiquitous form of chaos which appears in open flows.

An important characteristic of such transiently chaotic systems is that tracers entering the mixing region are typically *trapped* there for long times. In fact, there is a set of tracer trajectories *never* leaving the mixing region. Among these non-escaping trajectories periodic orbits can be found with a period which is an integer multiple of the flow's period T. Such periodic orbits are best visualized on a *stroboscopic map*, which is a series of snapshots taken at integer multiples of the flow's period T. On the stroboscopic map the periodic tracer orbits of period mT trace out a series of m different points.

Of course, such permanently trapped orbits are quite exceptional, and they are all unstable. This means that most typically the tracers having entered the mixing region leave it after some — typically long — time. The non-escaping orbits, however, form a fractal set (a fractal "cloud" of points on the stroboscopic map), called the *chaotic saddle*. It is called "saddle" because, similar to a saddle-point, the unstable trajectories forming it can be reached from a set of exceptional initial tracer positions that converge to the non-escaping trajectories as time goes by. All other trajectories, however, are escaping into the outflow region sooner or later.

The union of all exceptional trajectories converging to the non-escaping orbits is the *stable manifold* of the chaotic saddle. The tracers starting from the close vicinity of the stable manifold are advected towards the chaotic saddle, and they follow some of the non-escaping orbits for a while. Once repelled from a non-escaping orbit, they might be trapped again by the stable manifold of another non-escaping orbit, exhibiting a kind of random walk among them. Finally, they leave the chaotic saddle along its *unstable manifold*. On the stroboscopic map, both the stable and unstable manifolds are complicatedly winding fractal curves [19–31] with some fractal dimension D_0.

The unstable manifold can be directly observed both in experiments [35] as well as in environmental flows [36–38]. A droplet of tracers initially overlapping with the stable manifold is advected towards the chaotic saddle where it gets trapped for a long time.

The particles starting further away from the stable manifold are washed out quickly by the background flow, and they do not exhibit the kind of random walk among the orbits of the chaotic saddle. The tracers starting close to the stable manifold, however, spend a long time in the mixing region, and finally leave it along the unstable manifold of the chaotic saddle. This means that it is the unstable manifold where the tracers accumulate after sufficiently long time, as illustrated schematically in Fig. 1. This observation implies that although the chaos is transient in the mixing region, the particles trace out a *permanent fractal pattern*. If the unstable manifold is visualized by placing a single droplet of dye into the flow, the unstable manifold will be faded out as times goes by.

Indeed, the number of particles present in the mixing region decays exponentially in time with the exponent κ, which is called the *escape rate* [34]:

$$N(t) = N(0)e^{-\kappa t} . \tag{2}$$

The reciprocal of κ can be considered to be the average lifetime of transient chaos [34], and $1/\kappa$ is the average time a tracer spends in the mixing region.

As an illustrative example, we consider the flow of a viscous fluid around a cylinder with a background velocity pointing along the x-axis. At intermediate background velocities (whose dimensionless measure, the Reynolds number is of the order of 10^2) no stationary velocity field is stable, instead, a strictly periodic behavior sets in with period T, see Fig. 2. Two vortices are created behind the cylinder within each period, one above and another one below the x-axis. These two vortices are delayed by a time

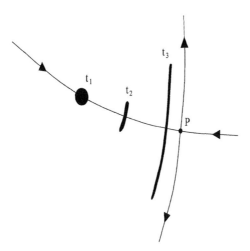

Fig. 1. Schematic illustration of a droplet of dye converging to the unstable manifold of the chaotic saddle. Point P illustrates a point of the saddle, while the two intersecting lines represent the stable and unstable manifolds. The droplet overlapping with the stable manifold at t_1 stretches as time goes on. The points on the stable manifold move towards P, while the other points are repelled from it along the unstable manifold.

$T/2$. The vortices first grow in size, then detach from the cylinder and start to drift downstream. This alternating separation of vortices from the upper and lower cylinder surface is called the *von Kármán vortex street* and is characterized by a strictly periodic velocity field of period T [39].

After a short length of travel, the vortices are destabilized and destroyed due to the viscosity of the fluid. Far away from the cylinder upstream and downstream the flow is practically stationary. The mixing region is thus located in the wake of the cylinder.

To obtain the velocity distribution one has to solve the two-dimensional viscous Navier–Stokes equations with no-slip boundary condition along a circle [20–22]. For simplicity we use here an analytic model for the streamfunction introduced in Ref. [22] motivated by a direct numerical simulation of the Navier–Stokes flow carried out by Jung and Ziemniak [21] at Reynolds number 250.

In Fig. 3 two particle trajectories are shown, with initial conditions deviating by an amount on the order of $10^{-3}R$. The trajectories diverge from each other rapidly, one leaves the wake at the top side, the other one at the bottom side of the cylinder. The typically exponential growth of the distance between initially close particles is a unique sign of the chaotic tracer motion, although the flow itself is strictly periodic, without chaoticity.

Fig. 4 shows a snapshot of the chaotic saddle and its stable and unstable manifolds in the wake of the cylinder. The chaotic saddle consist of a countable infinite number of periodic orbits and an uncountable number of non-periodic orbits. Tracers inserted on any black dot in Fig. 4a stay in the wake forever.

Particles inserted exactly on the stable manifold (Fig. 4b) converge to trajectories of the chaotic saddle after infinitely long time. If a tracer is, however, inserted off the

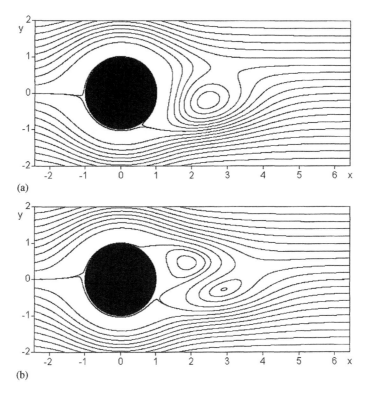

Fig. 2. Snapshots taken at times $t = 0 \, (\text{mod} \, T)$ and $t = T/4 \, (\text{mod} \, T)$ on the streamlines of the von Kármán vortex street. The fluid flows from left to right. During the first half time period $(T/2)$ a vortex is born at the top side of the cylinder, while the vortex at the bottom side dies out due to viscosity. The streamlines at times $t = T/2 \, (\text{mod} \, T)$ and $t = 3T/4 \, (\text{mod} \, T)$ are the mirror images of these figures with respect to the $y = 0$ axis.

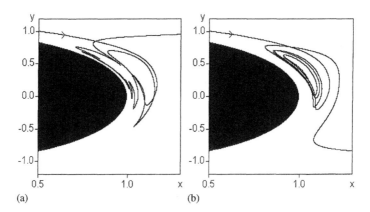

Fig. 3. Two particles initially close to each other trace out completely different paths. The cylinder is elongated in the horizontal direction for better visualization.

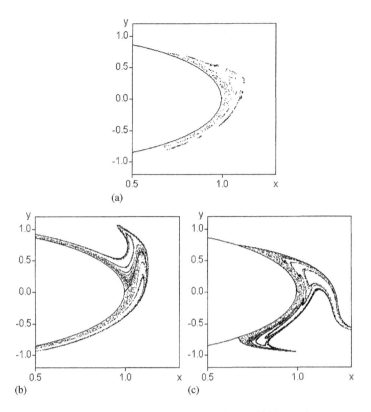

Fig. 4. The chaotic saddle (a) and its stable (b) and unstable (c) manifolds are shown on snapshots taken at $t = 0 \, (\mathrm{mod} \, T)$. While the chaotic saddle is a fractal cloud of points, its stable (unstable) manifold is a complicated curve reaching in the far upstream (downstream) region. The cylinder is elongated in the horizontal direction for better visualization.

stable manifold, but very close to it, it leaves the wake along the unstable manifold (Fig. 4c).

The unstable manifold is traced out by an ensemble of trajectories initially over-lapping with the stable manifold. This is illustrated in Fig. 5, where the tracers still present in the mixing region are shown after some time. As tracers spending a long time in the wake of the cylinder, that is, being trapped in the mixing region, finally leave it along the unstable manifold, it is natural to expect that any kind of transport processes occurring mainly take place along this fractal set. This can indeed be seen in laboratory experiments [15]. The above-mentioned property of the tracer dynamics implies that classical flow visualization techniques based on dye evaporation or streak-lines trace out fractal curves (unstable manifolds) which are *different from streamlines* or any other characteristics of the Eulerian velocity field (for several flow visualization photographs of this type see Ref. [39]).

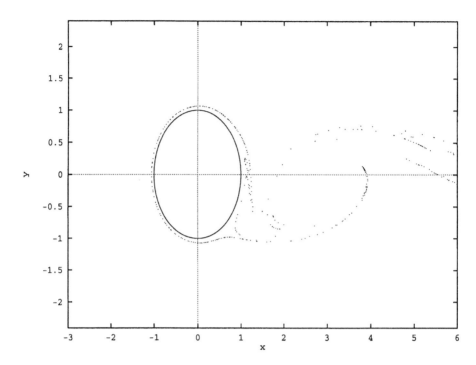

Fig. 5. The unstable manifold of the chaotic saddle is traced out by the tracers injected into the flow in front of the cylinder. The coverage is not perfect due to the finite number of particles. Initially 300×300 particles were inserted into the flow in the region $x \in [-2.55, -2.45]$, $y \in [-0.1, 0.1]$, and, for computational simplicity, their time-evolution was observed on a grid of size $1/300$. The snapshot was taken at $t = 2T$, where T is the period of the flow.

3. Advection of active tracers in open flows

In this section we consider the effect of the chaotic advection on the active processes on the flow's surface described by *kinetic reactions* [40,41]. Tracers injected into the flow are advected passively, they do not influence the flow. If, however, they come closer to each other than a given reaction distance σ, they interact with each other, creating thus the *product* particles. For our discussions we consider the *autocatalytic* reaction: $A + B \rightarrow 2B$. We insert a tiny seed of reacting B tracers into the flow covered with A particles. This way the reaction events occur on the surface between the areas covered by A and B particles: the A-type tracers become B within a distance σ. This distance can be considered as a *reaction range*. For computational simplicity the instantaneous reactions occur at integer multiples of a time lag τ, during which only advection occurs.

Fig. 6 shows the spreading of reagents B in the wake of the cylinder after a long time. The initial position of the tracers is the same as in Fig. 5. After a short transient (of about $4T$) a steady state is reached, which implies a constant production rate of B tracers in the wake. Note that the active tracers also trace out the unstable manifold,

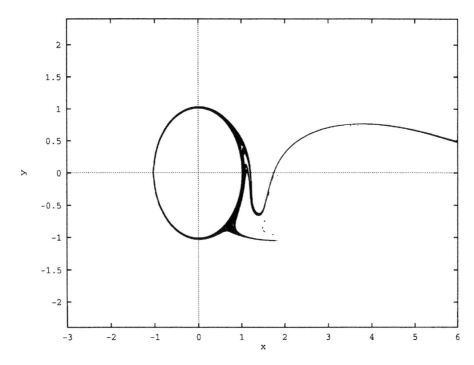

Fig. 6. The unstable manifold of the chaotic saddle is traced out by the autocatalytic tracers (B, black) injected into the flow in front of the cylinder. The coverage is much more efficient than in Fig. 5 due to the reactions. Initially 300×300 B particles were inserted into the flow in the region $x \in [-2.55, -2.45]$, $y \in [-0.1, 0.1]$, the rest of the fluid surface was covered by A. For computational simplicity, the time evolution was observed on a grid of size $1/300$. The snapshot was taken at $t = 20T$, where T is the period of the flow. The model parameters were $\sigma = 1/150$, $\tau = T/5$.

but the coverage is much wider due to the autocatalytic reactions. This means that the reactions occur *on the surface of a fattened-up fractal.*

Based on this observation, a simple theory [42,43] can give the area $\mathscr{A}_B(t)$ covered by the reacting tracers in the mixing region at time t. By taking the limit $\tau \to 0$, $\sigma \to 0$ but keeping σ/τ finite, a time continuous reaction equation can be obtained:

$$\dot{\mathscr{A}}_B = -\kappa \mathscr{A}_B + g\frac{\sigma}{\tau}\mathscr{A}_B^{-\beta}. \tag{3}$$

Here g is a constant, and

$$\beta = \frac{D_0 - 1}{2 - D_0} \tag{4}$$

is a nontrivial, positive exponent depending only on the fractal dimension D_0 of the unstable manifold. If the reactions occur along a simple line, that is, the surface between the A and B particles is not a fractal, we have $D_0 = 1$. This implies, via $\beta = 0$, that (3) describes a classical surface reaction [44] with reaction front velocity σ/τ in the presence of the escape (κ) of the products.

The negative exponent $-\beta$ in Eq. (3) implies that the reactions are enhanced due to the fractal boundary between the different reagents. In fact, the less reagent is present in the mixing region, the more effective the reactions become because of the larger resolved surface. This leads to a balance between the escape due to advection (first term) and the production due to reactions (second term in (3)). Since in our case the long-time advection dynamics is concentrated on a fractal, the unstable manifold of the passive flow, we obtain essential deviations from traditional chemical or population dynamics theories derived for a well-stirred environment. These observations might be of relevance for other applications of statistical physics where the support of the underlying kinetic theory is not a smooth subset of the full phase space.

Via direct substitution, one can check that the solution to Eq. (3) is

$$\mathscr{A}_B(t) = \left(\frac{g\sigma}{\kappa\tau} - K e^{-\kappa/(2-D_0)t} \right)^{2-D_0} , \tag{5}$$

where K is an integration constant, and it is related to the initial area $\mathscr{A}_B(0)$ via

$$K = \frac{g\sigma}{\kappa\tau} - [\mathscr{A}_B(0)]^{1/(2-D_0)} . \tag{6}$$

One can see from solution (5) that in the long-time limit $t \to \infty$, the area will be expressed as $(\varepsilon^*)^{2-D_0}$, i.e., as a coverage of the fractal unstable manifold with stripes of a *non-zero* average width $\varepsilon^* = g\sigma/\kappa\tau$. In case of no chemical reactions ($\sigma = 0$), we obtain the usual exponential emptying dynamics with κ being the escape rate, just as expected. The appearance of a novel term in the chemical rate equation is a macroscopic consequence of an underlying kinetic theory. In contrast to the classical theory, however, the chemically active tracers occupy a fractal subset of the full phase space (the unstable manifold) only.

The balance mentioned above can serve as a possible answer for a long-standing question called the *paradox of plankton* [45]. In well-mixed environments, classical studies [46,47] predict that all competing species die out except the most perfectly adapted ones for all the limiting factors. As the number of different limiting factors is rather small (on the order of 10), it is quite surprising that the number of different competing plankton populations is quite large. The keyword in the above problem is "well-mixed environment". We have seen that in open flows the advected tracers are *not* well-mixed, instead, they form a structured spatial distribution of tracers. It is thus natural to expect that in open flows, when the activity (in this biological sense the competition) is restricted to the surface of a fattened-up fractal, the number of competing species can be larger than the number of limiting factors.

The competition in our case is modeled by two autocatalytic reactions $A + B \to 2B$ and $A + C \to 2C$ using the same resource A, which is the only limiting factor. Both species have different replication abilities σ_B and σ_C, and mortality rates δ_B and δ_C. The mortality rate is the probability that an organism dies out during the time lag τ. In a well-mixed environment, the traditional theory implies that only species *B or C* survives the competition, the one with superior reproduction abilities. In open flows, as illustrated in Fig. 7, the coexistence of the competing species *B and C* can be observed. Both species are present in the wake of the cylinder, thus both of them are pulled along the unstable manifold. Here their activity is enhanced, which leads to increased access for both species to the background material A for which they compete.

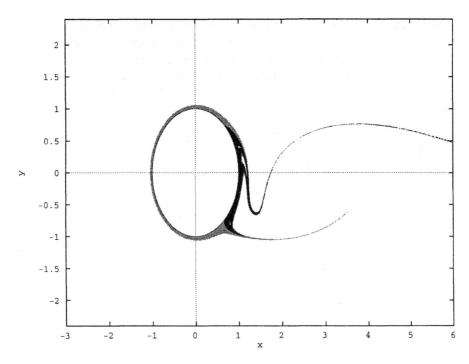

Fig. 7. The unstable manifold of the chaotic saddle is covered by the competing species B (light-grey) and C (dark-grey) at time $t = 20T$, after reaching the steady state. The background material A is shown in white. Initially 300×300 B particles were inserted into the flow in the region $x \in [-2.55, -2.45]$, $y \in [0, 0.1]$, and 300×300 C particles in the region $x \in [-2.55, -2.45]$, $y \in [-0.1, 0]$. The grid size chosen was $1/300$. The model parameters were $\sigma_B = 1/150$, $\sigma_C = 1/300$, $\delta_B = 0.5$, $\delta_C = 0.0001$, and $\tau = 0.2T$.

Thus, due to the fact that the mixing of the different species is *not perfect*, coexistence along the unstable manifold is ensured [48] in a wide range of parameter *differences* characterizing the activity of the species.

4. Discussion

We have seen that particles advected by open flows trace out complicated fractal patterns. The particles spend a long time here, and they possess a largely increased surface to perform any kind of (chemical or biological) activity. Active tracers (chemical reagents or biological species) thus fatten-up the unstable manifold, and the major part of the activity takes place on a fractal set. Such processes have been observed e.g. in atmospheric chemistry, like the ozone depletion at the polar vortex. Here the filamental spatial distribution of $ClONO_2$ at the vortex edge [36,38] might be identified as the result of a reaction $ClO + NO_2 \rightarrow ClONO_2$ along a fractal. Similarly, in aquatic systems, the evolution of plankton populations was also reported to possess filamental spatial distribution [49–52].

Although the environmental flows are often thought of as *closed* ones (in the sense that there is no escape), they can still produce filamental spatial distribution, cf. [53,54]. One reason for this is that on the time-scale of the active environmental processes, and in a fixed frame of observation, these large-scale flows can be considered to be open with a net current flowing through the observation region. Additionally, there is no significant feedback into the region of observation on the time-scale of the active processes.

When the activity takes place along a non-trivial fractal boundary, a new kind of surface reaction equation (3) has been derived. It contains important chaos parameters, like the escape rate κ and the fractal dimension D_0. These parameters, however, depend uniquely on the parameters of the hydrodynamics, like the Reynolds number. This equation, based on microscopic properties of the advection dynamics, gives a global, macroscopic description of the product depending only measurable quantities. This can lead to estimation of observable quantities that could provide a verification of the theory.

Acknowledgements

This research has been supported by the NSF-MRSEC at University of Maryland, by the NSF-DMR, by ONR(physics), CNPq/NSF-INT, by the DOE, by the US-Hungarian Science and Technology Joint Fund under Project numbers 286 and 501, by the Hungarian Science Foundation T019483, T025793, T029789, and F029637, by the Hungarian-British Intergovernmental Science and Technology Cooperation Program GB-66/95, and by the Hungarian Ministry of Culture and Education under grant number 0391/1997.

References

[1] H. Aref, J. Fluid Mech. 143 (1984) 1.
[2] H. Aref, S. Balachandar, Phys. Fluids 29 (1986) 3515.
[3] J. Chaiken, R. Chevray, M. Tabor, Q.M. Tan, Proc. Roy. Soc. London A 408 (1986) 165.
[4] J.M. Ottino, The Kinematics of Mixing: Stretching, Chaos and Transport, Cambridge University Press, Cambridge, 1989.
[5] A. Crisanti, M. Falcioni, G. Paladin, A. Vulpiani, La Riv. Nuovo Cimento 14 (1991) 207.
[6] S. Wiggins, Chaotic Transport in Dynamical Systems, Springer, Berlin, 1992.
[7] H. Aref (Ed.), Chaos applied to fluid mixing, (special issue) Chaos Solitons Fractals 4(6) (1994).
[8] R.B. Rybka et al., Phys. Rev. E 48 (1993) 757.
[9] T.H. Solomon, J.P. Gollub, Phys. Rev. A 38 (1988) 6280.
[10] T.H. Solomon, E.R. Weeks, H.L. Swinney, Physica D 76 (1994) 70.
[11] R.T. Pierrehumbert, Chaos Solitons Fractals 4 (1994) 1091.
[12] V.V. Meleshko, G.J.F. van Heijst, Chaos 4 (1994) 977.
[13] V.V. Meleshko et al., Phys. Fluids A (1992) 2779.
[14] J.B. Kadtke, E.A. Novikov, Chaos 3 (1993) 543.
[15] J.C. Sommerer, E. Ott, Science 259 (1993) 281.
[16] J.C. Sommerer, Physica D 76 (1994) 85.
[17] D. Beigie, A. Leonard, S. Wiggins, Chaos Solitons Fractals 4 (1994) 749.

[18] H. Aref, S.W. Jones, S. Mofina, I. Zawadski, Physica D 37 (1989) 423.
[19] V. Rom-Kedar, A. Leonard, S. Wiggins, J. Fluid. Mech. 214 (1990) 347.
[20] K. Shariff, T.H. Pulliam, J.M. Ottino, Lect. Appl. Math. 28 (1991) 613.
[21] C. Jung, E. Ziemniak, J. Phys. A 25 (1992) 3929.
[22] C. Jung, T. Tél, E. Ziemniak, Chaos 3 (1993) 555.
[23] E. Ziemniak, C. Jung, T. Tél, Physica D 76 (1994) 123.
[24] Á. Péntek, T. Tél, Z. Toroczkai, J. Phys. A 28 (1995) 2191.
[25] Á. Péntek, T. Tél, Z. Toroczkai, Fractals 3 (1995) 33.
[26] Á. Péntek, Z. Toroczkai, T. Tél, C. Grebogi, J.A. Yorke, Phys. Rev. E 51 (1995) 4076.
[27] G. Károlyi, A. Péntek, T. Tél, Z. Toroczkai, Chaotic tracer dynamics in open hydrodynamical flows, in: E. Infeld, R. Żelazny, A. Gałkowski (Eds.), Proceedings of International Conference on Nonlinear Dynamics, Chaotic and Complex Systems, Zakopane, Poland, 1995, Cambridge UP, Cambridge, 1997, pp. 24–38.
[28] J.A. Kennedy, J.A. Yorke, Topology Appl. 80 (1997) 201.
[29] M.A. Sanjuan et al., Chaos 7 (1997) 125.
[30] M.A. Sanjuan et al., Phys. Rev. Lett. 78 (1997) 1892.
[31] G. Károlyi, T. Tél, Phys. Rep. 290 (1997) 125.
[32] A. Provenzale, Annu. Rev. Fluid. Mech. 31 (1999) 55.
[33] A. Babiano, J.H.E. Cartwright, O. Piro, A. Provenzale, preprint, 1999.
[34] T. Tél, in: Hao Bai-lin (Ed.), Directions in Chaos, Vol. 3, 1990, pp. 149–211; STATPHYS'19, World Scientific, Singapore, 1996, pp. 346–362.
[35] J.C. Sommerer, H.-C. Ku, H.E. Gilreath, Phys. Rev. Lett. 77 (1996) 5055.
[36] S. Edouard, B. Legras, V. Zeitlin, J. Geophys. Res. 101 (1996) 16 771.
[37] M.G. Balluch, P.H. Haynes, J. Geophys. Res. 102 (1997) 23 487.
[38] A. Mariotti, C.R. Mechoso, B. Legras, Ozone filaments from the southern polar night vortex, J. Geophys. Res. (1999) submitted for publication.
[39] M. Van Dyke, An Album of Fluid Motion, The Parabolic Press, Stanford, 1982.
[40] G. Metcalfe, J.M. Ottino, Phys. Rev. Lett. 72 (1994) 2875.
[41] G. Metcalfe, J.M. Ottino, Chaos, Solitons Fractals 6 (1995) 425.
[42] Z. Toroczkai, G. Károlyi, Á. Péntek, T. Tél, C. Grebogi, Phys. Rev. Lett. 80 (1998) 500.
[43] G. Károlyi, Á. Péntek, Z. Toroczkai, T. Tél, C. Grebogi, Phys. Rev. E 59 (1999) 5468.
[44] L.D. Landau, E.M. Lifschitz, Fluid Mechanics, Pergamon Press, Oxford, 1987.
[45] G.E. Hutchinson, Am. Nat. 95 (1961) 137.
[46] G.F. Gause, A.A. Witt, Am. Nat. 69 (1935) 596.
[47] G. Hardin, Science 131 (1960) 1292.
[48] I. Scheuring, G. Károlyi, Á. Péntek, T. Tél, Z. Toroczkai, A model for resolving the plankton paradox: coexistence in open flows, Freshwater Biol. (1999), in press.
[49] W.H. Thomas, C.H. Gibson, J. Appl. Phycol. 2 (1990) 71.
[50] W.H. Thomas, C.H. Gibson, Deep Sea Res. 37 (1990) 1583.
[51] C.H. Gibson, W.H. Thomas, J. Geophys. Res. 100 (1995) 24 841.
[52] S.A. Spall, K.J. Richards, A numerical model of mesoscale frontal instabilities and plankton dynamics, Deep-Sea Res., in press.
[53] S. Edouard, B. Legras, B. Lefevre, R. Eymard, Nature 384 (1996) 444.
[54] Z. Neufeld, C. López, P.H. Haynes, Phys. Rev. Lett. 82 (1999) 2606.

ELSEVIER

Physica A 274 (1999) 132–139

www.elsevier.com/locate/physa

Application of statistical physics to politics

Serge Galam *

*Laboratoire des Milieux Désordonnés et Hétérogènes[1], Universite Paris 6, Tour 13 - Case 86,
4 place Jussieu, 75252 Paris Cedex 05, France*

Abstract

The concept and technics of real space renormalization group are applied to study majority rule voting in hierarchical structures. It is found that democratic voting can lead to totalitarianism by keeping in power a small minority. Conditions of this paradox are analyzed and singled out. Indeed majority rule produces critical thresholds to absolute power. Values of these thresholds can vary from 50% up to at least 77%. The associated underlying mechanism could provide an explanation for both former apparent eternity of communist leaderships and their sudden collapse. © 1999 Elsevier Science B.V. All rights reserved.

1. Setting the limits

Modern theory of critical phenomena is based on the fundamental concepts of universality and irrelevant variables [1]. These two concepts mean that different physical systems, like for instance a magnet and a liquid, behave the same way when passing from one macroscopic state to another macroscopic state. Well-known examples are the magnet becoming a para-magnet, the liquid, a gas, and maybe even the creation of the universe from nothing with the big bang. Most of the microscopic properties of the physical compounds involved turn out to be irrelevant for describing the macroscopic change which in turn appears to be universal. While the number of physical systems undergoing phase transitions is infinite, all associated phase transitions can be described in terms of only a small finite number of universality classes. Only a few parameters, like space dimensionality, determine which universality class the system belongs to. The abstract and general nature of the statistical physics framework makes it tempting to extend such notions to non-physical systems, and in particular to social systems,

---

* Fax: +33-1-44273854.
E-mail address: galam@ccr.jussieu.fr (S. Galam)
[1] Laboratoire associé au CNRS (UMR no. 7603).

0378-4371/99/$ - see front matter © 1999 Elsevier Science B.V. All rights reserved.
PII: S 0378-4371(99)00320-9

for which, in many cases, there exists an interplay between microscopic properties and macroscopic realities.

Nevertheless, the two fields of physical sciences and social sciences are rather different. However, the process of going in parallel from one atom and one human to, respectively, several atoms in bulk and a social group has much in common. More precisely, it is the hypothesis behind the present approach. It is worth stressing that we are not claiming that our models will explain all aspects of human behavior. Like any modeling effort, it is appropriate only to some classes of phenomena in social science and not to others.

However, such an approach should be carefully controlled. To just map a physical theory built for a physical reality, onto a social reality, could be at best a nice metaphor, but without predictability, and at worst a misleading and wrong social theory. Physics has been successful in describing macroscopic behavior using properties of the constituent microscopic elements. The task here, is to borrow from physics those technics and concepts used to tackle the complexity of aggregations. The challenge is then to build a collective theory of social behavior along similar lines, but within the specific constraints of the psycho-social reality. The constant danger is for the theorist to stay in physics, using a social terminology and a physical formalism. The contribution from physics should thus be restricted to qualitative guidelines for the mathematical modeling of complex social realities. Such a limitation does not make the program less ambitious.

2. Real space: from physics to politics

In this paper we present an application of statistical physics to political sciences [2]. We apply the concept and technics of real space renormalization group [1] to study the majority rule voting process within hierarchical structures. In particular, we focus on the conditions for a given political party to get for sure, full power at the hierarchy top level.

We find that majority rule voting produces critical threshold to total power. Having an initial support above the critical threshold guarantees full power at the top. The value of the critical threshold to power is a function of the voting structure, namely the size of voting groups and the number of voting hierarchical levels. Using these results a new explanation is given to the sudden and abrupt fall of eastern former communist parties.

Here we apply the real space renormalization group scheme of collective phenomena in physics to a radically different reality, namely a social one with emphasis on other technical aspects than the ones usually used in physics. While there is an abstract and formal tool to study phase transitions, here we associate a political reality to each step of the renormalization group transformation. On this basis it is worth stressing that we apply statistical physics to political sciences not as a qualitative metaphor but indeed as a guide to build a quantitative model to study the effect of the majority rule voting on the democratic representation of groups within a hierarchical organization.

Like in physics we start from local cells constituted by a small number of degrees of freedom, here by individuals. Similar to Ising spins, to keep the analysis simple, people can choose only between two political tendencies A and B. Associated proportions of A and B support in the system, a political group, a firm, a network, a society, are supposed to be known. We denote them by p_0 for the overall A-support and $1 - p_0$ for the B-support. We are assuming that each member does have an opinion.

Once formed, each cell elects a representative, either an A or a B using a majority rule. These elected people (the equivalent of the super spin rescaled to an Ising one in real space renormalization group) constitute the first hierarchical level of the organization called level-1. New cells are then formed at level-1 from these elected people. They in turn elect new representatives to build level-2. This process is repeated again and again.

In physics the rescaled degree of freedom is fictitious while here it is a real person. We are not using a theoretical scheme to embody some complex features but instead we are building up a real organization where each voting step is real. Moreover, at odd with physics the number of iterations in the renormalization process is finite and the focus is on the stable fixed points.

The rest of the paper is as follows. In the next section we present and study the case of 3-person cells. A critical threshold to power is singled out. It equals 50%. A minority is found to self-eliminate within a few voting levels. A making-sense bias in the voting rule is then introduced in Section 3. A one vote bonus is added to the tendency already in power in cases of A–B equality. For 4-person cells, this bias shifts the critical threshold to power from 50% to 77%. Size effects are analyzed in Section 4. Analytic formulas are then derived. In particular, given an initial support p_0 for the A, the number of voting levels necessary for their self-elimination is obtained. Section 5 puts the results in a more practical perspective. The last section contains a discussion on both former apparent eternity of communist leaderships and their sudden collapse. Some perspectives are outlined.

3. The simplest fair case

We start from a population distributed among two tendencies A and B with respectively p_0 and $1 - p_0$ proportions within the system. It could be either a political group, a firm, or a whole society. At this stage each member does have an opinion. From now on, we will use the political language.

Cells are constituted by randomly aggregating groups of three persons. It could be by home localization or working place. Each cell then elects a representative using majority rule. To have an A elected requires either three A or two A in the cell. Otherwise, it is a b who is elected. These elected persons constitute the first level of the hierarchy denoted level-1. The same process of cell forming can be repeated within level-1 from the elected persons. Making the cell to vote produces an additional level, namely level-2. We can go on in the same way from a level-n to a level-$(n+1)$. The probability to have an A elected at level $(n+1)$ is then

$$p_{n+1} \equiv P_3(p_n) = p_n^3 + 3p_n^2(1 - p_n),$$ (1)

where p_n is the proportion of elected A persons at level-n.

We call $P_3(p_n)$ the voting function. It has three fixed points $p_d = 0$, $p_{c,3} = \frac{1}{2}$ and $p_t = 1$. The first one corresponds to the disappearance of A. The last one p_t represents the totalitarian situation where only A are present. Both are stable. In contrast, p_c is unstable. It determines the threshold to full power. Starting from $p_0 < \frac{1}{2}$ repeating voting leads towards 0 while the flow is in the direction of 1 for $p_0 > \frac{1}{2}$.

Therefore, majority rule voting produces the self-elimination of any proportion of the A-tendency as long as $p_0 < \frac{1}{2}$. However, the democratic self-elimination to be completed requires a sufficient number of voting levels.

At this stage the instrumental question is to determine the number of levels required to ensure full leadership to the initial larger tendency. The analysis will turn relevant to reality only if this level numbers are small. Most organizations have only a few levels, and always less than 10.

For instance, starting from $p_0 = 0.43$ we get successively $p_1 = 0.40$, $p_2 = 0.35$, $p_3 = 0.28$, $p_4 = 0.20$, $p_5 = 0.10$, $p_6 = 0.03$ down to $p_7 = 0.00$. Therefore, seven levels are sufficient to self-eliminate 43% of the population.

Though the aggregating voting process eliminates a tendency it stays democratic since it is the leading tendency (more than 50%) which after all gets the total leadership of the organization. It is worth noting the symmetry with respect to A and B tendencies.

4. The simplest killing case

In the real world, things are not as fair as above and often it turns out very hard, if not impossible, to change an organization leadership. We will now illustrate this situation.

Considering yet the simplest case we constitute groups of four persons instead of three. The salient new feature is the 2A–2B configuration for which there exists no clear majority. In most social situations it is well admitted that to change a policy required a clear cut majority. In case of no decision, things will stay as they are. It is a bias in favor of the already existing. Often this bias is achieved giving for instance, one additional vote to the committee president.

Along this line, the voting function becomes non-symmetrical. Assuming B were in power, for A to be elected at level $n+1$ we have

$$p_{n+1} \equiv P_4(p_n) = p_n^4 + 4p_n^3(1 - p_n),$$ (2)

where p_n is the proportion of elected A persons at level-n. In contrast, for a B it is,

$$1 - P_4(p_n) = p_n^4 + 4p_n^3(1 - p_n) + 2p_n^2(1 - p_n)^2,$$ (3)

where the last term embodies the bias in favor of B. From Eqs. (2) and (3) the stable
fixed points are still 0 and 1. However, the unstable one is drastically shifted to

$$p_{c,4} = \frac{1 + \sqrt{13}}{6} \tag{4}$$

which makes the threshold to power to A about 77%. Moreover, the process of
self-elimination is accelerated. For instance from $p_0 = 0.69$ we have the series $p_1 =$
0.63, $p_2 = 0.53$, $p_3 = 0.36$, $p_4 = 0.14$, $p_5 = 0.01$, and $p_6 = 0.00$. It shows that
using an a priori reasonable bias in favor of B turns a majority rule democratic voting
to a totalitarian outcome. Indeed to get to power A must pass over 77% of support
which is almost out of the realm of possibility. The above series shows how 63% of
a population disappears from the leadership within only five voting levels.

5. Larger voting groups

Most real organizations work with voting cells larger than three or four. To account
for this size variable we generalize the above scheme to cells with r voting persons.
We then have to determine the voting function $p_{n+1} = P_r(p_n)$. Using a majority rule
it becomes

$$P_r(p_n) = \sum_{l=r}^{l=m} \frac{r!}{l!(r-l)!} p_n^l (1 + p_n)^{r-l}, \tag{5}$$

where $m = (r+1)/2$ for odd r and $m = (r+1)/2$ for even r which thus accounts for
the B-bias.

The two stable fixed points $p_d = 0$ and $p_t = 1$ are preserved by enlarging the group
size. However, while the unstable one stays $p_{c,r} = \frac{1}{2}$ for odd sizes, it varies with r for
even sizes. It starts at $p_{c,4} = (1 + \sqrt{13})/6$ with the limit $p_{c,r} \to \frac{1}{2}$ when $r \to \infty$.

We can then calculate analytically the critical number of levels n_c at which $p_{n_c} = \varepsilon$
with ε being a very small number. It determines the level of confidence of the prediction
to have no A elected. One way to evaluate n_c is to expand the voting function $p_n =
P_r(n-1)$ around the unstable fixed point $p_{c,r}$:

$$p_n \approx p_{c,r} + (p_{n-1} - p_{c,r})\lambda_r, \tag{6}$$

where $\lambda_r \equiv dP_r(p_n)/dp_n|_{p_{c,r}}$ with $P_r(p_c) = p_{c,r}$. Rewriting the last equation as,

$$p_n - p_{c,r} \approx (p_{n-1} - p_{c,r})\lambda_r \tag{7}$$

we can then iterate the process to get

$$p_n - p_{c,r} \approx (p_0 - p_{c,r})\lambda_r^n, \tag{8}$$

from which we get the critical number of levels n_c at which $p_n = \varepsilon$. Taking the ln on
both side of Eq. (8) gives

$$n \approx -\frac{\ln(p_c - p_0)}{\ln \lambda_r} + n_0, \tag{9}$$

where $n_0 \equiv \ln(p_{c,r} - \varepsilon)/\ln \lambda_r$ is valid as it is not too far from $p_{c,r}$. However, it turns out to be a rather good estimate even down to the stable fixed point 0 by making it equal to 1 while taking the integer part of Eq. (9). For a more accurate calculation of n_0 see [3,4].

6. The magic formula

Most organizations do not change their structure at every election or decision event. They are set once and then do not change any longer. The number of hierarchical levels is thus fixed and constant. Therefore, to make our analysis useful we have to invert the question "how many levels are needed to eliminate a tendency" onto "given n levels what is the necessary overall support to get full power".

It is worth keeping in mind that situations for, respectively, A and B tendencies are not always symmetric. Here we stress the dynamics of voting with respect to A. To implement this operative question, we invert Eq. (7) to obtain

$$p_0 = p_{c,r} + (p_n - p_{c,r})\lambda_r^{-n}, \tag{10}$$

two critical thresholds appears now. The first one, the disappearence threshold $p_{d,r}^n$ which gives the value of support under which A disappears for sure at the top level leadership. It is given by Eq. (10) putting $p_n = 0$,

$$p_{d,r}^n = p_{c,r}(1 - \lambda_r^{-n}). \tag{11}$$

Putting in parallel $p_n = 1$ gives the second threshold $p_{f,r}^n$ above which A gets full and total power. Using Eq. (11), we get

$$p_{f,r}^n = p_{d,r}^n + \lambda_r^{-n} \tag{12}$$

which shows the appearence of a new region for $p_{d,r}^n < p_0 < p_{f,r}^n$. In that region A neither disappears totally nor gets full power (p_n is neither 0 nor 1). It is therefore a coexistence region where some democracy is prevailing since results of the election process are only probabilistic. No tendency is sure of winning making alternating leadership a reality. However, as seen from Eq. (12), this democratic region shrinks as a power law λ_r^{-n} of the number n of hierarchical levels. Having a small number of levels thus puts higher the threshold to a total reversal of power but simultaneously lowers the threshold for non-existence.

Again, the above formulas are approximate since we have neglected corrections in the vicinity of the stable fixed points. However, they give the right qualitative behavior. Actually $p_{d,r}^n$ fits to $n+1$ and $p_{f,r}^n$ to $n+2$. For more accurate formulas see [3,4].

To get a practical feeling of what Eqs. (11) and (12) mean, let us illustrate them for the case $r = 4$ where we have $\lambda = 1.64$ and $p_{c,4} = (1 + \sqrt{13})/6$. Considering 3, 4, 5, 6 and 7 level organizations, $p_{d,r}^n$ is equal to, respectively, 0.59, 0.66, 0.70, 0.73 and 0.74. In parallel, $p_{f,r}^n$ equals 0.82, 0.80, 0.79, 0.78 and 0.78. These series emphasize drastically the totalitarian character of the voting process.

7. Some prospectives

Up to now we have treated very simple cases to single out main trends produced by democratic voting aggregating over several levels. In particular, we have singled out the existence of critical thresholds to full power. Moreover, these thresholds are not necessarily symmetric for both the tendencies in competition. In the biased 4-cell case it is around 0.77% for A.

Such asymmetries are indeed always present in most realistic situations in which more than two groups are competing. Let us consider for instance the case of three competing groups A, B and C. Assuming a 3-cell case, now the ABC configuration is unsolved using majority rule as it was for the preceding two tendencies AABB 4-cell configuration. For the AB case we made the bias in favor of the group already in power, like giving an additional vote to the committee president.

For multi-group competitions typically the bias results from the parties agreement. Usually the two largest parties, say A and B are hostile while the smallest one C would compromise with either one of them. Then the ABC configuration gets a C elected. In such a case, we need 2A or 2B to elect, respectively, an A or a B. Otherwise a C is elected. Therefore, the elective function for A and B are the same as for the AB 3-cell model. It means that the critical threshold to full power to A and B is 50%. In otherwords for initial A and B support of less than 50% C gets full power. The required number of levels is obtained from the above formulas.

It is possible to generalize to as many groups as wanted. The analysis becomes more heavy but the mean features of voting flows towards a fixed point are preserved.

8. Conclusion

To conclude, we comment on some possible new explanation to the recently generalized auto-collapse of eastern communist parties. Up to this historical and drastic event, communist parties seemed eternal with the same leaderships ever. Once they collapsed most explanations were related to both an opportunistic change within the various organizations together with the end of the soviet army threat.

Maybe the explanation is different and related to our hierachical model. Indeed communist organizations are based on the structutral concept of democratic centralism which is nothing other than a tree-like hierachy with a rather high critical threshold to power. Suppose it was of the order of 80% like in our 4-cell case. We could then consider that the internal opposition to the orthodox leadership did grow a lot and massively over several decades to eventually reach and pass the critical threshold with the associated sudden rise of the internal opposition. Therefore, the sudden collapse of eastern communist parties might be as a result of a very long and solid phenomena. Such an explanation does not oppose additional constraints but emphasizes the internal mechanism within these organizations.

At this stage it is important to stress that modeling social and political phenomena is not stating some absolute truth but instead singling out some basic trends within very complex situations.

References

[1] Sh-k Ma, Modern Theory of Critical Phenomena, The Benjamin Inc., Reading, MA, 1976.
[2] S.M. de Oliveira, P.M.C. de Oliveira, D. Stauffer, Non-Traditional Applications of Computational Statistical Physics: Sex, Money, War, and Computers, Springer, Berlin, in press.
[3] S. Galam, Majority rule, hierarchical structure and democratic totalitarianism, J. Math. Psychol. 30 (1986) 426.
[4] S. Galam, Social paradoxes of majority rule voting and renormalization group, J. Stat. Phys. 61 (1990) 943–951.

ELSEVIER

Physica A 274 (1999) 140–148

www.elsevier.com/locate/physa

Application of statistical physics to the Internet traffics

Misako Takayasu[a,*], Kensuke Fukuda[b], Hideki Takayasu[c]

[a]*Faculty of Science and Technology, Keio University, 890-12 Kashimada, Saiwai-ku, Kawasaki 211-0958, Japan*
[b]*NTT Network Innovation Laboratories, 3-9-11, Midori-cho, Musashino, Tokyo 180-8585, Japan*
[c]*Sony Computer Science Lab., Takanawa Muse Bldg., 3-14-13 Higashi-gotanda, Shinagawa-ku, Tokyo 141-0022, Japan*

Abstract

Fluctuations of information flow in a public domain of the Internet are analyzed in view of statistical physics. It is shown that the characteristics of the massive traffics can be well described by the framework of dynamic phase transition between sparse and congested phases. © 1999 Elsevier Science B.V. All rights reserved.

1. Introduction

An interdisciplinary new field of science is growing with the growth of the Internet. Computer scientists, mathematicians and statistical physicists are now working on this emerging field focusing especially on the fractal properties of the fluctuations of information traffics.

In 1994, Leland et al. reported a statistical self-similarity in the number fluctuation of packets in an Ethernet cable inside a laboratory [1]. In the same year Csabai independently discovered another self-similarity in the fluctuation of round trip times of the "ping"-command which indirectly reflect the level of congestion along a path in the Internet [2]. These facts cast serious doubts on the most basic assumption of traditional traffic theory that packet arrivals can be approximated by a Poisson process. Intensive observations have followed these pioneering findings and such fractal properties have been confirmed by many observations [3–7].

---

* Corresponding author.
E-mail address: misako@future.st.keio.ac.jp (M. Takayasu)

0378-4371/99/$ - see front matter © 1999 Elsevier Science B.V. All rights reserved.
PII: S 0378-4371(99)00398-2

In 1996, M. Takayasu et al. discovered a phase transition behavior between sparse and congested phases by carefully observing the "ping"-fluctuations [8].

In a sparse phase the power spectrum is close to a white noise that is consistent with the classical Poisson assumption. The $1/f$ type power spectrum is observed near the phase transition point which is realized at a moderately congested path. It is argued that the phase transition behaviors including the $1/f$ power spectrum can be nicely modeled by a mathematically well-known stochastic process called the contact process [9] on the Cayley tree. The basic assumption of applying the contact process is that congestion at a router may propagate contagiously to neighboring routers. Spatio-temporal behaviors of congestion level in an actual Internet path have recently been observed by generalizing the "ping"-experiment and the existence of contagious propagation of congestion among neighboring routers has been proved [10,11].

Numerical simulation of packet traffic on an artificial network not only clarifies that the phase transition is a continuous phase transition accompanied with critical behaviors, but it also demonstrates that the critical point is technologically very important by the following meaning [12]: Below the critical point the flow is sparse and more information can be injected, while above the critical point the possibility of loosing packets due to overflow increases rapidly, namely, the total flow rate is the largest and the reliability is high at the critical point.

Theoretically, it is pointed out that even the simplest information traffic system consisted of a random information input and a buffer shows a phase transition behavior typically when the buffer capacity is infinite [13]. The sparse phase corresponds to the case that the mean input is smaller than the maximum output and the averaged quantity of information accumulated at the buffer is finite. As the mean input rate is increased the accumulation at the buffer increases on an average, and at a critical point the averaged accumulation diverges. The critical point condition is given by the simple relation that the mean input rate is equal to the maximum output rate, and the power spectrum of output flow fluctuation with independent random input is shown to follow an inverse square law. It should be noted that even the mean value of the accumulated information is infinite at the critical point the actual accumulation level fluctuates following a Brownian motion and sometimes the buffer becomes empty.

This phase transition is a local phenomenon that can occur in any buffer system because it is due to the general nonlinear response of the buffer. The global phase transition phenomenon discovered by the "ping"-experiment is due to propagation of congestion among routers and it gives the $1/f$ fluctuation at the critical point, namely, these two phase transitions should belong to different categories. The whole system of the Internet can be viewed as a huge ensemble of "phase transition elements" and the system has a rich variety of phase transition behaviors.

In this paper we observe actual information flow fluctuation at a public domain of the Internet and show that a new type of phase transition behaviors can be confirmed from real-time sequences of flow density.

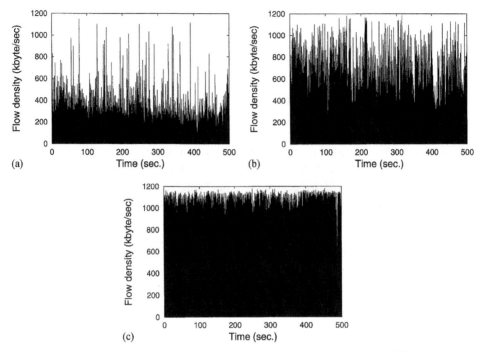

Fig. 1. Examples of flow density fluctuation. (a) Low mean density, 200 kbytes/s. (b) Middle mean density, 500 kbytes/s. (c) High mean density, 1000 kbytes/s.

2. Observation of information flow density fluctuations

In order to observe fluctuations of information flow in the Internet we set a personal computer as a monitoring host on an Ethernet cable that connects the gateway of Keio University and the WIDE Internet backbone that belongs to public domain. The data-link layer of the campus backbone is 20 Mbps ATM, and the gateway is connected to the network operating center of the WIDE backbone by Ethernet of 10 Mbps. Our monitoring host is connected to this segment via a non-intelligent hub. There is no other traffic source in the Ethernet cable in our experiment, namely we collected only information traffics passing through these links. The measurement host saves all data that go through this link both upstream and downstream into its hard-disk via the "tcpdump" command. One observation lasts 4 h and we made time sequential data of information flow density with unit time 0.1 s.

Examples of temporal fluctuations of information flow are shown in Fig. 1(a)–(c) with period 500 s, whose mean flow densities are about 200, 500 and 1000 kbytes/s, respectively. In order to characterize the difference of these periods more clearly, we plot the distributions of information flow density with time unit 0.1 s in Fig. 2. In the low-density case the distribution has a single peak around 100 kbyte/s. As the mean density increases, the distribution's width expands widely. The distribution becomes nearly flat around the intermediate density of 500 kbyte/s. The distribution for higher

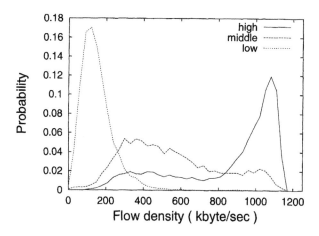

Fig. 2. Probability distributions of flow density fluctuations for the three cases in Fig. 1. (a) The dotted line. (b) The broken line. (c) The solid line.

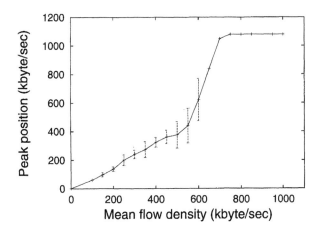

Fig. 3. The peak flow density of probability distribution as a function of the mean flow density.

mean flow density again has a single peak but its position is at a high flow density. As a total we observe 252 sets of 500 s periods and confirmed that the shape of probability density for each box changes smoothly as a function of the mean flow density interpolating the three curves in Fig 2.

The peak position and the peak's-width at half-height of the probability density are plotted in Figs. 3 and 4 as functions of the mean flow density in each box. When the mean flow density is less than 500 kbyte/s, the peak position is nearly proportional to the mean flow density and the peak width increases gradually. This region corresponds to the sparse phase. Around 600 kbytes/s the distribution becomes nearly flat, consequently the peak width becomes the largest. When the mean flow density is larger than 700 kbyte/s, the peak position is nearly fixed at a value a little smaller

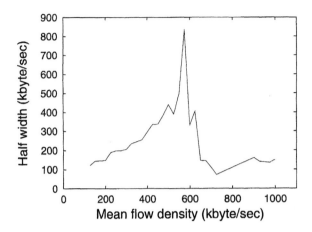

Fig. 4. The peak-width at half-height of the probability distribution.

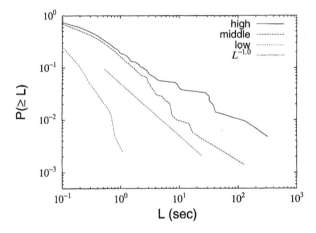

Fig. 5. Distribution of congestion duration time for the three cases in Fig. 1. (a) The dotted line. (b) The broken line. (c) The solid line. The dotted straight line gives the slope −1.

than the capacity of the Ethernet cable and the half-width of the distribution also nearly converges to a certain value. The changes of probability densities can be viewed as a dynamic phase transition with the control parameter given by the mean flow density. The critical point is estimated to be a little smaller than 600 kbytes/s from Fig. 4 by the value that maximize the width of the distribution.

In Fig. 5 the distribution of time interval, L, that the flow densities are larger than a given threshold flow density continuously for L s. As an evidence of the critical behavior we can find a power-law distribution at the mean density of 600 kbytes/s. The estimated slope of the distribution at the critical state is −1 which is consistent with the $1/f$ power spectrum [15]. Other evidences of phase transition such as divergence of autocorrelation time can also be confirmed from the time sequential data [14]. These

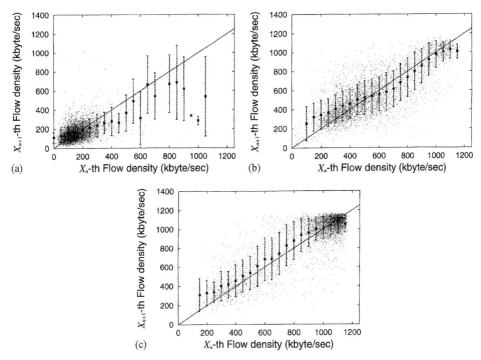

Fig. 6. Map analysis of the fluctuations in Fig. 1(a), (b) and (c), respectively. The dotted straight line shows the 45° line.

are typical properties of dynamic phase transition often called the critical slowing down of relaxation speed.

3. A map analysis of information fluctuation

In this section we analyze the short-time correlation of information flow fluctuation by introducing a map analysis that is a plot of X_n vs. X_{n+1}, where X_n denotes the flow density at time step n.

In Fig. 6(a)–(c) the flow density maps are plotted with time step 0.1 s for the data corresponding to Fig. 1(a), (b) and (c), respectively. In each figure there are 4999 points and each error bar shows the standard deviation around the mean value for each flow density. The solid line shows the relation $X_n = X_{n+1}$.

In the case of sparse state, Fig. 6(a), both X_n and X_{n+1} values tend to concentrate around the peak point at 150 kbytes/s. For $X_n > 150$ the data points are generally lower than the line of $X_n = X_{n+1}$. The large variations in the higher flow densities are due to the shortage of data. In the case, near the critical point, Fig. 6(b), the data points are widely scattered around the line, $X_n = X_{n+1}$, in the middle range of flow density, $400 < X_n < 1000$. For smaller flow densities, $X_n < 400$, the mean value of data points are above the line, and for larger flow densities, $X_n > 1000$, the data points are below the line. In the case of congested state, Fig. 6(c), the mean values are generally above the line for $X_n < 1000$ and many data points are concentrated around the point of $X_n = X_{n+1} = 1000$.

Summarizing these observation facts on the relation of X_n vs. X_{n+1}, we can assume the following first-order autoregressive map with an external random force term, f_n, as the simplest approximation,

$$X_{n+1} = g(X_n) + f_n .$$ (1)

Here, the map function $g(X)$ is empirically constructed from the measured data by introducing a piecewise linear function with two bending points for $g(X)$:

$$g(X) = \begin{cases} a_1 X + b_1 & \text{for } 0 < X < c_1 , \\ X + b & \text{for } c_1 \leqslant X < c_2 , \\ a_2(X - 1.0) + b_2 & \text{for } c_2 \leqslant X \leqslant 1.0 , \end{cases}$$ (2)

where the maximum flow is normalized to 1.0. The parameters, c_1 and c_2, indicate the bending points, and b_1 and b_2 are the minimum and maximum values of the mean values. The parameter b is the most important parameter that controls the mean flow density. For $b < 0$ the generated traffic has a low mean flow density fixed point corresponding to the sparse phase, and for $b > 0$ the generated traffic belongs to the congested phase. The critical point is given by the condition $b = 0$ which means that all the traffic points over the range of $c_1 < X < c_2$ are the fixed points, thus the variance of the fluctuation becomes very large. The values of a_1 and a_2 indicate the line slopes in the small and large flow density range and are determined by requiring the continuity of $g(X)$. As a best fit with the observation we set the parameters $c_1 = 0.25$, $c_2 = 0.75$, $b_1 = 0.1$ and $b_2 = 0.9$.

The generated fluctuations capture all the basic phase transition properties of the real fluctuations qualitatively as for the probability density shapes, the peak point shifts, the width broadening, the rapid increase of autocorrelation time and the power-law distribution of interval of successive high-density flows. However, from the quantitative viewpoint the critical exponents do not fit perfectly. For example, the slope of the log–log plot of the successive high-density interval corresponding to Fig. 5 at the critical point is -0.5 instead of -1.0 directly reflecting the white noise property of the external noise term. Obviously this simple empirical map model needs revision especially on the external noise term.

4. Discussion

We have seen that the Internet can be viewed as a huge composite of phase transition elements and we can find a rich variety of phase transition behaviors. Microscopically each router can show a phase transition behavior due to the buffer's non-linear property. Macroscopically there occurs another type of phase transition due to contagious propagation of congestion among routers. By observing information flow density at a point we can find phase transition behaviors from the fluctuations in the time sequential data as described in the second section.

In order to confirm the validity of this data analysis method we performed several other observations at different observation points. In any case all the results are consistent with those described in this paper. An interesting finding is that the estimated critical flow densities are about 60% of the maximum flow density in general. It is technologically very important to estimate the critical density because packet loss probability becomes dominant at flow density higher than the critical point.

The map analysis captures the qualitative properties of the phase transition very well. It is a transition of fixed points from a low density to a high density, and the map is tangent to the $45°$ line at the critical point. However, as we mentioned the simple map model cannot reproduce the real fluctuation in a quantitative level. The model, especially the external force term, should be revised taking into account the real mechanism of this phase transition behavior.

As the physical mechanism of the present phase transition is not elucidated yet, we are now studying the properties of Ethernet connection by numerical simulation. There is an interesting rule of avoiding collisions of information packets in the Ethernet communication called the back-off. The back-off rules are very complicated, but in a rough sense the rules are given as follows; when two routers shearing an Ethernet cable are trying to send information packets, the one that has just sent a packet successfully has a priority to emit the next packet and the other router must wait. By this effect there occurs a new type of phase transition between two phases; one phase is a low-density mixed flow phase and the other is a high-density oscillatory phase in which active router switches nearly periodically. We are now investigating detail properties of this phase transition by numerical simulation.

Acknowledgements

We wish to thank Y. Watanabe for providing the measurement environment at the public domain of the WIDE Internet backbone for our academic study of Internet traffic. Our research is partially supported by the Japan Society for the Promotion of Science, "Research for the Future" Program: JSPS-RFTF96P00503.

References

[1] W.E. Leland, M.S. Taqqu, W. Willinger, D.V. Willson, IEEE/ACM Trans. Networking 2 (1994) 1.
[2] I. Csabai, J. Phys. A 27 (1999) 417.
[3] V. Paxson, S. Floyd, IEEE/ACM Trans. Networking 3 (1995) 226.
[4] P. Pruthi, A. Erramilli, Proceedings of IEEE ICC'95, 1995, p. 445.
[5] A. Erramili, O. Narayan, W. Willinger, IEEE/ACM Trans. Networking 4 (1996) 209.
[6] W. Willinger, M.S. Taqqu, R. Sherman, D.V. Willson, IEEE/ACM Trans. Networking 5 (1997) 71.
[7] M.E. Crovella, A. Bestavros, IEEE/ACM Trans. Networking 5 (1997) 835.
[8] M. Takayasu, H. Takayasu, T. Sato, Physica A 233 (1996) 924.
[9] T.M. Liggett, Interacting Particle Systems, Springer, New York, 1985.

[10] K. Fukuda, H. Takayasu, M. Takayasu, Fractals 7 (1999) 23.
[11] K. Fukuda, H. Takayasu, M. Takayasu, Adv. Performance Anal. 2 (1999) 45.
[12] A.Yu. Tretyakov, H. Takayasu, M. Takayasu, Physica A 253 (1998) 315.
[13] M. Takayasu, A. Tretyakov, K. Fukuda, H. Takayasu, in: D.E. Wolf (Ed.), Traffic and Granular Flow '97, Springer, Berlin, 1998, p. 57.
[14] M. Takayasu, H. Takayasu, K. Fukuda, Physica A, to appear.
[15] M. Takayasu, Physica A 197 (1993) 371.

ELSEVIER

Physica A 274 (1999) 149–157

www.elsevier.com/locate/physa

Applications of statistical mechanics in number theory

Marek Wolf

Institute of Theoretical Physics, University of Wrocław, Pl. Maxa Borna 9, PL-50-204 Wrocław, Poland

Abstract

The links between statistical physics and number theory are discussed. First the attempts to prove the Riemann Hypothesis by means of the suitable spin model and the Lee–Yang theorem about zeros of the partition function are shortly reviewed. Next, the analogies between random walks and prime numbers are mentioned. In the last section the partition function of the system whose energies are defined by the distances between consecutive primes is calculated. The arguments are given that such a "prime numbers gas" behaves like a set of noninteracting harmonic oscillators. © 1999 Elsevier Science B.V. All rights reserved.

1. Introduction

There are a lot of links between number theory and physics, see e.g. two fat books [1,2]. Very well known are applications of number theory in chaos, both classical and quantum. As an example the Fibonnaci numbers can be mentioned: there is an ubiquity of places in the theory of chaos, where they appear (see [3]). Other papers where some mathematical facts about primes were applied to the study of quantum chaos can be found in [4–7]. The Hardy–Littlewood conjecture on the gaps between (not necessary consecutive) primes inspired some works on the correlations between periodic orbits, see [8,9]. There are also a few papers where primes served as a toy model for testing the methods used by physicists. For example in [10] the multifractality of primes was investigated, while in [11] the appropriately defined Lyapunov exponents for the distribution of primes were calculated numerically. $1/f$ noise in the distribution of primes was found in [12]. Quite recently the Wiener–Khintchine formula relating the spectral densities and the autocorellation function was applied in [13] to the problem of distribution of the pairs of primes.

E-mail address: mwolf@ift.uni.wroc.pl (M. Wolf)

In this lecture I will confine myself only to the influence of statistical mechanics on the number theory. In Section 2 I will review application of the concepts and ideas from the statistical mechanics to the Riemann $\zeta(s)$ function and the Riemann Hypothesis. In the next section I will speak about the links between random walks and prime numbers. In the last section I will discuss some of my own results which allow to treat prime numbers as a gas of noninteracting harmonic oscillators.

2. The Riemann hypothesis and statistical mechanics

The Riemann hypothesis is probably the most famous open problem in mathematics. The Riemann's zeta function $\zeta(s)$ for a complex variable s is defined for $\mathrm{Re}\,s > 1$ by

$$\zeta(s) = \sum_{n=1}^{\infty} \frac{1}{n^s} = \prod_p \left(1 - \frac{1}{p^s}\right)^{-1} \tag{1}$$

and for other values of s by its analytic continuation. One of such representation is given by

$$\zeta(s) = \frac{1}{(1 - 2^{1-s})\Gamma(s)} \int_0^{\infty} \frac{t^{s-1}}{1 + \exp t} \, dt, \quad s \neq 1. \tag{2}$$

For negative even integer values of s, i.e. $s = -2, -4, -6, \ldots$ $\zeta(s)$ is zero, while all other non-trivial zeros of $\zeta(s)$ must lie in the strip $0 < \mathrm{Re}\,s < 1$. It has been conjectured by Riemann that all non-trivial zeros of $\zeta(s)$ actually lie on the critical line $\mathrm{Re}\,s = \frac{1}{2}$. This statement is known as the Riemann hypothesis (hereafter referred to as RH). It is very important in the number theory and other branches of mathematics as well because there are thousands of theorems proved under the assumption of the validity of RH.

The $\zeta(s)$ function fulfills the following functional equation [27]

$$2\Gamma(s)\cos\left(\frac{\pi}{2}s\right)\zeta(s) = (2\pi)^s \zeta(1-s). \tag{3}$$

It is analogous to the Kramers–Wannier [14] duality relation for the partition function of the two-dimensional Ising model with parameter J expressed in units of kT (i.e. equal to interaction constant divided by kT)

$$Z(J) = 2^N (\cosh(J))^{2N} (\tanh(J))^N Z(\tilde{J}), \tag{4}$$

where N denotes the number of spins and $\tilde{J}$ is related to J via $e^{-2\tilde{J}} = \tanh(J)$, see e.g. [15]. This analogy is a starting point for a series of papers [16–20] where attempts to find the appropriate spin model possessing the partition function $Z(\beta)$ expressed by the $\zeta(s)$ function were undertaken. Knauf succeeded to find such spin system that its partition function $Z(\beta)$ is equal to the ratio of two zeta functions: $Z(\beta) = \zeta(\beta-1)/\zeta(\beta)$. To localize the zeros of the partition function $Z(\beta)$ the Lee–Yang theorem [21] on the zeros could be used, what in turn should give the Riemann conjecture that all nontrivial zeros of $\zeta(s)$ are of the form $1/2 + i\sigma$. I would like to point out another paper written by a physicist aimed at the proof of the RH. Namely in [22] Susumu Okubo tried to find the hamiltonian whose eigenvalues coincide with the imaginary parts of the zeros

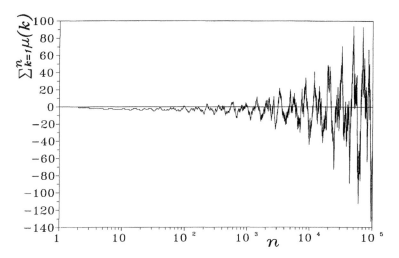

Fig. 1. Plot of $M(n) = \sum_{k=1}^{n} \mu(k)$ for N up to 10^5.

of $\zeta(s)$. That is, if the zeros of $\zeta(s)$ can be interpreted as the eigenvalues of $1/2+\mathrm{i}T$, where T is a Hermitian operator on some Hilbert space, then since the zeros of a Hermitian operator are real, the Riemann hypothesis follows. This idea was originally put forth by Polya and Hilbert at the beginning of the century. The problem is that Okubo was not able to show that his hamiltonian is self-adjoint. In the similar spirit are the papers written by Allain Connes [23,24], the review of these attempts can be found in [25].

3. Random walks and prime numbers

There were a few papers published on the relation between random walks and number theory. In 1973 P. Bilingsley published the paper "Prime numbers and Brownian motion" [26], where he defined the set of random walks (RW) employing the factorization of integers into the primes. There exists even the link between RW and RH. Namely let $\mu(n)$ denote the the Möbius function:

$$\mu(m) = \begin{cases} 1 & \text{if } m = 1 , \\ 0 & \text{if } m \text{ is divisible by a square of a prime} , \\ (-1)^k & \text{otherwise} , \end{cases} \quad (5)$$

where k is the number of prime divisors of m. For example $\mu(25) = 0$, $\mu(14) = 1$, $\mu(30) = -1$. Let $M(n)$ denote the cumulative values of the Möbius function: $M(n) = \sum_{k=1}^{n} \mu(k)$. The values of $\mu(n)$ are equiprobably equal to 1 and -1 and in Fig. 1 I have plotted the graph of $M(n)$ for $n < 10^5$. The resemblance of $M(n)$ to the symmetrical random walk led Mertens over 100 years ago to make the conjecture that $M(n)$ grows not faster than the mean displacement of the symmetrical random walk, i.e. that

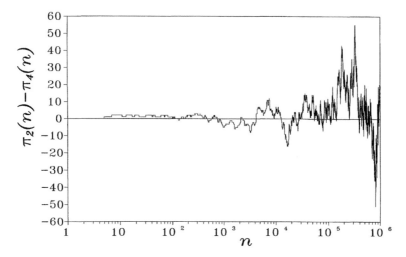

Fig. 2. The plot of $W(n) = \pi_2(n) - \pi_4(n)$ up to $n = 10^6$.

$|M(n)| < \sqrt{n}$. Later it was shown that the Mertens conjecture imply the Riemann Hypothesis, but the converse is not true. In contrast the modified Mertens conjecture $M(n) = \mathcal{O}(n^{1/2+\varepsilon})$ is equivalent to the Riemann Hypothesis, see [27]. The fact that $M(n)$ behaves like a one dimensional random walk was pointed out in [28] and used to show that RH is "true with probability 1". However in 1985 Andrew Odlyzko and Hermann te Riele [29] have disproved the Mertens conjecture and they raised even the possibility that $\lim\sup_{n\to\infty} |M(n)|n^{-1/2} = \infty$. Finally let us notice that the Möbius function has the physical interpretation, namely in [30] it was shown that $\mu(n)$ can be interpreted as the operator $(-1)^F$ giving the number of fermions in quantum field theory. In this approach the property $\mu(m) = 0$ when m is not squarefree is equivalent to the Pauli exclusion principle.

In [31] I have defined the following random walk on the primes. Among the primes the subset of Twin and Cousins primes is distinguished: Twins are such numbers $\{p, p'\}$ that both p and $p' = p+2$ are prime and primes $\{p, p'\}$ such that both p and $p' = p+4$ are prime and are called Cousins. A common belief is that there is infinity of twins and cousins, but this still remains unproved. The random walk is constructed as follows: If we meet the pair of Twins, then the random walker makes the step say up and if the pair of Cousin is encountered, then the step down is performed. If $\pi_2(x)$ will denote the number of Twins up to n and $\pi_4(n)$ the number of Cousins up to n then the random walk $W(n)$ is given by $W(n) = \pi_2(n) - \pi_4(n)$, see Fig. 2. In [31] I gave numerical evidence that such a deterministic walk possess the properties of the ordinary one dimensional random walk. In particular the number of returns to the origin in n steps is proportional to $\sqrt{n}$ — obeys the square-root law. Also the mean fluctuation function $F(l)$ displays perfect power law dependence $F(l) \sim l^{1/2}$ indicating that the defined RW is not correlated. The function $F(l)$ is defined by the equation:

$$F^2(l) = \langle (\Delta W(l))^2 \rangle - \langle \Delta W(l) \rangle^2, \qquad (6)$$

where $\Delta W(l) = W(l + l_0) - W(l_0)$ and the average is performed over all starting points l_0 in the random walk [32,33]. For the usual random walk the function $F(l)$ is described by the power law: $F(l) \sim l^\alpha$ with the exponent $\alpha = \frac{1}{2}$ and for the above random walk I found in [31] it is $\alpha = 0.49$. The set of returns to the origin (i.e. a set of such n that $W(n) = 0$) has the fractal dimension 0.51, what is very close to the theoretical value $\frac{1}{2}$.

4. Prime numbers as a gas of harmonic oscillators

Bernard Julia has created statistical theory of numbers in [16]. He considered a gas of independent bosonic oscillators of the energies being the logarithms of consecutive primes: $E_n = \ln(p_n)$ and showed that such a system possess the canonical partition function coinciding with the Riemann function. In this section I will discuss the same gas but with energies defined in a different way.

We consider the "prime number gas" in which the role of atoms is played by the primes and the "energy of excitation" of a given "atom" — prime p_n is equal to the distance to the next prime p_{n+1}. Explicitly we have $p_1 = 2, p_2 = 3, p_3 = 5, \ldots,$ $p_{15} = 47, p_{16} = 53, \ldots$ and hence $E_1 = 1, E_2 = 2, E_3 = 2, \ldots, E_{15} = 6, \ldots$. There is no interaction between these "atoms". We will show that such a system behaves like an ensemble of harmonic oscillators.

Let $h_N(d) = number\ of\ pairs\ p_n, p_{n+1} < N\ with\ d = p_{n+1} - p_n$ counts the number of consecutive primes separated by the gap $d = p_{n+1} - p_n$ and smaller than N. Here the letter h stays as an abbreviation for "histogram". In the paper [34] it was conjectured that $h_N(d)$ decreases exponentially with d:

$$h_N(d) \sim 2c_2 \frac{\pi^2(N)}{N} \prod_{p \mid d,\, p > 2} \frac{p - 1}{p - 2} e^{-d\pi(N)/N} . \tag{7}$$

Here the gaps d can be arbitrary even numbers $d = 2, 4, 6, \ldots$, the constant

$$c_2 \equiv \prod_{p > 2} \left(1 - \frac{1}{(p - 1)^2}\right) = 0.66016\ldots \tag{8}$$

is called "Twin constant" and $\pi(N)$ denotes the total number of primes smaller than N. I have counted on a computer the number of gaps between consecutive primes up to $N = 2^{44} \approx 1.76 \times 10^{13}$. During the computer search the data representing the function $h_N(d)$ were stored at values of N, forming the geometrical progression with the ratio 4, i.e. at $N = 2^{20}, 2^{22}, \ldots, 2^{42}, 2^{44}$. Such a choice of the intermediate thresholds as powers of 2 was determined by the employed computer program, because the primes were coded as bits. The resulting curves are plotted on the semi-logarithmic axes in Fig. 3. The straight lines are the least-square fits of the assumed exponential decrease of $h_N(d)$ with d to the actual values. A lot of regularities can be seen in this figure. In particular, the characteristic oscillating pattern of points is caused by the product

$$J(d) = \prod_{p|d,\, p>2} \frac{p-1}{p-2} \tag{9}$$

appearing in (7). This product is also responsible for the phenomenon of "champions" — the most often occurring gaps between consecutive gaps, see [37]. Putting in (7) the Gauss formula $\pi(N) \sim N/\ln(N)$ we get that

$$h_N(d) \sim 2c_2 \frac{N}{\ln^2(N)} \prod_{p|d,\, p>2} \frac{p-1}{p-2} e^{-d/\ln(N)} . \tag{10}$$

Because the number $h_N(d)$ is directly connected to the probability of picking up randomly the pair of consecutive primes separated by d we recognize in the above formula the Boltzmann factor for the system of the excitations given by d and the temperature identified with $\ln(N)$:

$$d \leftrightarrow E_n ,$$

$$\ln(N) \leftrightarrow kT ,$$

$$e^{-d/\ln(N)} \leftrightarrow e^{-E_n/kT} .$$

Here $d = 2, 4, \ldots$ thus values of d are equally distributed which suggests that (10) leads to the partition function for the harmonic oscillator. Calculating the partition function of the finite system of primes $p_n < N$ we have to take into account that the gaps d are bounded for finite N. Namely the maximal gap $G(N)$ between two consecutive primes is roughly (the Cramer conjecture, see [35,36]):

$$G(N) \sim \ln^2(N) . \tag{11}$$

Hence we have the cutoff at the maximal gap which depends on N. For the partition function $Z(N)$ of the "gas of primes" we get the following result:

$$Z(N) = \sum_{d=2}^{G(N)} e^{-d/\ln(N)} = \frac{1 - (1/N)e^{2/\ln(N)}}{e^{2/\ln(N)} - 1} . \tag{12}$$

For large N we can skip the term proportional to $1/N$ and in this way we get the result

$$Z(N) = \frac{1}{e^{2/\ln(N)} - 1} , \tag{13}$$

what corresponds exactly to the harmonic oscillator with parameters $\hbar\omega = 4$, because for the latter we have

$$Z(\beta) = \sum_{n=0}^{\infty} e^{-\beta\hbar\omega(n+\frac{1}{2})} = \frac{1}{e^{\beta\hbar\omega/2} - 1} . \tag{14}$$

The "prime gas" considered above is somehow strange because N should be identified with the volume of the system and simultaneously $\ln(N)$ serves as a temperature $\ln(N) = kT$. Thus, we can say that the state equation of the above gas is $kT = \ln(V)$ and there is no pressure.

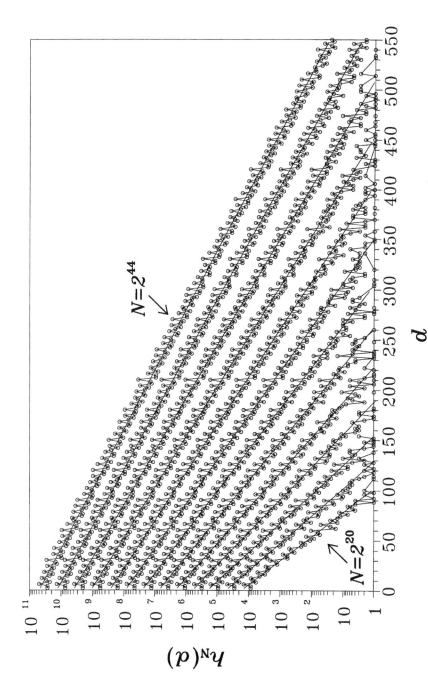

Fig. 3. The plot showing the dependence of the histogram $h_N(d)$ on d at $N = 2^{20}, 2^{22}, \ldots, 2^{44}$. There is a logarithmical scale on the y-axis, while on the x-axis there is a linear scale. The values of $h_N(d)$ obtained from the computer search are represented by small circles. The points oscillate around the straight lines with period 6. Let us mention "local" spikes at $d = 30 (= 2 \cdot 3 \cdot 5), 60, \ldots$ and at $d = 210$. They are called "champions", see [37].

Another corroboration of the above conclusion that the "gas of primes" with the "energy of excitation" of a given prime p_n equal to the distance to the next prime p_{n+1} behaves like a harmonic oscillator can be found in the paper of Mussardo [38,39]. In [38] Mussardo raised the question: what is a quantum mechanical system whose energy eigenvalues can be one-to-one mapped into the sequence of prime numbers. He showed that it is possible to construct in the semiclassical way (the WKB method) the potential whose eigenvalues will coincide after rescaling with the prime numbers. The potential $W(x)$ is obtained after inverting the following infinite series:

$$x(W) = \frac{\hbar}{\sqrt{2m}} \sum_{m=1}^{\infty} \frac{\mu(m)}{m} \int_2^W \frac{\mathscr{E}^{(1-m)/m}}{\ln \mathscr{E} \sqrt{W - \mathscr{E}}} \, d\mathscr{E} \,. \tag{15}$$

The potential $W(x)$ can be calculated numerically and it turns out that for large x harmonic oscillator $W(x) \sim x^2$ is obtained what can be expected in the WKB approximation. However, everybody will agree that primes are not such a simple system like a harmonic oscillator. The explanation is as follows: In the above calculations we have used a lot of approximate expressions and thus the result (12) is not strictly precise; for example, the error term is unknown in the main conjecture (7). Mussardo in his derivation has also made a lot of crude approximations. Thus we can say that the primes behave like a harmonic oscillator *in the 0th order approximation*. In higher orders there are corrections which are responsible for all subtleties and mysteries in the distribution of prime numbers.

5. Instead of summary

I would like to follow the advise of Prof. D. Mermin [40] to the beginner speakers and I will skip the summary. He wrote: *The ubiquitous heavy-handed concluding summary should be omitted; a talk should tell such a good story that a summary is uncalled for. Imagine "War and Peace' ending with a summary"*. I agree with this as well as with many other statements in this article.

References

[1] J.M. Luck, P. Moussa, M. Waldschmidt (Eds.), Number Theory and Physics, Springer, Berlin, 1990.
[2] M. Waldschmidt, P. Moussa, J.-M. Luck, C. Itzykson (Eds.), From Number Theory to Physics, Springer, Berlin, 1992.
[3] Schuster, Deterministic Chaos: An Introduction, Physik Verlag, Weinheim, 1989.
[4] M. Berry, Some quantum-to-classical asymptotics, in: A. Voros et al. (Eds.), Chaos and Quantum Physics, Elsevier, Amsterdam, 1991.
[5] R. Aurich et al., Universal signatures of quantum chaos, Phys. Rev. Lett. 73 (1994) 1356.
[6] R. Aurich et al., in: G. Casati et al. (Eds.), Quantum Chaos, CXIX 'Enrico Fermi' International School of Physics, North-Holland, Amsterdam, 1993.
[7] P. Sarnak, Artihmetic quantum chaos, Israeli Mathematical Conference Proceedings, Vol. 8, 1995, p. 183.
[8] N. Argaman et al., Correlations in the actions of periodic orbits derived from quantum chaos, Phys. Rev. Lett. 71 (1993) 4326–4329.

[9] E. Bogomolny, P. Lebœuf, Statistical properties of the zeros of zeta functions — beyond the Riemann case, Nonlinearity 7 (1994) 1155–1167.

[10] M. Wolf, Physica 160A (1989) 24.

[11] Z. Gamba, J. Hernando, L. Romanelli, Phys. Lett. A 145 (1990) 106.

[12] M. Wolf, $1/f$ noise in the distribution of primes, Physica A 241 (1997) 439.

[13] H. Gopalkrishna Gadiyar, R. Padma, Ramanujan — fourier series, the Wiener–Khintchine formula and the distribution of prime pairs, Physica A 269 (1999) 503–510.

[14] H.A. Kramers, G.H. Wannier, Phys. Rev. 60 (1941) 252.

[15] R.P. Feynman, Statistical Mechanics, W.A. Benjamin, New York, 1972.

[16] B. Julia, Statistical theory of numbers, in: J.M. Luck, P. Moussa, M. Waldschmidt (Eds.), Number Theory and Physics, Springer, Berlin, 1990, p. 276.

[17] B. Julia, Physica A 203 (1994) 425.

[18] A. Knauf, J. Statist. Phys. 73 (1993) 423.

[19] A. Knauf, Comm. Math. Phys. 153 (1993) 77.

[20] A. Knauf, Comm. Math. Phys. 196 (1998) 703.

[21] T.D. Lee, C.N. Yang, Phys. Rev. 87 (1952) 404, 410.

[22] S. Okubo, J. Phys. A 31 (1998) 1049–1057.

[23] A. Connes, C. R. Acad. Sci. 323 (1996) 1231.

[24] J.-B. Bost, A. Connes, Selecta Math. (New Series) 1 (1995) 411.

[25] P.B. Cohen, Dedekind zeta functions and quantum statistical mechanics, Preprint ESI 617 (1998), available at http://www.esi.ac.at.

[26] P. Billingsley, Prime numbers and Brownian motion, Amer. Math. Mon. 80 (1973) 1099–1115.

[27] e.g. E. Titchmarch, The Theory of the Riemann Zeta-Function, Oxford University Press, Oxford, 1986.

[28] I.J. Good, R.F. Churchouse, Math. Comput. 22 (1968) 857.

[29] A.M. Odlyzko, H.J.J. te Riele, J.f.d. reine und angewandte Math. 357 (1985) 138.

[30] D. Spector, Commun. Math. Phys. 127 (1990) 239.

[31] M. Wolf, Physica A 250 (1998) 335.

[32] C.-K. Peng, S.V. Buldyrev, A.L. Goldberger, S. Havlin, F. Sciortino, M. Simons, H.E. Stanley, Nature 356 (1992) 168.

[33] A.A. Tsonis, J.B. Elsner, J. Stat. Phys. 81 (1995) 869–880.

[34] M. Wolf, Some conjectures on the gaps between consecutive primes, available at www.ift.uni.wroc.pl/~mwolf.

[35] H. Cramer, On the order of magnitude of difference between consecutive prime numbers, Acta Arith. 2 (1937) 23–46.

[36] D. Shanks, On maximal gaps between successive primes, Math. Comput. 18 (1964) 464.

[37] A. Odlyzko, M. Rubinstein, M. Wolf, Experim. Math. 8 (1999) 107.

[38] G. Mussardo, The quantum mechanical potential for the prime numbers, Preprint ISAS/EP/97/153, available as cond-mat/9712010.

[39] R. Matthews, New Scientist, 10 January 1998, p. 18.

[40] D. Mermin, Phys. Today, November 1992, p. 9.

ELSEVIER

Physica A 274 (1999) 158–170

www.elsevier.com/locate/physa

Computational test of kinetic theory of granular media

M.D. Shattuck[a], C. Bizon[b], J.B. Swift[a], Harry L. Swinney[a,*]

[a] Center for Nonlinear Dynamics and Department of Physics, The University of Texas at Austin, Austin, TX 78712, USA
[b] Colorado Research Associates, 3380 Mitchell Lane, Boulder, CO 80301 USA

Abstract

Kinetic theory of granular media based on inelastic hard sphere interactions predicts continuum equations of motion similar to Navier–Stokes equations for fluids. We test these predictions using event-driven molecular dynamics simulations of uniformly excited inelastic hard spheres confined to move in a plane. The event-driven simulations have been previously shown to quantitatively reproduce the complex patterns that develop in shallow layers of vertically oscillated granular media. The test system consists of a periodic two-dimensional box filled with inelastic hard disks uniformly forced by small random accelerations in the absence of gravity. We describe the inelasticity of the particles by a velocity-dependent coefficient of restitution. Granular kinetic theory assumes that the velocities at collision are uncorrelated and close to a Maxwell–Boltzmann distribution. Our two-dimensional simulations verify that the velocity distribution is close to a Maxwell–Boltzmann distribution over 3 orders of magnitude in velocity, but we find that velocity correlations, of up to 40% of the temperature, exist between the velocity components parallel to the relative collision velocity. Despite the velocity correlations we find that the calculated transport coefficients compare well with kinetic theory predictions. © 1999 Elsevier Science B.V. All rights reserved.

1. Introduction

Transport and processing of granular materials is important in industries ranging from food preparation to pharmaceuticals to coal processing. However, theoretical understanding of granular flows has lagged significantly behind that of liquid and gas flows. No basic theory of granular flows comparable to the Navier–Stokes equations for fluids has attained widespread acceptance [1], and it has been argued that such a theory is not possible [2]. This lack of understanding leads to significant waste in

---

* Corresponding author. Fax: +1-512 471 1558.
E-mail address: swinney@chaos.ph.utexas.edu (H.L. Swinney)

0378-4371/99/$ - see front matter © 1999 Elsevier Science B.V. All rights reserved.
PII: S 0378-4371(99)00433-1

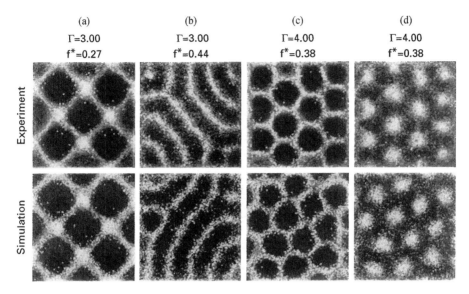

Fig. 1. Comparison of standing wave patterns obtained in experiment and an event-driven molecular dynamics simulation: (a) squares, (b) stripes, (c) and (d) alternating phases of hexagons [4]. All patterns oscillate at $f/2$. The layer depth is 5.42 times the particle diameter σ. The experiments use lead spheres sieved between 0.5 and 0.6 mm in a container which is 100σ on each side.

solids processing [3]. An increased understanding of granular flows could improve this situation.

Recent experiments [5–7] show that the patterns formed in vertically oscillated granular layers are strikingly similar to patterns seen in fluid systems (see Fig. 1).This similarity suggests that in this and other rapidly flowing granular systems, equations of motion similar to Navier–Stokes may apply. Such equations have been derived from kinetic theory for granular media flows under the assumption of binary hard sphere interactions in the limit of small energy loss in collisions [8–14]. In particular, for granular flows Jenkins and Richman [8,9] derived mass, momentum, and energy conservation equations:

$$\frac{\partial n}{\partial t} + \nabla \cdot (n\mathbf{v}) = 0 \,, \tag{1}$$

$$n\frac{\partial \mathbf{v}}{\partial t} + n\mathbf{v} \cdot \nabla \mathbf{v} = -\nabla \cdot \underline{\mathbf{P}} \,, \tag{2}$$

$$n\frac{\partial T}{\partial t} + n\mathbf{v} \cdot \nabla T = -\nabla \cdot \mathbf{q} - \underline{\mathbf{P}} : \underline{\mathbf{E}} - \gamma \,, \tag{3}$$

where n is the number density, $\mathbf{v}$ is the vector velocity with components v_i, T the granular temperature (which is not the thermodynamic temperature but by analogy to molecular gases is the variance of the velocity distribution), and $E_{ij} = \frac{1}{2}(\partial_i v_j + \partial_j v_i)$ are the elements of the symmetrized velocity gradient tensor $\underline{\mathbf{E}}$. The constitutive relation for the pressure tensor $\underline{\mathbf{P}}$ is Newton's stress law,

$$\underline{\mathbf{P}} = (P - 2\lambda \mathrm{Tr}\underline{\mathbf{E}})\underline{\mathbf{I}} - 2\mu(\underline{\mathbf{E}} - (\mathrm{Tr}\underline{\mathbf{E}})\underline{\mathbf{I}}) \,, \tag{4}$$

where Tr denotes trace and $\underline{\mathbf{I}}$ is the unit tensor. For the heat flux $\mathbf{q}$, the constitutive relation is Fourier's heat law,

$$\mathbf{q} = -\kappa \nabla T \,. \tag{5}$$

For a two-dimensional system the equations close [9] with the equation of state, which is the ideal gas equation of state with a term that includes dense gas and inelastic effects,

$$P = (4/\pi\sigma^2)vT[1 + (1+e)G(v)] \,, \tag{6}$$

and the predicted values for the transport coefficients. These are the temperature loss rate per unit volume γ, the bulk viscosity λ, the shear viscosity μ, and the thermal conductivity κ,

$$\gamma_0 = \frac{16vG(v)}{\sigma^3}(1 - e^2)\left(\frac{T}{\pi}\right)^{3/2} \,, \tag{7}$$

$$\lambda_0 = \frac{8vG(v)}{\pi\sigma}\sqrt{\frac{T}{\pi}} \,, \tag{8}$$

$$\mu_0 = \frac{v}{2\sigma}\left[\frac{1}{G(v)} + 2 + \left(1 + \frac{8}{\pi}\right)G(v)\right]\sqrt{\frac{T}{\pi}} \,, \tag{9}$$

$$\kappa_0 = \frac{2v}{\sigma}\left[\frac{1}{G(v)} + 3 + \left(\frac{9}{4} + \frac{4}{\pi}\right)G(v)\right]\sqrt{\frac{T}{\pi}} \,, \tag{10}$$

where the subscript 0 denotes predicted values, σ is the hard disk diameter and $v = n\pi\sigma^2/4$ is the area fraction. $G(v)$ is a correction for positional correlations, which are important as the density increases. $G(v)$ is determined from $g(r, v)$, the radial distribution function, which is the probability of having a pair of particles whose relative distance is in the interval $r, r + dr$ at a density v, normalized by the same probability for an ideal gas at the same density and evaluated at $r = \sigma : G(v) \equiv vg(\sigma, v)$. $g(r, \sigma)$ gives the increased probability of collisions due to excluded volume in dense gases. For elastic hard disks, $G(v)$ is often described by a formula derived by Carnahan and Starling [15],

$$G_{CS}(v) = \frac{v(16 - 7v)}{16(1 - v)^2} \,. \tag{11}$$

Eq. (11) works well for elastic particles if the solid fraction is below 0.675, where a phase transition occurs [16], and is often used in modeling granular media [9]. $G(v) \equiv vg(\sigma, v)$ defines G in terms of the radial distribution function $g(r, v)$, evaluated at $r = \sigma$, while (11) is a particular model for G, denoted by the subscript CS.

The equations of motion (1)–(3) differ slightly from the dense gas Navier–Stokes equations by the addition of a temperature loss rate term (7) in the energy equation (3) and in the equation of state for the pressure (6), which depends on the ratio of the normal relative velocity v_n of a colliding pair after a collision to v_n before a collision (i.e., the coefficient of restitution e). Eq. (7) was derived for a coefficient of restitution which is independent of v_n, but in real materials e is a function of v_n [17]. In our simulation, we allow e to vary as [4]

$$e(v_n) = \begin{cases} 1 - Bv_n^{\beta}, & v_n < v_a, \\ \varepsilon, & v_n > v_a, \end{cases} \tag{12}$$

where v_n is the component of relative velocity along the line joining particle centers (normal to the contact surface), $B = (1 - \varepsilon)(v_a)^{-\beta}$, $\beta = \frac{3}{4}$, ε is a constant, chosen to be 0.7, and v_a is effectively set to unity as it is the velocity scale used to nondimensionalize all quantities to follow. By use of (12) a new volumetric loss rate γ_e is determined [18].

$$\gamma_e = \frac{4vG\sqrt{T}}{\sigma^3 \pi^{3/2}} \left[(1 - e_0^2)(v_a^2 + 4T)\exp(-v_a^2/4T) + 4I \right], \tag{13}$$

where

$$I = 2^{1+\beta} A T^{1+\beta/2}(\Gamma(2 + \tfrac{1}{2}\beta) - \Gamma(2 + \tfrac{1}{2}\beta, v_a^2/4T))$$

$$- A^2 2^{2\beta} T^{1+\beta}(\Gamma(2 + \beta) - \Gamma(2 + \beta, v_a^2/4T)), \tag{14}$$

$\Gamma(a)$ is the gamma function, and $\Gamma(a, b)$ is the incomplete gamma function. In the limit that $v_a \to 0$, $\gamma_e \to \gamma_0$.

A further assumption used in deriving (7) for γ is molecular chaos (i.e., no velocity–velocity correlations). Molecular chaos assumes that the velocity–velocity distribution function $\Theta_2(\mathbf{v}_1, \mathbf{v}_2)$ equals the product of the single-particle velocity distribution of particle 1, $\Theta(\mathbf{v}_1)$, and the single-particle velocity distribution of particle 2, $\Theta(\mathbf{v}_2)$. However, γ can be determined without regard to the form of $\Theta_2(\mathbf{v}_1, \mathbf{v}_2)$ as

$$\gamma = \frac{(1 - e^2)G}{\sigma} \frac{\langle v_n^2 \rangle_c}{\langle v_n \rangle_c} nT, \tag{15}$$

where e is the constant coefficient of restitution, and

$$\langle h(\mathbf{v}_1, \mathbf{v}_2) \rangle_c = \int_{\text{all collisions}} h(\mathbf{v}_1, \mathbf{v}_2)\Theta_2(\mathbf{v}_1, \mathbf{v}_2)\, d\mathbf{v}_1\, d\mathbf{v}_2. \tag{16}$$

$h(\mathbf{v}_1, \mathbf{v}_2)$ is any function of the collision velocities. If the distribution of relative normal velocity at collision is equal to that predicted by molecular chaos, $P(v_n) = (1/2T)v_n \exp(-v_n^2/4T)$, then $\langle v_n^2 \rangle_c = 4T$ and $\langle v_n \rangle_c = \sqrt{\pi T}$, so that $\gamma = \gamma_0$ is recovered.

While (1)–(10) have been available for 15 years, there have been few experimental [19–21] or numerical [18,22–24] tests. In this paper we will directly test the validity of these equations, using event-driven molecular dynamics simulations. Experiments such as those in vibrated layers of granular materials provide an unprecedented opportunity to study granular fluid-like behavior. However, experiments alone do not provide information on the microscopic underpinnings of the kinetic theory description, due to the difficulty of internal measurements in three-dimensional systems. To overcome this difficulty we have developed an event-driven molecular dynamics simulation capable of quantitatively reproducing laboratory experiments [4] in vertically oscillated granular media, including wavelength-changing secondary instabilities [25]. The simulation is based on assumptions similar to those for granular kinetic theory. In particular, particles obey Newton's laws between binary instantaneous collisions (hard sphere model) that conserve momentum but dissipate energy. However, unlike the kinetic theory models, the energy dissipation can be large, and there is no restriction on the velocity distribution.

In Section 2 we describe the simulation technique and discuss its quantitative verification with experiments in vertically oscillated shallow granular media. In Section 3 we describe tests of the assumptions and results of kinetic theory.

2. Numerical model

The purpose of simulations of our model is to test granular kinetic theory assumptions and predictions. However, to determine if granular kinetic theory applies to real granular materials we must first validate our simulation by comparison with experiments. For this purpose we have chosen a vertically oscillated thin granular layer that shows the kind of fluid-like behavior to which kinetic theory is likely to apply. In this system a thin layer of granules is vertically oscillated by an electromagnetic shaker. Different patterns emerge for various shaking amplitudes and frequencies [5–7] (see Fig. 1).

2.1. Vertically oscillated granular layer

We have developed an event-driven numerical simulation of hard spheres interacting through momentum conserving, energy dissipating collisions [4]. In this type of simulation [26,27] time advances from collision to collision with ballistic motion between collisions. A sorted list of the time-to-next-collision for each particle is used to determine when the next collision will occur. The simulation performs the collision using an operator which maps the velocities of each particle before the collision to their values after the collision. Collisions conserve momentum but not energy. The collision duration is assumed zero, therefore limiting the particle interactions to binary collisions. Energy is lost in collisions through a normal coefficient of restitution e defined by the ratio of the outgoing normal relative velocity v_n to the incoming v_n. Therefore, energy is lost at a rate of $1 - e^2$ per collision. e is a function of v_n given by (12), which is consistent with real materials [28] and prevents the numerical instability of inelastic collapse [17].

To validate the simulation, we conducted experiments in a vertically oscillated cell $100\sigma \times 100\sigma$ with $60\,000 \pm 8$ lead spheres ($\sigma = 0.55$ mm) [4]. Experiments and simulations under the same conditions are compared in Fig. 1 using the nondimensional frequency $f^* = f\sqrt{H/g}$, where f is the frequency of oscillation, H is the height of the layer, and g is the acceleration of gravity and the nondimensional maximum oscillation acceleration amplitude is $\Gamma = A(2\pi f)^2/g$, where A is the maximum stroke amplitude.

Three collisional particle properties — the coefficient of friction μ, the value of the constant portion of normal coefficient of restitution ε (see (12)), and the cutoff for the rotation coefficient of restitution β_0 — must be determined for the simulation. For the comparison to experiments, β_0 is taken from the literature [29], and ε and μ are determined by adjusting their values until the pattern wavelength in the simulation and experiment matched in two specific runs. For the comparisons to the kinetic theory, rotation is ignored and therefore β_0 and μ are not needed. This is consistent with the kinetic theory, which also ignores rotation.

The results of the simulation for various control parameters are shown in Fig. 1. Patterns obtained in the simulation and experiment at the same values of the control parameters show a striking correspondence. Further, the pattern wavelengths for various f^* in experiment and simulation agree well, even when comparing the simulation in a cell 100σ wide with experiments in a large container with a diameter of 982σ.

2.2. Randomly accelerated forcing of particles in a box with no gravity

For comparison to kinetic theory we restrict our full three-dimensional simulation (which includes particle rotations) to two dimensions and to particles that do not rotate. We use a periodic cell which is 52.6σ on each side; the number of particles determines the density of the granular gas. The particles interact through the velocity-dependent coefficient of restitution described by (12). In elastic hard sphere models, transport properties, velocity distributions, velocity correlations, and $G(v)$ can all be determined from simulations at thermal equilibrium. For inelastic hard sphere models, because of the energy loss, the equilibrium state is for all particles to be at rest. Therefore, we constantly add energy in a stochastic manner (random accelerated forcing) to achieve a steady state at finite temperature. The situation is opposite that in simulations on non-equilibrium systems of elastic particles, where the constant energy input from the driving must be removed through an artificial means [30].

In our model each particle moves under a uniform acceleration:

$$\mathbf{a}_i = a_0 \hat{\mathbf{r}}_i . \tag{17}$$

The magnitudes of all particle accelerations, a_0, are the same, but the directions, $\hat{\mathbf{r}}_i$, are randomly and uniformly chosen. When a collision occurs, two particles are given new $\hat{\mathbf{r}}_i$. In order to conserve total momentum, we hold the total acceleration of the particles at zero by giving exactly opposite accelerations to pairs of particles. Initially, each particle is paired with another, and these are given opposite accelerations. Later, when one particle is chosen and its acceleration randomized, its partner particle is also given a new acceleration, opposite to the first particle's. The random acceleration model is chosen because it closely models thin flat disks on an air table [18,31]. This correspondence with an experimentally realizable system can be exploited in the future to directly test kinetic theory against a physical system. This type of forcing has several technical implications for the simulation which are described in Ref. [18].

Spatially uniform application of the accelerated forcing model to an inelastic gas of hard disks produces a homogeneous steady state. From this state we extract the velocity distribution function, velocity–velocity correlations, $G(v)$, and the temperature loss rate γ for various constant densities and temperatures. To extract the shear viscosity and the thermal conductivity we introduce a spatially dependent heating to produce either a velocity gradient to measure viscosity or a temperature gradient to measure thermal conductivity.

3. Tests of granular kinetic theory

In this section we test granular kinetic theory against our validated event-driven molecular dynamics simulation of a two-dimensional granular gas heated by random acceleration. Comparison of other types of heating as well as more detailed analysis of our results can be found in Ref. [18].

3.1. Test of assumptions

3.1.1. Velocity distribution functions

To obtain (1)–(10) it is assumed that the deviations of the velocity distribution $\Theta(v)$ of an inelastic hard sphere gas from a Maxwell–Boltzmann distribution can be expanded in a polynomial in the velocity. The lowest-order term of the expansion is unity, corresponding to a $\Theta(v)$ equal to the Maxwell–Boltzmann distribution, which is the distribution for an unheated elastic gas. To test this assumption for the steady-state that is produced by accelerated forcing, we determine the velocity distribution for various temperatures and densities and find that $\Theta(v)$ is close to the Maxwell–Boltzmann distribution, as seen in Fig. 2. The deviations which do exist become stronger as the density and temperature increase. Because of the velocity-dependent coefficient of restitution (12), increasing temperature has the same effect as decreasing the average coefficient of restitution due to lower average collision velocities. These deviations tend to flatten the distribution, increasing the probability in the tails and slightly in the peak, and decreasing the probability in between [18]. Similar types of deviations, but much stronger, have been observed in experiments on a dilute, vertically oscillated granular layer [21].

3.1.2. Velocity–velocity correlations

A further assumption of kinetic theory is that of molecular chaos — that the particle velocities involved in a collision are uncorrelated. Strong velocity correlations have been reported in driven granular media [32–34], and so we measure velocity–velocity correlation functions to test the kinetic theory assumption. Given two particles, labeled 1 and 2, $\hat{\mathbf{k}}$ is a unit vector pointing from the center of 1 to the center of 2. Particle 1's

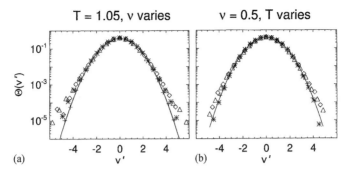

Fig. 2. Velocity distribution function $\Theta(v')$ for accelerated forcing, compared to Maxwell–Boltzman distributions (solid curves). The velocities are scaled with the square root of the temperature T, so that $v' = v/\sqrt{T}$ and $\Theta(v') = P(v')\sqrt{T}$, where $P(v')$ is the probability distribution of v'. In (a) the average temperature is approximately 1.05, and the solid fraction is varied (+: $\nu = 0.1$, *: $\nu = 0.4$, $\Diamond$: $\nu = 0.6$, $\triangle$: $\nu = 0.8$). In (b) ν is fixed at 0.5 and the temperature is varied (+: $T = 3.0 \times 10^{-5}$, *: $T = 1.1 \times 10^{-2}$, $\Diamond$: $T = 1.05$, $\triangle$: $T = 256$).

velocity then has a component $v_1^{\|}$ parallel to and $v_1^{\perp}$ perpendicular to $\hat{\mathbf{k}}$; likewise for particle 2. We define two correlation functions

$$\langle v_1^{\|} v_2^{\|} \rangle = \sum v_1^{\|} v_2^{\|} / N_r \,, \tag{18}$$

$$\langle v_1^{\perp} v_2^{\perp} \rangle = \sum v_1^{\perp} v_2^{\perp} / N_r \,, \tag{19}$$

where the sums are over N_r particles such that the distance between the two particles is within δr of r. If particle velocities are uncorrelated, $\langle v_1^{\|} v_2^{\|} \rangle$ and $\langle v_1^{\perp} v_1^{\perp} \rangle$ will both give zero.

The parallel and perpendicular velocity correlations are plotted in Fig. 3 for particles driven with randomly accelerated forcing. Strong long-range velocity correlations are apparent. These correlations are not small, reaching as much as 40% of the temperature; typically, the perpendicular correlations are about one-half of the parallel correlations. Further, these correlations are long range — they extend the full length of the system. The parallel correlations drop to zero at $L/2$, while the perpendicular correlations reach zero around $r = 10\sigma$, and have a negative value but zero derivative at $L/2$. The long-range nature of the correlation is not due to the size of the computational cell. Similar cell-filling correlations were observed in cells 4, 16, and 64 times larger in area [24].

3.1.3. Equation of state

The virial theorem of mechanics as applied to hard spheres can be used to calculate the equation of state [35,36],

$$PV = NT + \frac{\sigma}{2t_m} \sum_c \hat{\mathbf{k}} \cdot \Delta \mathbf{v}_i \,, \tag{20}$$

where N is the total number of particles, V is the total volume, and the sum is over all collisions that occur during the measurement time t_m, $\Delta \mathbf{v}_i$ is the change in the

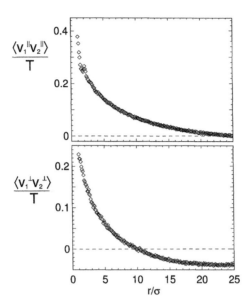

Fig. 3. Velocity correlations as a function of particle separation at $v = 0.5$, $T = 1.05$ for accelerated forcing in a box of length 52.6σ. The curve is constructed from 100 frames separated in time by 100 collisions per particle, and $\delta r = \sigma/10$.

velocity of the ith particle due to the collision, and $\hat{\mathbf{k}}$ is the unit vector pointing from particle center to particle center. In this form, measurement of pressure reduces to measurement of the average particle energy and the average change in the normal velocity at collision.

Using (20) to measure pressure, and assuming the equation of state (6), we produce a measurement of $G(v)$, denoted $G_s(v)$, where the subscript s stands for simulation. This measured value of G is compared to the Carnahan and Starling value $G_{CS}(v)$ from (11) and is shown in Fig. 4. Accurate characterization of $G(v)$ is important, because it occurs in the expressions for transport coefficients. $G_{CS}(v)$ consistently overestimates $G_s(v)$ above $v = 0.25$ with increasing error until $v > 0.675$, where elastic particles undergo a phase transition to an ordered state [16], and the two curves begin to converge and finally cross at $v = 0.8$.

3.2. Test of transport coefficient predictions

3.2.1. Volumetric temperature loss rate

We calculate the volumetric temperature loss rate γ for the simulation and compare to the kinetic theory value including the correction for the velocity-dependent coefficient of restitution γ_e (13), as shown in Fig. 5. The agreement is quite good but there is a systematic overestimate as T or v is increased. This deviation can be explained by velocity correlations discussed above. Velocity correlations change the values of $\langle v_n^2 \rangle_c$ and $\langle v_n \rangle_c$, which determine γ through (15). The ratio $\langle v_n^2 \rangle_c$ to $\langle v_n \rangle_c$, normalized by

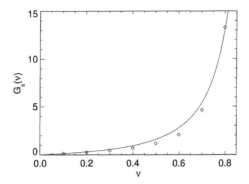

Fig. 4. (a) $G_s(v)$ for inelastic hard discs driven by accelerated forcing for $T = 1.05$ (open symbols). The solid curve is the Carnahan and Starling relation $G_{CS}(v)$, given by (11).

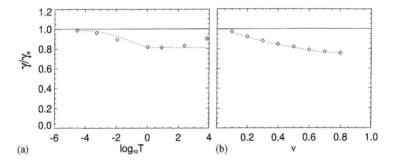

Fig. 5. Ratio of the temperature loss rate γ from molecular dynamic simulations to the prediction of kinetic theory (13) γ_e as a function of (a) temperature with $v = 0.5$ and (b) density with $T = 1.05$. The dotted lines show $(\sqrt{\pi}/4\sqrt{T})\langle v_n^2 \rangle_c / \langle v_n \rangle_c$ (see (15)).

the molecular chaos value $4\sqrt{T/\pi}$ tracks the values obtained in the simulation closely; see Fig. 5.

3.2.2. Thermal conductivity and shear viscosity

In order to measure the thermal conductivity and shear viscosity of the heated inelastic gas we introduce spatially dependent heating. Spatial inhomogeneity in the magnitude of the forcing leads to a stationary inhomogeneous temperature field, allowing measurement of heat flux and thermal conductivity κ; spatial inhomogeneity in the mean of one forcing component leads to a stationary inhomogeneous velocity field, allowing measurement of the momentum flux and the shear viscosity μ. The results of these calculations are shown in Fig. 6. While κ agrees well at low temperatures, there is a sizable error of 50% compared to the value predicted by the kinetic theory proposed by Jenkins and Richman [9]. One problem with this result is that in order to extract κ, Fourier's Law (5) must be assumed, providing no independent test of this relation. Analysis based on other closures of the Boltzmann equation predict a term in the heat flux proportional to the density gradient [37]. If such a term had a sizeable magnitude and were ignored, it would cause a reduction in the observed κ.

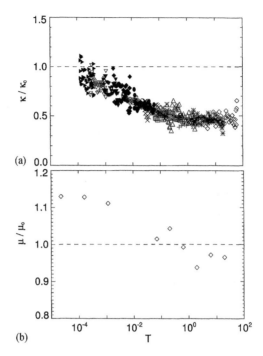

(a)

(b)

Fig. 6. (a) Ratio of the thermal conductivity κ measured from simulations to κ_0 from kinetic theory (10). Each symbol denotes a different run; for each run, the average solid fraction is 0.75. (b) Viscosity, normalized by the kinetic theory value μ_0, as a function of T, for $v = 0.6$.

The determination of μ, however, allows us, for each run at fixed T, to test Newton's viscosity law (4),

$$P_{yz} = -\mu \frac{\partial v_y}{\partial z}, \qquad (21)$$

where the viscosity μ is a constant of proportionality. We find that the linear relation of (21) holds and the slope of these curves then provide values for μ which are compared to the kinetic theory result from (9) and shown in Fig. 6(b). The maximum deviation is less than 15%. Unlike the loss rate and thermal conductivity, the shear viscosity is underestimated by kinetic theory at lower temperatures. However, the trend of decreasing transport with increasing T is the same.

4. Conclusions

We find that granular kinetic theory assumptions and predictions agree well with event-driven molecular dynamics simulations of a two-dimensional inelastic granular gas. Our simulation has been validated quantitatively with laboratory experiments of vertically oscillated thin granular layers. Further, the simulation has shown that the local velocity distribution in a vibrated layer is close to a Maxwell–Boltzmann distribution [38]. This suggests that the kinetic theory should apply to that system as well. In particular, kinetic theory assumes that velocities are distributed according to a Maxwell–Boltzmann distribution, and that at collisions there is no velocity correlation. Both of these are determined directly in our simulations. We find a Maxwell–Boltzmann distribution over 3 orders of magnitude in velocity (Fig. 2), but we find that velocity correlations of up to 40% exist between the velocity components parallel to the relative collision velocity (Fig. 3). Kinetic theory also predicts the form of the closure relations and the values of the transport coefficients in those relations. We determined the volumetric loss rate coefficient, which agrees with theory within 20% (Fig. 5); the origin of this difference is the substantial velocity correlation mentioned above. We find that the shear stress depends linearly on the shear velocity gradients, yielding the predicted Newtonian stress law with a shear viscosity that deviates less than 15% in the worst case from the value determined by kinetic theory (Fig. 6(b)). Using our current technique we cannot determine if Fourier's cooling law is obeyed, but under the assumption that it is, a thermal conductivity can be calculated; it has a maximum deviation from theory of 50% (Fig. 6(a)). We have also found that the value of the radial distribution function evaluated at the particle diameter, which is used to correct for spatial correlations that develop as the density is increased, is systematically over-estimated by the value predicted by Carnahan and Starling above a volume fraction v of about 0.25 (Fig. 4). Even for elastic particles, kinetic theory is not expected to work to arbitrarily high solid fractions; as density increases, deviations from kinetic theory are expected. Further as the inelasticity of particles increases, velocity correlations increase, reducing collisional transport. However, despite these difficulties the theory agrees remarkably well with the results of simulations.

Acknowledgements

This work was supported by the Engineering Research Program of the Office of Basic Energy Sciences of the US Department of Energy, and NSF — International Program (Chile).

References

[1] H.M. Jaeger, S.R. Nagel, R.P. Behringer, Phys. Today 49 (1996) 32.
[2] L.P. Kadanoff, Rev. Mod. Phys. 71 (1999) 435.
[3] E.W. Merrow, Chem. Eng. Progr. 81 (1985) 14.
[4] C. Bizon, M.D. Shattuck, J.B. Swift, W.D. McCormick, H.L. Swinney, Phys. Rev. Lett. 80 (1998) 57.
[5] F. Melo, P. Umbanhowar, H.L. Swinney, Phys. Rev. Lett. 72 (1994) 172.
[6] F. Melo, P.B. Umbanhowar, H.L. Swinney, Phys. Rev. Lett. 75 (1995) 3838.
[7] P. Umbanhowar, F. Melo, H.L. Swinney, Nature 382 (1996) 793.
[8] J.T. Jenkins, M.W. Richman, Arch. Rat. Mech. Anal. 87 (1985) 355.
[9] J.T. Jenkins, M.W. Richman, Phys. Fluids 28 (1985) 3485.
[10] J.T. Jenkins, M. Shahinpoor, in: J.T. Jenkins, M. Satake (Eds.), Mechanics of Granular Materials: New Models and Constitutive Relations, Elsevier, Amsterdam, 1983, p. 339.

[11] C.K.K. Lun, S.B. Savage, D.J. Jeffrey, N. Chepurniy, J. Fluid Mech. 140 (1983) 223.

[12] C.K.K. Lun, S.B. Savage, Acta Mech. 63 (1986) 15.

[13] C.K.K. Lun, J. Fluid Mech. 233 (1991) 539.

[14] A. Goldshtein, M. Shapiro, J. Fluid Mech. 282 (1995) 75.

[15] N. F. Carnahan, K.E. Starling, J. Chem. Phys. 51 (1969) 635.

[16] B.J. Alder, T.E. Wainwright, Phys. Rev. 127 (1962) 359.

[17] D. Goldman, M.D. Shattuck, C. Bizon, W.D. McCormick, J.B. Swift, H.L. Swinney, Phys. Rev. E 57 (1998) 4831.

[18] C. Bizon, M.D. Shattuck, J.B. Swift, H.L. Swinney, Phys. Rev. E (1999), in press.

[19] T.G. Drake, J. Fluid Mech. 225 (1991) 121.

[20] J. Delour, A. Kudrolli, J.P. Gollub, preprint, 1998.

[21] J.S. Olafsen, J.S. Urbach, Phys. Rev. Lett. 81 (1998) 4369.

[22] C.S. Campbell, Annu. Rev. Fluid Mech. 2 (1990) 57.

[23] E.L. Grossman, T. Zhou, E. Ben-Naim, Phys. Rev. E 55 (1997) 4200.

[24] C. Bizon, M.D. Shattuck, J.B. Swift, H.L. Swinney, in: J. Karkheck (Ed.), Dynamics: Models and Kinetic Methods for Nonequilibrium Many-Body Systems, NATO ASI Series E: Applied Sciences, Kluwer Academic publishers, Dordrecht, 1999, in press.

[25] J.R. de Bruyn, C. Bizon, M.D. Shattuck, D. Goldman, J.B. Swift, H.L. Swinney, Phys. Rev. Lett. 81 (1998) 1421.

[26] M. Marín, D. Risso, P. Cordero, J. Comput. Phys. 109 (1993) 306.

[27] B.D. Lubachevsky, J. Comput. Phys. 94 (1991) 255.

[28] W. Goldsmith, Impact, Edward Arnold Ltd., London, 1960.

[29] O.R. Walton, in: M.C. Roco (Ed.), Particulate Two-Phase Flow, Butterworth-Heinemann, Boston, 1993, pp. 884–911.

[30] D.J. Evans, G.P. Morriss, Statistical Mechanics of Nonequilibrium Liquids, Academic Press, San Diego, 1990.

[31] L. Oger, C. Annic, D. Bideau, R. Dai, S.B. Savage, J. Stat. Phys. 82 (1996) 1047.

[32] M.R. Swift, M. Boamfǎ, S.J. Cornell, A. Maritan, Phys. Rev. Lett. 80 (1998) 4410.

[33] J.A.G. Orza, R. Brito, T.P.C. van Noije, M.H. Ernst, Int. J. Mod. Phys. C 8 (1998) 953.

[34] T.P.C. van Noije, M.H. Ernst, E. Trizac, I. Pagonabarraga, Phys. Rev. E 59 (1999) 4326.

[35] J.O. Hirschfelder, C.F. Curtiss, R.B. Byrd, Molecular Theory of Gases and Liquids, Wiley, New York, 1954.

[36] D.C. Rapaport, The Art of Molecular Dynamics Simulation, Cambridge University Press, Cambridge, 1980.

[37] J.J. Brey, J.W. Duffy, A. Santos, J. Stat. Phys. 87 (1997) 1051.

[38] C. Bizon, Ph.D. Thesis, University of Texas, Austin, 1998.

ELSEVIER

Physica A 274 (1999) 171–181

www.elsevier.com/locate/physa

Granular flow, collisional cooling and charged grains

D.E. Wolf*, T. Scheffler, J. Schäfer[1]

Theoretische Physik, Gerhard-Mercator-Universität Duisburg, D-47048 Duisburg, Germany

Abstract

The non-Newtonian character of granular flow in a vertical pipe is analyzed. The time evolution of the flow velocity and the velocity fluctuations (or granular temperature) is derived. The steady-state velocity has a power law dependence on the pipe width with an exponent between 3/4 and 3/2. The flow becomes faster the more efficient the collisional cooling is, provided the density remains low enough. The dependence of collisional cooling on the solid fraction, the restitution coefficient and a possible electric charge of all grains is discussed in detail. © 1999 Elsevier Science B.V. All rights reserved.

PACS: 45.70.Mg; 83.50.Pk; 05.40.+j

Keywords: Granular flow; Collisional cooling; Charged grains

1. Introduction

Granular materials like dry sand are classical examples of many particle systems which differ significantly from solids and liquids in their dynamical behavior. They seem to have some properties in common with solids, like the ability of densely packed grains to sustain shear. Other properties, like the ability to flow through a hopper, remind one of a fluid. Upon looking more closely, however, it turns out that such similarities are only superficial. Force localization or arching in granular packings and their plastic yield behaviour are distictly different from solids, cluster instabilities and nonlinear internal friction make granular flow very different from that of ordinary, Newtonian fluids. These differences manifest themselves often in extraordinarily strong

---

* Corresponding author. Fax: +49-203 379 3163.

E-mail address: d.wolf@uni-duisburg.de (D.E. Wolf)

[1] Present address: Procter & Gamble European Services GmbH, D-65823 Schwalbach am Taunus, Germany.

fluctuations, which may cause accidents, when ignored in technological applications. This is why understanding the statistical physics of granular materials is important and has been very fruitful (see e.g. Refs. [1–4]). In this paper a few selected examples will be presented, which on the one hand document the progress in understanding brought about by applying concepts from statistical physics, and on the other hand point out some areas where important and difficult questions invite future research.

One of the simplest geometries displaying the non-Newtonian character of granular flow is an evacuated vertical tube through which the grains fall. The experimental investigation is difficult, however, because the flow depends sensitively on electric charging and humidity [5]. Nevertheless, the ideal uncharged dry granular medium falling in vacuum through a vertical pipe is an important reference case and a natural starting point for computer simulations.

Using this idealized example we shall show that the properties of granular flow can be explained, if two essential physical ingredients are understood: the interaction of the granular flow with the container walls and the phenomenon of collisional cooling. This technical term draws an analogy between the disordered relative motion of the agitated grains and the thermal motion of gas molecules. In a granular "gas", unlike a molecular gas, the relative motion of the grains is reduced in every collision due to the irreversible loss of kinetic energy to the internal degrees of freedom of the grains. This is called collisional cooling.

Agitated dry grains are usually electrically charged due to contact electrification. Its effect on the dynamical behavior of granular materials has hardly been studied so far. The reason is certainly not lack of interest, as intentional charging is the basis of several modern applications of granular materials in industrial processes. The reason is that well-controlled experiments with electrically charged grains are difficult, as is the theory, because of the long-range nature of the Coulomb interaction.

One such application is the electrostatic separation of scrap plastics into the raw materials for recycling. A similar technique is used to separate Potassium salts, a raw material for fertilizers, from rock salt. In a "conditioning process" chemically different grains get oppositely charged. Then they fall through a condenser tower, where they are deflected in opposite directions and hence separated. Such a dry separation has the advantage of avoiding the environmental damage, the old-fashioned chemical separation method would cause. Another application is powder varnishing. In order to avoid the harmful fumes of ordinary paints, the dry pigment powder is charged monopolarly and attracted to the grounded piece of metal to be varnished. Once covered with the powder, the metal is heated so that the powder melts and forms a continuous film.

Recently, a rather complete understanding has been reached on how monopolar charging affects collisional cooling [6,7]. However, little is known about the influence of the charges on the grain–wall interaction statistics. The monopolar case is much simpler than the bipolar one: if all grains repel each other, collisional cooling cannot lead to the clustering instability observed for neutral grains [8–10]. The case, where the grains carry charges of either sign, is much more difficult, because the clustering instability might even be enhanced. It has not been investigated yet.

2. Why the laws of Hagen, Poisseuille and Bagnold fail for granular pipe flow

Since the flow through a vertical pipe is such a basic example, it has been addressed many times, including some classical work from the last century. Here we remind the reader of some of the most elementary ideas and results concerning pipe flow, and at the same time show, where they fail. The general situation is much more complex, as we are going to point out in the subsequent sections.

Force balance requires that the divergence of the stress tensor σ_{ij} compensates the weight per unit volume in the steady state of the flowing material:

$$\partial_x \sigma_{zx} + \partial_z \sigma_{zz} = -mgn, \tag{1}$$

where m denotes the molecular or grain mass, n the number density of molecules or grains and g the gravitational acceleration. The partial derivative in the vertical (the z-) direction vanishes because of translational invariance along the pipe. ∂_x denotes the partial derivative in the transversal direction.[2]

In Newtonian or simple liquids the stress tensor is assumed to be proportional to the shear rate

$$\sigma_{zx} = \eta \partial_x v_z, \tag{2}$$

where the proportionality constant η is the viscosity. Inserting this into (1) immediately gives the parabolic velocity profile of Hagen–Poisseuille flow, $v_z = v_{\max} - (mgn/2\eta)x^2$. No-slip boundary conditions then imply that the flow velocity averaged over the cross section of the pipe scales like $\bar{v} \propto W^2$.

According to kinetic theory the viscosity η is proportional to the thermal velocity. In lowest order the thermal motion of liquid molecules is independent of the average flow velocity. It is given by the coupling of the liquid to a heat bath. For a granular gas, the thermal velocity must be replaced by a typical relative velocity δv of the grains. Due to collisional cooling δv would drop to zero, if there was no flow. This is the most important difference between liquid and granular flow. It shows, that for a granular gas the collision rate between the grains, and hence the viscosity cannot be regarded as independent of the average flow velocity in lowest order. Bagnold [11] argued that the typical relative motion should be proportional to the absolute value of the shear rate, $\eta \propto \delta v \propto |\partial_x v_z|$, so that

$$\sigma_{zx} \propto |\partial_x v_z| \partial_x v_z. \tag{3}$$

Inserting this into (1) leads to the result, that the average flow velocity must scale with the pipe diameter as $\bar{v} \propto W^{3/2}$.

However, Bagnold's argument ignores, that there is a second characteristic velocity in the system, which is $\sqrt{gd}$, where d is the diameter of the grains. It enters due to the nonlinear coupling between the flow velocity and the irregular grain motion, as we are going to point out in the next section. Hence, for granular flow through a vertical

---

[2] For the sake of transparency the equations are given for the two-dimensional case in this section.

pipe, the viscosity is a function of both the average flow velocity and $\sqrt{gd}$. This will change the scaling of $\bar{v}$ with the diameter of the pipe, of course.

Very little is known about the flow of dry granular materials at high solid fractions, where the picture of gas-like dynamics, which we employed so far, no longer applies. Hagen studied the discharge from a silo [12] and postulated, that the flow rate is not limited by plastic deformations inside the packing but by arching at the outlet. He assumes that the only dimensionful relevant parameters for outlets much larger than the grain size are g and the width W of the outlet, for which we use the same notation as for the pipe diameter. Therefore, he concludes, that up to dimensionless prefactors

$$\bar{v} \propto \sqrt{gW} \, . \tag{4}$$

He confirmed this experimentally for the silo geometry, where the outlet is smaller than the diameter of the container. It is tempting to expect that this also holds for pipe flow at high solid fractions. However, in our computer simulations we never observed such a behavior, although we studied volume fractions, which were so high, that the addition of a single particle would block the pipe completely. Without investigating dense granular flow any further in this paper, we just want to point out that Hagen's dimensional argument seems less plausible for a pipe than for a silo, because important arching now occurs at any place simultaneously along the pipe, and the dynamics of decompaction waves [13] and plastic deformations far from the lower end of the pipe may well depend on the dimensionless ratio W/d, for instance. This spoils the argument leading to (4), of course.

3. General equations of momentum and energy balance

A vertical pipe can be viewed essentially as a one-dimensional system, if one averages all dynamical quantities over the cross section. In the following section we derive the time evolution of such cross-sectional averages of the velocity and velocity fluctuations, assuming that they are constant along the pipe. This assumption ignores the spontaneous formation of density waves, which is legitimate if the pipe is sufficiently short. Then a homogeneous state can be maintained. It need not be stationary, though, and the equations we shall derive describe its temporal evolution. The physical significance of this study is based on the assumption that short sections of a long pipe are locally homogeneous and close to the corresponding steady state.

The translational invariance along the pipe implies that the average velocity only has an axial component. Its time evolution is given by the competition of a gain term, which is the gravitational acceleration g, and a loss term due to the momentum transfer to the pipe wall. Here we focus on the behavior at low enough densities, where the dynamics are dominated by collisions rather than frictional contacts. Then the momentum transfer to the pipe wall is proportional to the number of grain-wall collisions, $\dot{N}_w$. In each such collision the axial velocity of the colliding particle changes by an average value $\Delta\bar{v}$. All grains are assumed to be equal for simplicity. Hence the average axial velocity changes by $\Delta\bar{v}/N$ in a wall collision. The momentum balance then reads

$$\dot{\bar{v}} = g - \dot{N}_w\Delta\bar{v}/N \ . \tag{5}$$

More subtle is the energy balance which gives rise to an equation for the root-mean-square fluctuation of the velocity, $\delta v = \sqrt{\langle\bar{v}^2\rangle - \langle\bar{v}\rangle^2}$. This can be regarded as the typical absolute value of relative velocities.

The rates of energy dissipation $\dot{E}_{diss}$ and of change of kinetic and potential energy, $\dot{E}_{kin}$ and $\dot{E}_{pot}$ must add up to zero due to energy conservation,

$$0 = \dot{E}_{diss} + \dot{E}_{kin} + \dot{E}_{pot} \ . \tag{6}$$

The change in kinetic energy per unit time is

$$\dot{E}_{kin} = Nm(\bar{v}\dot{\bar{v}} + \delta v\dot{\delta v}) \ , \tag{7}$$

where N is the total number of particles in the pipe and m their mass. The potential energy (in the absence of Coulomb interactions between the grains) changes at a rate

$$\dot{E}_{pot} = -Nmg\bar{v} \ . \tag{8}$$

If only the irreversible nature of binary grain collisions is taken into account the energy dissipation rate is proportional to the number of binary collisions per unit time, $\dot{N}_g$, times the loss of kinetic energy in the relative motion of the collision partners,

$$\dot{E}_{diss} = \dot{N}_g\Delta E \ , \tag{9}$$

with

$$\Delta E = \Delta(m\delta v^2/2) = m\delta v\Delta(\delta v) \ . \tag{10}$$

Solving (6) for $\dot{\delta v}$ and replacing $\dot{\bar{v}} - g$ using (5) gives

$$\dot{\delta v} = (\bar{v}/\delta v)\dot{N}_w\Delta\bar{v}/N - \dot{N}_g\Delta(\delta v)/N \ . \tag{11}$$

As for the average velocity, (5), the typical relative velocity has a gain and a loss term. The gain term has a remarkable symmetry to the loss term in (5), which is completely general. Only the second term in (11) may be different, if additional modes of energy dissipation like collisions with the walls or friction are included. The gain term in (11) subsumes also the production of granular temperature in the interior of the pipe due to the finite shear rate, which is remarkable, as it expresses everything in terms of physics at the wall.

Once the loss terms of the balance equations, (5) and (11), are known, the time evolution of the average velocity and the velocity fluctuations can be calculated, because the gain terms are given. In this sense, it is sufficient to have a statistical description of collisional cooling (which gives the loss term in (11)) and of the momentum transfer of the granular gas to a wall (which gives the loss term in (5)) in order to describe granular flow in a vertical pipe. It turns out, that collisional cooling is easier, because it cannot depend on the average velocity due to the Galilei invariance of the grain–grain interactions, whereas the momentum transfer to the walls depends on both, $\bar{v}$ and δv.

4. Collisional cooling

We shall now specify $\dot{N}_g$ and $\Delta(\delta v)$. The time between two subsequent collisions of a particle can be estimated by the mean free path, λ, divided by a typical relative velocity, δv. Hence the number of binary collisions per unit time is proportional to

$$\dot{N}_g \propto N\delta v/\lambda .\tag{12}$$

Here we assumed that the flow is sufficiently homogeneous, and that the local variations of λ and δv are unimportant.

In each collision the relative normal velocity gets reduced by a factor, the restitution coefficient $e_n < 1$. For simplicity we assume that the restitution coefficient is a constant. Correspondingly a fraction of the kinetic energy of relative motion is dissipated in each collision

$$\Delta E = (1 - e_n^2)\frac{m}{2}\delta v^2 ,\tag{13}$$

with the grain mass m. According to (10), $\Delta(\delta v) = (1 - e_n^2)\delta v/2$. Putting this together, the dissipation rate (9) is [14]

$$\dot{E}_{\text{diss}} = k_g N\frac{m}{d}\delta v^3 .\tag{14}$$

The dimensionless proportionality constant k_g contains the dependence on the solid fraction $v \propto d/\lambda$ and the restitution coefficient e_n and can be calculated analytically, if one assumes that the probability distribution of the particles is Gaussian [15].

From these considerations one can draw a very general conclusion for the steady state values of $\bar{v}$ and δv. In the steady state the kinetic energy is constant, so that (6) together with (8) and (14) imply

$$\frac{\bar{v}_s}{\delta v_s^3} = \frac{k_g}{gd} .\tag{15}$$

Whenever the dissipation is dominated by irreversible binary collisions and the flow is sufficiently homogeneous, the steady flow velocity in a vertical pipe should be proportional to the velocity fluctuation to the power 3/2. The proportionality constant does not depend on the width of the pipe.

We tested this relation by two-dimensional event-driven molecular dynamics simulations [4]. The agreement is surprisingly good, given the simple arguments above, even quantitatively. However, it turns out that the proportionality constant in (15) has a weak dependence on the width of the pipe, which can be traced back to deviations of the velocity distribution from an isotropic Gaussian: the vertical velocity component has a skewed distribution with enhanced tail towards zero velocity [16].

5. Interaction of the granular flow with the wall

The collision rate $\dot{N}_w$ with the walls of the vertical pipe can be determined by noting that the number of particles colliding with a unit area of the wall per unit time for low density n is given by $|v_\perp| n$. As the typical velocity perpendicular to the pipe wall, $|v_\perp|$ is proportional to δv, one obtains

$$\dot{N}_w \propto N\delta v/W \, . \tag{16}$$

This is the place where the pipe width W enters into the flow dynamics.

To specify, by how much the vertical velocity of a grain changes on average, when it collides with the wall, is much more difficult, as it depends on the local geometry of the wall. In our simulations the wall consisted of a dense array of circular particles of equal size. When a grain is reflected from such a wall particle, a fraction of the vertical component of its velocity will be reversed. Instead of averaging this over all collision geometries, we give some general arguments narrowing down the possible functional form of $\Delta\bar{v}$. If we assume that the velocity distribution is Gaussian, all moments of any velocity component must be functions of $\bar{v}$ and δv. This must be true for $\Delta\bar{v}$, as well. For dimensional reasons it must be of the form

$$\Delta\bar{v} = \bar{v} f\left(\frac{\delta v}{\bar{v}}\right) \tag{17}$$

with a dimensionless function f. The physical interpretation of this is the following: the loss term in the momentum balance can be viewed as an effective wall friction. As long as the granular flow in the vertical pipe approaches a steady state, the friction force must depend on the velocity $\bar{v}$. The ratio $\delta v/\bar{v}$ can be viewed as a characteristic impact angle, so that the function f contains the information about the average local collision geometry at the wall. In principle, all dimensionless parameters characterizing the system may enter the funcion f, that is, apart from v also the restitution coefficient e_n and the ratios W/d and $gd/\bar{v}^2$. However, it is hard to imagine, that the width W of the pipe or the gravitational acceleration g influences the local collision geometry. Therefore we shall assume that f does not depend on W/d or $gd/\bar{v}^2$. On the other hand, it is plausible, that the restitution coefficient enters f. It will influence the spatial distribution of particles and also accounts for the correlation of the velocities, if some particle is scattered back and forth between the wall and neighboring particles inside the pipe, and hence hits the wall twice or more times without a real randomization of its velocity. Due to positional correlations among the particles, f should also depend on the solid fraction v. One can easily imagine that the average collision geometry is different in dense and in dilute systems.

Lacking a more precise understanding of the function f, we make a simple power law ansatz for it and write the loss term of (5) as

$$\begin{aligned}
\frac{\dot{N}_w}{N}\Delta\bar{v} &= \frac{1}{W}\,\delta v\bar{v}\,k_w\left(\frac{\delta v}{\bar{v}}\right)^\beta \\
&= k_w\,W^{-1}\delta v^{1+\beta}\bar{v}^{1-\beta} \, .
\end{aligned} \tag{18}$$

The dimensionless parameters k_w and β will be functions of v and e_n.

6. Time evolution and steady state

With these assumptions, the equations of motion (5) and(11) for granular flow through a vertical pipe become

$$\dot{\bar{v}} = g - k_w \, W^{-1} \delta v^{1+\beta} \bar{v}^{1-\beta} \,, \tag{19}$$

$$\dot{\delta v} = k_w \, W^{-1} \delta v^{\beta} \bar{v}^{2-\beta} - k_g \, d^{-1} \delta v^2 \,. \tag{20}$$

As the time evolution should not be singular for $\bar{v} = 0$ or $\delta v = 0$, the values of β are restricted to the interval

$$0 \leqslant \beta \leqslant 1 \,. \tag{21}$$

The meaning of the exponent β becomes clear, if we calculate the steady-state velocity from (19) and(15). The result is

$$\bar{v}_s = \sqrt{gd} \, k_g^{\gamma-1/2} k_w^{-\gamma} \left(\frac{W}{d} \right)^{\gamma} \,. \tag{22}$$

The exponent γ, which determines the dependence of the average flow velocity on the pipe diameter, is related to β by

$$\gamma = \frac{3}{2(2 - \beta)} \,. \tag{23}$$

Due to (21) we predict that in granular pipe flow

$$3/4 \leqslant \gamma \leqslant 3/2 \,, \tag{24}$$

as long as the flow is sufficiently homogeneous and the main dissipation mechanisms are binary collisions. Note that the exponent is always smaller than 2, which would be its value for Hagen–Poisseuille flow of a Newtonian fluid. $\gamma = 3/2$ is the prediction of Bagnold's theory, but in our simulations we found also values as small as 1, depending on the values of the solid fraction and the restitution coefficient [16].

The stationary value δv_s directly follows from (22) and (15). One obtains the same formula as (22) with γ replaced by $\gamma/3$.

7. Collisional cooling for monopolar charged grains

In this section we summarize our recent results [6,7], how the dissipation rate (14) is changed if all grains carry the same electrical charge q (besides having the same mass m, diameter d and restitution coefficient e_n). For simplicity, we assume that the charges are located in the middle of the insulating particles. The results are valid for grains in a three-dimensional space, $D = 3$.

Whereas the hard sphere gas has no characteristic energy scale, the Coulomb repulsion introduces such a scale,

$$E_q = q^2/d \,. \tag{25}$$

It is the energy barrier that two collision partners have to overcome, when approaching each other from infinity. It has to be compared to the typical kinetic energy stored in the relative motion of the particles, which by analogy with molecular gases is usually expressed in terms of the "granular temperature"

$$T = \delta v^2 / D .\tag{26}$$

If $E_q \ll mT$, one expects that the charges have negligible effect on the dissipation rate. Using (26) and the expression

$$v = \frac{\pi}{6} n d^3 \quad \text{with} \quad n = N/V\tag{27}$$

for the three-dimensional solid fraction, the dissipation rate (14) can be written in the form

$$\dot{E}_{\text{diss}}/V = k\, n^2 d^2 m T^{3/2}\tag{28}$$

with the dimensionless prefactor

$$k = k_g \pi \sqrt{3}/2v .\tag{29}$$

The advantage of writing it this way is that the leading n- or v-dependence is explicitly given. In the dilute limit $v \to 0$ the dissipation rate should be proportional to n^2, i.e. to the probability that two particles meet in an ideal gas.

Since the remaining factors in (28) are uniquely determined by the dimension of the dissipation rate, this equation must hold for charged particles as well. However, in this case the prefactor k will not only depend on e_n and v, but also on the dimensionless energy ratio E_q/mT. We found [6,7] that the following factorization holds:

$$k = k_0(e_n) g_{\text{chs}}(v, E_q/mT) ,\tag{30}$$

where

$$k_0 = 2\sqrt{\pi}(1 - e_n^2)\tag{31}$$

is the value of k for $v = E_q/mT = 0$. g_{chs} denotes the radial distribution function for charged hard spheres (chs) at contact, normalized by the one for the uncharged ideal gas.

For $v < 0.2$ and $E_q/mT < 8$, our computer simulations show that

$$g_{\text{chs}}\left(v, \frac{E_q}{mT}\right) \approx g_{\text{hs}}(v) \exp\left(-\frac{E_q}{mT} f(v)\right)\tag{32}$$

is a very good approximation. Here,

$$g_{\text{hs}} = \frac{2 - v}{2(1 - v)^3} \geqslant 1\tag{33}$$

is the well–known Enskog correction for the radial distribution function of (uncharged) hard spheres (hs) [17]. This factor describes that the probability that two particles collide is enhanced due to the excluded volume of all the remaining particles. The second, Boltzmann-like factor describes that the Coulomb repulsion suppresses collisions. The effective energy barrier $E_q f(v)$ decreases with increasing solid fraction, because two particles which are about to collide not only repel each other but are also pushed together by being repelled from all the other charged particles in the system. A two-parameter fit gives

$$f(v) \approx 1 - c_0 \, v^{1/3} + c_1 \, v^{2/3} \tag{34}$$

with

$$c_0 = 2.40 \pm 0.15 \quad \text{and} \quad c_1 = 1.44 \pm 0.15 \,. \tag{35}$$

Very general arguments [6] lead to the prediction that $c_1 = (c_0/2)^2$, which is confirmed by (35).

We expect deviations from (32) for larger v and E_q/mT, because the uncharged hard sphere system has a fluid–solid transition close to $v \approx 0.5$, and the charged system may become a Wigner crystal for any solid fraction, provided the temperature gets low enough.

8. Conclusion

We presented four main results. The steady-state velocity of granular flow in a vertical pipe should have a power law dependence on the diameter W of the pipe with an exponent γ ranging between 3/4 and 3/2, depending on the solid fraction and the restitution coefficient of the grains. This result was derived ignoring possible electric charges of the grains and assuming that the flow is sufficiently homogeneous and the main dissipation mechanism are binary collisions. This illustrates the genuinely non-Newtonian character of granular flow.

Second, the dependence of the steady-state velocity on the solid fraction v, the restitution coefficient e_n and – in the case of monopolar charging – the ratio between Coulomb barrier and kinetic energy, E_q/mT, is contained in the factor $k_g^{\gamma-1/2}k_w^{-\gamma}$ in (22). In the dilute limit $v \to 0$ as well as in the limit of nearly elastic particles $e_n \to 1$, the coefficient k_w, which describes how sensitive the momentum transfer to the wall depends on the local collision geometry, should remain finite, whereas k_g vanishes like $v(1 - e_n^2)$ according to (29) and (30). As $\gamma - 1/2 > 0$, the flow through a vertical pipe becomes faster the higher the solid fraction and the less elastic the collisions between the grains are (in the limit of low density and nearly elastic collisions). The physical reason for this is that in denser and more dissipative systems collisional cooling is more efficient, reducing the collisions with the walls and hence their braking effect. This remarkable behavior has been confirmed in computer simulations [16].

Third, monopolar charging leads to a Boltzmann-like factor in k_g or the dissipation rate, respectively, which means that for low granular temperature the dissipation rate becomes exponentially weak. The higher the density the less pronounced is this effect, because the effective Coulomb barrier $E_q f(v)$ hindering the collisions becomes weaker.

Finally, we derived the evolution equations for the flow velocity and the velocity fluctuation for granular flow through a vertical pipe, (19) and (20). These equations apply to the situation of homogeneous flow, which can only be realized in computer simulations of a sufficiently short pipe with periodic boundary conditions. In order to generalize these equations for flow that is inhomogeneous along the pipe one should replace the time derivatives by $\partial_t + \bar{v}(z,t)\partial_z$. In addition, a third equation, the continuity equation, is needed to describe the time evolution of the density of grains along the pipe. Such equations have been proposed previously [18–20] in order to study the kinetic waves spontaneously forming in granular pipe flow. Our equations are different.

Acknowledgements

We thank the Deutsche Forschungsgemeinschaft for supporting this research through grant No. Wo 577/1-3. The computer simulations supporting the theory presented in this paper were partly done at the John von Neumann Institut für Computing (NIC) in Jülich.

References

[1] H.J. Herrmann, J.-P. Hovi, S. Luding (Eds.), Physics of dry granular media, Kluwer, Dordrecht 1998.
[2] R.P. Behringer, J.T. Jenkins (Eds.), Powders and Grains '97, A.A. Balkema, Rotterdam, 1997.
[3] H.M. Jaeger, S.R. Nagel, R.P. Behringer, Rev. Mod. Phys. 68 (1996) 1259.
[4] D.E. Wolf, in: K.H. Hoffmann, M. Schreiber (Eds.), Computational Physics: Selected Methods, Simple Exercises, Serious Applications, Springer, Berlin 1996, pp. 64–95.
[5] T. Raafat, J.P. Hulin, H.J. Herrmann, Phys. Rev. E 53 (1996) 4345.
[6] T. Scheffler, D.E. Wolf, Phys. Rev. E, submitted for publication
[7] T. Scheffler, J. Werth, D.E. Wolf, in: P. Entel, D.E. Wolf (Eds.), Structure and Dynamics of Heterogeneous Systems, World Scientific, Singapore, 1999.
[8] I. Goldhirsch, G. Zanetti, Phys. Rev. Lett. 72 (1993) 1619.
[9] S. McNamara, W.R. Young, Phys. Fluids A 4 (1992) 496.
[10] S. McNamara, W.R. Young, Phys. Rev. E 53 (1996) 5089.
[11] R.A. Bagnold, Proc. Roy. Soc. A 225 (1954) 49.
[12] G. Hagen, Monatsberichte Preuß, Akad. d. Wiss., Berlin, 1852, pp. 35–42.
[13] J. Duran, T. Mazozi, S. Luding, E. Clément, J. Raichenbach, Phys. Rev. E 53 (1996) 1923.
[14] P.K. Haff, J. Fluid Mech. 134 (1983) 401.
[15] C.K.L. Lun, S.B. Savage, D.J. Jeffrey, N. Chepurniy, J. Fluid Mech. 140 (1984) 223.
[16] T. Scheffler, J. Schaefer, D. E. Wolf, to be published.
[17] N.F. Carnahan, K.E. Starling, J. Chem. Phys. 51 (1969) 635.
[18] J. Lee, M. Leibig, J. Phys. I France 4 (1994) 507.
[19] A. Valance, T. Le Pennec, Eur. Phys. J. B 5 (1998) 223.
[20] T. Riethmüller, L. Schimansky-Geier, D. Rosenkranz, T. Pöschel, J. Stat. Phys. 86 (1997) 421.

ELSEVIER

Physica A 274 (1999) 182–189

www.elsevier.com/locate/physa

Application of statistical mechanics to collective motion in biology

Tamás Vicsek[a,*], András Czirók[a], Illés J. Farkas[a], Dirk Helbing[b]

[a]*Department of Biological Physics, Eötvös University, Pázmány Péter Sétány 1A,
H-1117 Budapest, Hungary*
[b]*II. Institute of Theoretical Physics, University of Stuttgart, Pfaffenwaldring 57/III,
70550 Stuttgart, Germany*

Abstract

Our goal is to describe the collective motion of organisms in the presence of fluctuations. Therefore, we discuss biologically inspired, inherently non-equilibrium models consisting of self-propelled particles. In our models the particles corresponding to organisms locally interact with their neighbours according to simple rules depending on the particular situation considered. Numerical simulations indicate the existence of new types of transitions. Depending on the control parameters both disordered and long-range ordered phases can be observed. In particular, we demonstrate that (i) there is a transition from disordered to ordered motion at a finite noise level even in one dimension and (ii) particles segregate into lanes or jam into a crystalline structure in a model of pedestrians. © 1999 Elsevier Science B.V. All rights reserved.

1. Introduction

The collective motion of organisms (birds, for example), is a fascinating phenomenon many times capturing our eyes when we observe our natural environment. In addition to the aesthetic aspects of collective motion, it has some applied aspects as well: a better understanding of the swimming patterns of large schools of fish can be useful in the context of large scale fishing strategies. Our interest is also motivated by the recent developments in areas related to statistical physics. During the last 15 years or so there has been an increasing interest in the studies of far-from-equilibrium systems typical in our natural and social environment. Concepts originated from the physics of phase transitions in equilibrium systems [1–3] such as collective behaviour, scale invariance and renormalization have been shown to be useful in the understanding of

---

* Corresponding author.
E-mail address: h845vic@ella.hu (T. Vicsek)

0378-4371/99/$ - see front matter © 1999 Elsevier Science B.V. All rights reserved.
PII: S 0378-4371(99)00317-9

various non-equilibrium systems as well. Simple algorithmic models have been helpful in the extraction of the basic properties of various far-from-equilibrium phenomena, like diffusion limited growth [4,5], self-organized criticality [6] or surface roughening [7]. Motion and related transport phenomena represent a further characteristic aspect of non-equilibrium processes [8], including traffic models [9,10], thermal ratchets [11] or driven granular materials [12,13].

Self-propulsion is an essential feature of most living systems. In addition, the motion of the organisms is usually controlled by interactions with other organisms in their neighbourhood and randomness plays an important role as well. In Ref. [14] a simple model of self propelled particles (SPP) was introduced capturing these features with a view toward modelling the collective motion [15] of large groups of organisms such as schools of fish, herds of quadrupeds, flocks of birds, or groups of migrating bacteria, correlated motion of ants [16] or pedestrians [17,18].

It turns out that modelling collective motion leads to interesting specific results inscr2 with the relevant dimensions (from 1–3 [19,14,20,21]). Here we discuss a one-dimesional and a quasi-one-dimensional version of the related models.

2. Ordered motion for finite noise in 1d

Since in 1d the particles cannot get around each other, some of the important features of the dynamics present in higher dimensions are lost. On the other hand, motion in 1d implies new interesting aspects (groups of the particles have to be able to change their direction for the opposite in an organized manner) and the algorithms used for higher dimensions should be modified to take into account the specific crowding effects typical for 1d (the particles can slow down before changing direction and dense regions may be built up of momentarily oppositely moving particles).

In a way the system we study can be considered as a model of organisms (people, for example), moving in a narrow channel. Imagine that a fire alarm goes on, the tunnel is dark, smoky, everyone is extremely excited. People are both trying to follow the others (to escape together) and behave in an erratic manner (due to smoke and excitement). Will they escape (move in a selected direction in an orderly manner) or become trapped (keep moving back and forth)? One way to study this and related applied questions is carrying out simulations.

Thus, we consider N off-lattice particles along a line of length L. The particles are characterized by their coordinate x_i and dimensionless velocity u_i updated as

$$x_i(t + \Delta t) = x_i(t) + v_0 u_i(t)\Delta t, \tag{1}$$

$$u_i(t + \Delta t) = G(\langle u(t)\rangle_{S(i)}) + \xi_i, \tag{2}$$

where ξ_i is a random number drawn with a uniform probability from the interval $[-\eta/2, \eta/2]$. The local average velocity $\langle u\rangle_{S(i)}$ for the ith particle is calculated over the particles located in the interval $[x_i - \Delta, x_i + \Delta]$, where we fix $\Delta = 1$. The function G tends to set the velocity in average to a prescribed value v_0: $G(u) > u$ for $u < 1$

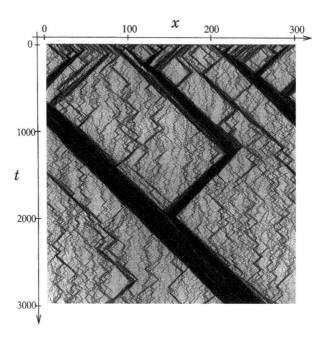

Fig. 1. The first 3000 time steps of the 1d SPP model [$L = 300$, $N = 600$, $\eta = 2.0$ (a) and $\eta = 4.0$ (b)]. The darker grey scale represents higher particle density.

and $G(u) < u$ for $u > 1$. In the numerical simulations [13] one of the simplest choices for G was implemented as

$$G(u) = \begin{cases} (u+1)/2 & \text{for } u > 0, \\ (u-1)/2 & \text{for } u < 0 \end{cases} \tag{3}$$

and random initial and periodic boundary conditions were applied.

In Fig. 1 we show the time evolution of the model for $\eta = 2.0$. In a short time the system reaches an ordered state, characterized by a spontaneous broken symmetry and clustering of the particles. In contrast, for $\eta = 4.0$ the system remains in a disordered state. Ordering for finite noise in 1d is a surprising phenomenon and is entirely due to the specific non-equilibrium nature of our system.

To understand the phase transitions observed in the various SPP models, efforts has been made to set up a consistent continuum theory in terms of v and ρ, representing the coarse-grained velocity and density fields, respectively. The first approach [22,23] has been made by Toner and Tu for $d > 1$ who investigated the following set of equations

$$\partial_t v + (v\nabla)v = \alpha v - \beta |v|^2 v - \nabla P + D_L \nabla(\nabla v) + D_1 \nabla^2 v + D_2 (v\nabla)^2 v + \xi$$

$$\partial_t \rho + \nabla(\rho v) = 0, \tag{4}$$

where the $\alpha, \beta > 0$ terms make v have a non-zero magnitude, $D_{L,1,2}$ are diffusion constants and f is an uncorrelated Gaussian random noise. The pressure P depends on the local density only, as given by the expansion

$$P(\rho) = \sum_n \sigma_n(\rho - \rho_0)^n \,.$$ (5)

The non-intuitive terms (4th and 6th in the RHS) in Eq. (4) were generated by the renormalization process.

Tu and Toner were able to treat the problem analytically and show the existence of an ordered phase in 2D, and also extracted exponents characterizing the density–density correlation function. They showed that the upper critical dimension for their model is 4 and the theory does not allow an ordered phase in 1d.

However, as we have shown, there exist SPP systems in one dimension which exhibit an ordered phase for low noise level. This finding motivated us to construct an alternative continuum model for 1d [19].

$$\partial_t u = f(u) + \mu^2 \partial_x^2 u + \alpha \frac{(\partial_x u)(\partial_x \rho)}{\rho} + \zeta \,,$$ (6)

$$\partial_t \rho = -v_0 \partial_x(\rho u) + D \partial_x^2 \rho \,,$$ (7)

where $u(x,t)$ is the coarse-grained dimensionless velocity field, $f(u)$ is an antisymmetric function with $f(u) > 0$ for $0 < u < 1$ and $f(u) < 0$ for $u > 1$, $\bar{\zeta} = 0$, and $\overline{\zeta^2} = \sigma^2/\rho\tau^2$. These equations are different both from the equilibrium field theories and from the nonequilibrium system defined through Eq. (4). The main difference comes from the nature of the coupling term $(\partial_x u)(\partial_x \rho)/\rho$. This term can be derived, but here we present a plausible interpretation of its origin.

The main point is that one would like to have a term in the equation for the velocity which would result in the slowing down (and eventually in the "turning back") of the particles under the influence of a larger number of particles moving oppositely. When two groups of particles move in the opposite direction, the density locally increases and the velocity decreases at the point they meet. Let us consider a particular case, when particles move from left to right and the velocity is locally decreasing while the density is increasing as x increases (particles are moving towards a "wall" formed between two oppositely moving groups). The term $(\partial_x u)$ is less, the term $(\partial_x \rho)$ is larger than zero in this case. Together they have a negative sign resulting in the slowing down of the local velocity. This is a consequence of the fact that there are more slower particles (in a given neighbourhood) in the forward direction than faster particles coming from behind, so the average action (a particle is trying to take the average velocity of its neighbours and simultaneously move with a preferred velocity) experienced by a particle in the point x slows it down. Thus, the $(\partial_x u)(\partial_x \rho)/\rho$ term does what we expect from it.

Given the equation, and the picture of domain walls formed between oppositely moving groups of particles we can use the following simple arguments to interpret long range ordering.

A small domain of (left moving) particles moving opposite to the surrounding (right moving) ones is *bound to interact* with more and more right moving particles and, as a result, the domain wall assumes a specific structure which is characterized by a buildup of the right moving particles on its left side, while no more than the originally present left moving particles remain on the right-hand side of the domain wall. This process "transforms" non-local information (the size of the corresponding domains) into a *local asymmetry of the particle density* which, through the instability of the walls [19], results in a leftward motion of the domain wall, and consequently, eliminates the smaller domain.

This can be demonstrated schematically as

$$> > > > > > > > > > > < < < < < < < < > > > > > > > > > > > > > >$$
$$A \qquad\qquad\qquad B$$

where by $>(<)$ we denoted the right (left) moving particles. In contrast to the Ising model the A and B walls are very different and have non-equivalent properties. In this situation the B wall will break into a B_1 and B_2, moving in opposite directions, B_1 moving to the left and B_2 moving to the right, leaving the area $B_1 - B_2$ behind, which is depleted of particles.

$$> > > > > > > > > > > < < < < < \qquad > > > > > > > > > >$$
$$A \qquad\qquad\qquad B_1 \quad B_2$$

At the A boundary the two type of particles slow down, while, due to the instability shown to be present in the system, the wall itself moves in a certain direction, most probably to the right. Even in the other, extremely rare case (i.e., when the A wall moves to the left), an elimination of the smaller domain $(A - B_1)$ takes place since the velocity of the domain wall A is smaller than the velocity of the particles in the "bulk" and at B_1 where the local average velocity is the same as the preferred velocity of the particles. Thus, the particles tend to accumulate at the domain wall A, which again, through local interactions leads to the elimination of the domain $A - B_1$.

3. Simulations of pedestrians in a corridor

In our studies of collective phenomena in systems of moving organisms we also considered humans walking in opposite directions along a corridor. Here we assumed that half of the people tended to go from left to right (and vice versa) with a constant velocity, unless they encountered another person. The interaction potential had a hard core (corresponding to the physical size of a pedestrian) and was assumed to decay as a power law within a cutoff radius of about twice of the hard core. All this can be represented in the form of equations of motion which we solved numerically [24].

In addition to the interest in the properties of systems involving walkers [17], there are several further motivations to study such a model. A system of light (rising) and heavy (sinking) particles in a vertical column of fluid or a system of oppositely charged colloidal particles in an electric field represent potential applications of our model. In fact, the system we study is a generalization to the continuum case of the two-species driven lattice gas proposed recently [25], with a number of relevant modifications arising from the adaptation to the off-lattice case. In a wider context, these models can be considered as simplified paradigms of systems consisting of entities with opposing interests (drives). In the present work, we consider the behaviour of a limited number of particles in a confined geometry, and our results are primarily valid for this "mesoscopic" situation. In the quickly growing literature on mesoscopic systems there are many examples of the potential practical relevance of phenomena occurring in various models for finite sizes [26–28].

We started our simulations with N (typically between 24 and 72) particles which were randomly distributed in the corridor without allowing overlaps. For half of the particles a driving into the $(-1,0)$ direction, and for the other half a driving into the $(1,0)$ direction was assigned. Integrating equation using periodic boundary conditions produced the following results: For small noise amplitudes η and moderate densities, our simulations lead to the formation of coherently moving linear structures (just as if the particles moved along traffic lanes). For relatively large N, jamming occurs, depending on the respective initial condition. Nevertheless, lane formation, if at all, happens relatively quickly and is very dominant and robust in our model (in contrast with simpler lattice gas versions we have tested).

The mechanism of lane formation can be understood as follows: Particles moving against the stream or in areas of mixed directions of motion will have frequent and strong interactions. In each interaction, the encountering particles move a little aside in order to pass each other. This sidewards movement tends to separate oppositely moving particles. Moreover, once the particles move in uniform lanes, they will have very rare and weak interactions. Hence, the tendency to break up existing lanes is negligible, when the fluctuations are small. Furthermore, the most stable configuration corresponds to a state with a minimal interaction rate [29].

Whereas lane formation was also observed in previous studies of related models with *deterministic dynamics only* [30], in the present, more realistic model we have discovered a surprising phenomenon when we increased the noise amplitude. If the fluctuations were large enough, they were able to prevent lane formation or even to destroy previously existing lanes, eventually causing a mutual blocking of the opposite directions of motion for given, sufficiently high densities. In this blocked state, the particles were arranged in a hexagonal lattice structure, very much like in a crystal. In particular, there are densities, at which we observe *bistability*. That is, we find lanes at small noise amplitudes, but in a crystallized state, if the noise amplitude is large.

Thus, one can drive the system from a "fluid" state with lanes of uniform directions of motion to a "frozen" state just by increasing the fluctuations. We call this transition "freezing by heating" [24]. Although this transition exists for the simplest version of the model, it is more pronounced if the gradients of the repulsive potentials are multiplied with factors

$$\{\lambda + (1 - \lambda)[1 + \cos(\varphi_i)]/2\}, \tag{8}$$

where φ_i denotes the angle between the direction of motion e_i of particle i and the direction of the object exerting the repulsive force. The parameter λ reflects the relative influence of forces "from behind", in order to mimic that the interactions may depend on the (relative) velocities of the particles, a situation quite common in realistic driven systems.

Why is freezing by heating new? Some glasses may crystallize when slowly heated. However, this is a well understood phenomenon: the amorphous state is metastable for those temperatures, and crystallization means an approach to the more stable state with smaller free energy (smaller internal energy). In our case, crystallization is achieved by spontaneously driving the system with the help of noise uphill towards higher internal energy. The system would like to maximize its efficiency [29], but instead it ends up with minimal efficiency due to noise-induced crystallization. The role of "temperature" or noise here is to destroy the energetically more favourable fluid state, which inevitably leads to jamming. Crystallization occurs in this jammed phase, because it yields the densest packing of particles. The corresponding transition seems to be related to the off-lattice nature of our model and is different from those reported for driven diffusive systems on a lattice. It should be noted that the transition we find is not sharp, which is in part a consequence of the mesoscopic nature of the phenomenon.

We would like to point out that "freezing by heating" is likely to be relevant to situations involving pedestrians under extreme conditions (panics). Imagine a very smoky situation, caused by a fire, in which people do not know which is the right way to escape. When panicking, people will just try to get ahead, with a reduced tendency to follow a certain direction. Thus, fluctuations will be very large, which can lead to fatal blockings.

Acknowledgements

This work was supported by OTKA F019299, FKFP 0203/1997, and the DFG by the Heisenberg scholarship He 2789/1-1.

References

[1] S.-K. Ma, Statistical Mechanics, World Scientific, Singapore, 1985.
[2] S.-K. Ma, Modern Theory of Critical Phenomena, Benjamin, New York, 1976.
[3] H.E. Stanley, Introduction to Phase Transitions and Critical Phenomena, Oxford University Press, Oxford, 1971.
[4] T.A. Witten, L.M. Sander, Phys. Rev. Lett. 47 (1981) 1400.
[5] T. Vicsek, Fractal Growth Phenomena, 2nd Edition, World Scientific, Singapore, 1992.
[6] P. Bak, C. Tang, K. Wiesenfeld, Phys. Rev. Lett. 59 (1987) 381.
[7] A-L. Barabási, H.E. Stanley, Fractal Concepts in Surface Growth, Cambridge University Press, Cambridge, 1995.
[8] M. Schreckenberg, D.E. Wolf (Eds.), Traffic and Granular Flow '97, Springer, New York, 1998.
[9] See, e.g., D. Helbing, B.A. Huberman, Nature 396 (1998) 738.
[10] D. Helbing, A. Hennecke, M. Treiber, Phys. Rev. Lett. 82 (1999) 4360, and references therein.
[11] M.O. Magnasco, Phys. Rev. Lett. 71 (1993) 1477.
[12] Y.L. Duparcmeur, H.J. Herrmann, J.P. Troadec, J. Phys. (France) I 5 (1995) 1119.
[13] J. Hemmingsson, J. Phys. A 28 (1995) 4245.
[14] T. Vicsek, A. Czirók, E. Ben-Jacob, I. Cohen, O. Shochet, Phys. Rev. Lett. 75 (1995) 1226.

[15] C.W. Reynolds, Computer Graphics 21 (1987) 25.
[16] E.M. Rauch, M.M. Millonas, D.R. Chialvo, Phys. Lett. A 207 (1995) 185.
[17] D. Helbing, J. Keitsch, P. Molnar, Nature 387 (1997) 47.
[18] D. Helbing, F. Schweitzer, P. Molnar, Phys. Rev. E. 56 (1997) 2527.
[19] A. Czirók, A.-L. Barabási, T. Vicsek, Phys. Rev. Lett. 82 (1999) 209.
[20] A. Czirók, H.E. Stanley, T. Vicsek, Spontaneously ordered motion of self-propelled particles, J. Phys A 30 (1997) 1375.
[21] A. Czirók, M. Vicsek, T. Vicsek, Collective motion of organisms in three dimensions, Physica A 264 (1999) 299.
[22] J. Toner, Y. Tu, Phys. Rev. Lett. 75 (1995) 4326.
[23] J. Toner, Y. Tu, Phys. Rev. E 58 (1998) 4828.
[24] D. Helbing, I. Farkas, T. Vicsek, preprint cond-mat/9904326.
[25] B. Schmittmann, K. Hwang, R.K.P. Zia, Europhys. Lett. 19 (1992) 19.
[26] Y. Imry, Introduction to Mesoscopic Physics, Oxford University Press, Oxford, 1997.
[27] N.G. van Kampen, Stochastic Processes in Physics and Chemistry, North-Holland, Amsterdam, 1981.
[28] M-L. Tan, I. Goldhirsch, Phys. Rev. Lett. 81 (1998) 3022.
[29] D. Helbing, T. Vicsek, New J. Phys. 1 (1999) no. 13.
[30] D. Helbing, P. Molnár, Phys. Rev. E 51 (1995) 4282.

ELSEVIER

Physica A 274 (1999) 190–199

www.elsevier.com/locate/physa

Formation of colony patterns by a bacterial cell population

M. Matsushita[a,*], J. Wakita[a], H. Itoh[a], K. Watanabe[a], T. Arai[a],
T. Matsuyama[b], H. Sakaguchi[c], M. Mimura[d]

[a]*Department of Physics, Chuo University, Kasuga, Bunkyo-ku, Tokyo 112-8551, Japan*
[b]*Department of Bacteriology, Niigata University School of Medicine, Asahimachidōri,
Niigata 951-8510, Japan*
[c]*Department of Mathematics, Faculty of Engineering, Tokushima University, Minami-Josanjima,
Tokushima 770-8506, Japan*
[d]*Department of Mathematical and Life Sciences, Hiroshima University,
Higashi-Hiroshima 739-8526, Japan*

Abstract

Bacterial species *Bacillus subtilis* is known to exhibit various colony patterns, such as diffusion-limited aggregation (DLA)-like, compact Eden-like, dense branching morphology (DBM)-like, concentric ring-like and disk-like, depending on the substrate softness and nutrient concentration. We have established the morphological diagram of colony patterns, and examined and characterized both macroscopically and microscopically how they grow. For instance, we have found that there seem to be two kinds of bacterial cells; active and inactive cells, the former of which drive colony interfaces outward. The active cells are particularly distinguished from the inactive ones at the tips of growing branches of a DBM-like colony as the characteristic fingernail structure. We have also found that the concentric ring-like colony is formed as a consequence of alternate repetition of advancing and resting of the growing interface which consists of active cells. Based on our observations, we have constructed a phenomenological but unified model which produces characteristic colony patterns. It is a reaction–diffusion type model for the population density of bacterial cells and the concentration of nutrient. The essential assumption is that there exist two types of bacterial cells; *active* cells that move actively, grow and perform cell division, and *inactive* ones that do nothing at all. Our model is found to be able to reproduce globally all the colony patterns seen in the experimentally obtained morphological diagram, and is phenomenologically quite satisfactory. © 1999 Elsevier Science B.V. All rights reserved.

---

* Corresponding author. Fax: + 81-3 3817 1792.
E-mail address: matusita@phys.chuo-u.ac.jp (M. Matsushita)

1. Introduction

Much attention has recently been paid to pattern formation in various fields from various viewpoints [1–5]. Above all, pattern formation in biological systems is the most interesting and exciting, but in general very complicated. Understanding it may be our ultimate goal for studying pattern formation. Pattern formation of populations of simple biological objects may be a good starting point because it is, under some conditions, dominated by purely physical conditions. We will here pay attention to the colony formation of bacteria, which are one of the simplest biological objects [6]. In particular, we will discuss experimental and theoretical attempts to elucidate the mechanism on a part of rich-in-variety bacterial colony formation [7].

There is a twofold advantage of using bacteria for the study of pattern formation in general. Firstly, one can easily vary their environmental or living conditions. For instance, by varying the nutrient concentration, one can to some extent control the growth or multiplication rate of bacteria. One can also control their motility by varying the agar concentration: cells can hardly move or show active motion on a hard agar plate, while on a soft agar plate they actively move around by themselves. One can thus turn on/off the cell division and/or active motion rather easily. This means that one can investigate the wide spectrum of bacterial behavior, from active to passive. One may then bridge over non-biological (physico-chemical) and biological pattern formation by studying bacterial colonies. Secondly, the cell size of individual bacteria is just good. In contrast to, e.g., dendritic crystal growth, one can easily observe both the macroscopic growth of colony patterns and the microscopic structure and motion of individual cells that constitute the colonies. Hence one may relate microscopic observations of bacterial cells to macroscopic colony formation. It is very important to bridge the gap between microscopic and macroscopic observations in order to understand pattern formation in general.

In order to study the morphological change due to environmental conditions, we vary only two parameters; concentrations of nutrient C_n and agar C_a in a thin agar plate as the incubation medium. The latter controls the medium hardness. Other parameters such as temperature ($35°C$) are kept constant. Also, throughout the present experiments we use only one bacterial species *Bacillus* (*B.*) *subtilis*. The experimental procedures are standard and were described elsewhere [8].

We have so far observed various colony patterns by varying C_n and C_a, and established a morphological diagram for a wild-type strain of *B. subtilis* [8–11]. Fig. 1 shows the diagram drawn in the ($C_a^{-1}, \log C_n$)-plane. This means that agar plates as living environments for bacteria become softer as the diagram is followed from left to right, and they become richer in nutrient from bottom to top.

We have found that quasi-two-dimensional colony patterns grown on agar surfaces are classified into five types in the diagram, i.e., A (diffusion-limited aggregation [12] (DLA)-like), B (Eden [13]-like), C (concentric ring-like), D (disk-like) and E (dense branching morphology [1–3,5] (DBM)-like). In this paper we first examine and characterize experimentally how bacterial colonies grow.

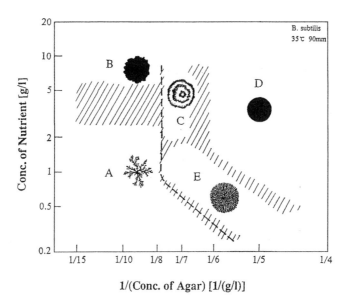

Fig. 1. Morphological diagram of colonies of B. subtilis.

Looking at the morphological diagram shown in Fig. 1, one may have a naive question whether the diversity of colony patterns observed in experiments is caused by different effects or governed by the same underlying principles. Various models [9,14–16] have so far been proposed to describe these colony patterns. They are, however, not very successful in that they can only reproduce some of the observed patterns such as DLA-like and DBM-like. Pondering over these models, one question naturally arises if there is any unified model with two parameters corresponding to C_n and C_a, which generates all these growth patterns. Thus, our second concern in this article is to propose, based on our observations, a unified model [17,18] to describe the morphological diversity of colony patterns shown in Fig. 1.

2. Growth diversity of colonies

So far we have shown that the bacterial species *Bacillus subtilis* exhibits qualitatively different two-dimensional colony patterns as a result of growth and cell division on the surface of thin agar plates by feeding on nutrients [8–11]. Colony patterns change drastically when the concentrations of agar and nutrient, C_a and C_n, are globally varied. They are classified qualitatively into five types, each of which is observed in the regions labeled A–E in Fig. 1.

In the region A (hard agar medium with poor nutrient) colony patterns exhibit tip-splitting growth with characteristically ramified structures. These patterns look similar to those observed in diffusion-limited processes in electrochemical deposition, dielectric breakdown and viscous fingering [1–3,5]. Also, the structures are clearly

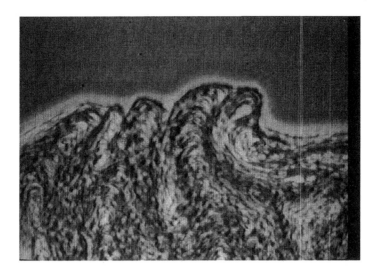

Fig. 2. A photograph showing the growing interface of a colony in the region B. Note that many cells connect end to end like long strings of sausages, bundles of which form the folded interface. $C_a = 10$ g/l and $C_n = 20$ g/l. The photo, whose width is 2 mm, was taken 3 days after inoculation.

reminiscent of two-dimensional diffusion-limited aggregation (DLA) clusters, which can be generated by a simple particle model for randomly branched patterns grown through diffusion-limited processes [1–3,5,12]. In fact, seemingly different phenomena mentioned above are all known to be explained by DLA. The growth of colonies in the region A has also been shown to be clearly described in terms of DLA [19].

Increasing C_n with high fixed C_a values corresponding to the change from the region A to the region B in Fig. 1, the branch thickness of the colony increases gradually and colony patterns eventually become compact and Eden-like in the region B. Although growing interface of a colony exhibits characteristic roughness just as Eden clusters [4,13], microscopic observations of the growing interface reveal that microscopic growth process is very different from that of the Eden model. We found that the bacterial cells are not individually separate but they connect with each other end to end, and hence they do not move actively on a solid agar surface. The colony interface consists, therefore, of bundles of many chains of bacterial cells and slowly advances forward (about 10 μm/min) by increasing their length along the interface and folding themselves (Fig. 2). As a result, it seems that the interface has strong, long-range lateral correlations. In fact, the self-affine analysis has been performed [10], giving the value of roughness exponent $\alpha \simeq 0.78$, very different from $\alpha = 0.5$ for the Eden model which has no long-range lateral correlations.

In the region C, a rather narrow region between regions B and D, colonies spread and rest alternately, leaving stationary concentric ring-like patterns [20]. It seems that the periodicity does not depend so strongly on the nutrient concentration C_n. It also seems that when they are in the period of migration, their cells become longer than when they are in the rest period. However, these properties are not so prominent as

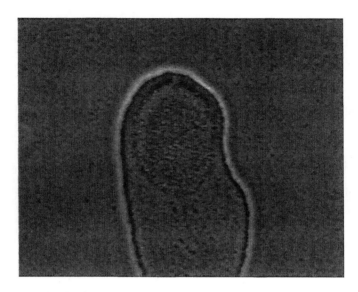

Fig. 3. Photograph of the tip of a growing branch in the region E. Note the fingernail structure consisting of active cells. $C_a = 5$ g/l and $C_n = 0.5$ g/l. Width of the photo is about 0.5 mm.

for *Proteus* [21]. The detailed experimental results will soon be published elsewhere [22].

For high C_n and low C_a (soft agar medium with rich nutrient) in the region D colony patterns drastically change to be spread homogeneously, which look macroscopically like a perfect disk. According to the microscopic observations [9], a colony consists of a monolayer of bacterial cells and they move around actively and more or less randomly inside the colony. It is, therefore, likely that their movement can be described in terms of diffusion [9]. There is no submacroscopic branching at all. The colony growth in this region D has been found to be consistent with the behavior of solutions of the two-dimensional Fisher equation [9].

Finally, in the region E between A and D, there emerge colony patterns clearly reminiscent of the so-called dense branching morphology (DBM) [1–3,5]. Although the branching is very dense, the advancing envelope looks characteristically smooth compared with DLA-like colonies in the region A. Microscopic observations [11] reveal that bacterial cells composing growing branch tips move around very actively, while those composing branches deep inside are inactive. In fact, a group of actively moving cells are clearly observed in the tip of each growing branch, as if they were a fingernail (Fig. 3). They drive the growth of the branch tip, leaving inactive cells behind. Interestingly, it was confirmed macroscopically that both averaged branch width and averaged branch gap decrease systematically when increasing nutrient concentration C_n, while their ratio remains unchanged with the approximate value of one [11].

Fig. 1 shows that the morphological change is more conspicuous from region A (DLA-like) to region E (DBM-like) than the gradual change from A to region B (Eden-like). The growth rate in region E is also much higher than that in region A.

It takes a colony about one month in A and about one week in B to grow to about 5 cm (half the diameter of a petri dish containing agar medium), whereas it takes only one day in C and E and half a day in D. This observation suggests a large difference in the way colonies grow. In fact, a thick broken line in Fig. 1 indicates the boundary (not clearcut, though) beyond which the active movement of bacterial cells is observed, at least around the growing interface of colonies.

3. Model for morphological diversity

Based on our observations of colony growth described in the previous section, let us here propose a unified model to describe all of the growth patterns shown in Fig. 1. The essential assumption in our modeling is the introduction of *internal state* [14] in each cell: If it becomes higher, the cell moves actively, grows and performs cell-division, while if it becomes lower, the cell becomes inactive. In order to take this situation into account in the simplest way conceivable, we assume that bacterial cells consist of two types; active and inactive cells. While former cells move around, grow and multiply, the latter do nothing at all. This assumption may be justified by the observation of a group of active cells driving the growth of branch tips and leaving inactive cells behind in the region E (Fig. 3).

Let $b(r,t)$ and $s(r,t)$ be the population densities of the active and inactive cells, respectively, where $b + s$ is the total population density, and $n(r,t)$ be the nutrient concentration. These quantities are assumed to be governed by the following reaction–diffusion equations:

$$\frac{\partial b}{\partial t} = \nabla(d(b,n)\nabla b) + \varepsilon g(n)b - a(b,n)b , \tag{1}$$

$$\frac{\partial s}{\partial t} = a(b,n)b , \tag{2}$$

$$\frac{\partial n}{\partial t} = \nabla^2 n - g(n)b . \tag{3}$$

Note that the equations are already reformulated in the non-dimensional form. Thus, d in the first term of Eq. (1) is the ratio of the diffusion coefficients of bacterial cells and nutrient. Let us, for the time being, assume that d is a constant parameter, independent of b and n. The coefficient $\varepsilon g(n)$ in the second term of Eq. (1) indicates the growth rate of bacteria, and $a(b,n)$ in the third term is the rate of change from the active to inactive cells. In Eq. (1) we neglect the reverse change from the inactive cells to the active ones for simplicity. This is acceptable because once cells become inactive well behind the growing front in a colony, they have little chance to become active again. We also put $g(n) = n$ for simplicity.

The coefficient $a(b,n)$ is determined due to the mechanism of cell's internal state. It is legitimate to assume that $a(b,n)$ decreases if the nutrient concentration increases: If the nutrient decreases, cells become inactive. Our observations indicate that when cells invade a virgin territory to advance the colony interface, they tend to gather and form a group like a raft and move together actively. On the other hand, if they happen to be separated from each other and b is quite small, each cell becomes inactive. We assume, therefore, that $a(b,n)$ is also a decreasing function of b. In order to meet these assumptions in the simplest form we take $a(b,n) = a_0(1+b)^{-1}(1+n)^{-1}$. It turned out that the detailed mathematical expression is not so important.

Eqs. (1) and (3) are closed for the population density of active cells b and the nutrient concentration n. Hence the population density of inactive cells s can be obtained from them through Eq. (2). Initial conditions we imposed are

$$b(\mathbf{r},0) = b_0(\mathbf{r}),\qquad\qquad(4)$$

$$n(\mathbf{r},0) = n_0,\qquad\qquad(5)$$

$$s(\mathbf{r},0) = 0,\qquad\qquad(6)$$

where $b_0(\mathbf{r})$ indicates a point-like distribution (with some randomness added) of the initial population density of bacterial cells and n_0 is the initial concentration of nutrient distributed homogeneously. It should be noted that the parameters (d, n_0) in the present model system correspond to (C_a^{-1}, C_n) in our experiments. In fact, as C_a is increased, the cell motility greatly decreases, while the nutrient diffusivity is not influenced so much, resulting in the decrease of their ratio d.

Let us now demonstrate spatial patterns produced by the total cells $b(\mathbf{r},t) + s(\mathbf{r},t)$ which were obtained numerically from (1)–(6) in two dimensions, when varying n_0 and d. In the following, we take $a_0 = 2400$ and $\varepsilon = 20$. For small values of both n_0 and d (nutrient-poor and hard environment), we had DLA-like patterns, an example of which is shown in Fig. 4(a). They look similar to the ones observed in region A in Fig. 1. In fact, they are self-similar and the fractal dimension is about 1.67, which is consistent with the value for two-dimensional DLA. Increasing d with more or less the same value of n_0, the patterns change to DBM-like ones, as seen in Fig. 4(b), which correspond to those observed in the region E in Fig. 1. On the other hand, for large n_0 and d (nutrient-rich and semi-soft environment), the patterns of $b + s$ take simply expanding disks (Fig. 4(d)), which are quite similar to those observed in the region D in Fig. 1. If d is decreased with n_0 large, the total cells exhibit periodic growth: They repeat advancing the growth front and halting alternately, leaving an immovable concentric ring-like pattern, as seen in Fig. 4(c). This corresponds to the repetitive growth observed in the region C in Fig. 1.

Summarizing the numerical results obtained so far, our reaction–diffusion model exhibits the morphological diagram of colony patterns in Fig. 5. It should be noted that this diagram reproduces experimental observations (Fig. 1) fairly well, except that there is no Eden-like, compact pattern.

As for Eden-like patterns in region B (hard agar medium with rich nutrient), experimental observations indicate that the bacterial cells hardly move, but cells grow moderately because of enough nutrient supply. Hence the resulting colony pattern expands due to a mechanism different from the others: as described in the previous

Fig. 4. Patterns produced by the total cells $b+s$. (a) DLA-like ($d=0.05$, $n_0=0.98$), (b) DBM-like ($d=0.12$, $n_0=0.855$), (c) concentric ring-like ($d=0.05$, $n_0=1.2$), (d) disk-like ($d=0.12$, $n_0=1.5$).

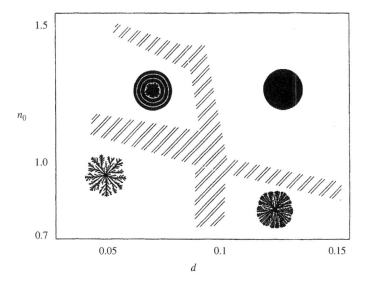

Fig. 5. Morphological diagram in the (d, n_0) plane obtained numerically from the present reaction–diffusion model.

Fig. 6. Eden-like pattern produced by the total cells $b + s$. Obtained from the modified version: $d = d_1 b$ ($d_1 = 0.04$, $n_0 = 4.0$).

section and shown in Fig. 2, the interface progresses because of collective growth of connected cells, instead of their individual active movements [10]. In order to model this situation in a simple way let us replace a constant parameter d with a bacterial population-dependent diffusion term $d(b) = d_1 b$ with a positive constant parameter d_1. The resulting modified system exhibits Eden-like patterns, an example of which is shown in Fig. 6.

4. Discussions and summary

We have first investigated the colony growth of a bacterial species *Bacillus subtilis*. By varying agar and nutrient concentrations, C_a and C_n, in the agar medium for the incubation, we have observed that colonies exhibit characteristically different patterns, which we classified into five types as shown in Fig. 1. We have also examined and characterized both macroscopically and microscopically how the colonies grow.

We have then proposed a unified model to explain the morphological diversity of bacterial colonies and demonstrated our numerical results. It should be noted that our model equations are quite simple, and to obtain such diverse patterns we varied only two parameters, d and n_0, which correspond to C_a^{-1} and C_n, respectively, in our experiments. Comparing numerical results shown in Figs. 4–6 with experimental ones shown in Fig. 1, our model is phenomenologically quite satisfactory. We believe that the present model may be a good one to start with, to analyze data obtained from more detailed experiments such as the dependence of spacing and periodicity on the nutrient concentration for the concentric ring-like colonies.

References

[1] D. Avnir (Ed.), The Fractal Approach to Heterogeneous Chemistry, Wiley, Chichester, 1989.

[2] E. Ben-Jacob, P. Garik, Nature 343 (1990) 523.

[3] T. Vicsek, Fractal Growth Phenomena, 2nd Edition, World Scientific, Singapore, 1992.

[4] A.-L. Barabási, H.E. Stanley, Fractal Concepts in Surface Growth, Cambridge University Press, Cambridge, 1995.

[5] P. Meakin, Fractals, Scaling and Growth Far From Equilibrium, Cambridge University Press, Cambridge, 1998.

[6] P. Singleton, Introduction to Bactria, 2nd Edition, Wiley, Chichester, 1992.

[7] J.A. Shapiro, M. Dworkin (Eds.), Bacteria as Multicellular Organisms, Oxford University Press, New York, 1997.

[8] M. Ohgiwari, M. Matsushita, T. Matsuyama, J. Phys. Soc. Jpn. 61 (1992) 816.

[9] J. Wakita, K. Komatsu, A. Nakahara, T. Matsuyama, M. Matsushita, J. Phys. Soc. Jpn. 63 (1994) 1205.

[10] J. Wakita, H. Itoh, T. Matsuyama, M. Matsushita, J. Phys. Soc. Jpn. 66 (1997) 67.

[11] J. Wakita, I. Ràfols, H. Itoh, T. Matsuyama, M. Matsushita, J. Phys. Soc. Jpn. 67 (1998) 3630.

[12] T.A. Witten, L.M. Sander, Phys. Rev. Lett. 47 (1981) 1400.

[13] See, e.g., F. Family, T. Vicsek (Eds.), Dynamics of Fractal Surfaces, World Scientific, Singapore, 1991.

[14] E. Ben-Jacob, O. Shochet, A. Tenenbaum, I. Cohen, A. Czirók, T. Vicsek, Nature 368 (1994) 46.

[15] K. Kawasaki, A. Mochizuki, M. Matsushita, T. Umeda, N. Shigesada, J. Theor. Biol. 188 (1997) 177.

[16] S. Kitsunezaki, J. Phys. Soc. Jpn. 66 (1997) 1544.

[17] M. Mimura, H. Sakaguchi, M. Matsushita, in preparation.

[18] M. Matsushita, J. Wakita, H. Itoh, I. Ràfols, T. Matsuyama, H. Sakaguchi, M. Mimura, Physica A 249 (1998) 517.

[19] M. Matsushita, H. Fujikawa, Physica A 168 (1990) 498.

[20] H. Fujikawa, Physica A 189 (1992) 15.

[21] O. Rauprich, M. Matsushita, C.J. Weijer, F. Siegert, S.E. Esipov, J.A. Shapiro, J. Bacteriol. 178 (1996) 6525.

[22] I. Ràfols, J. Wakita, H. Itoh, T. Matsuyama, M. Matsushita, in preparation.

ELSEVIER

Physica A 274 (1999) 200–215

www.elsevier.com/locate/physa

Application of statistical mechanics to stochastic transport

J. Łuczka

Department of Theoretical Physics, Institute of Physics, Silesian University, 40-007 Katowice, Poland

Abstract

The problem of transport of Brownian particles in spatially periodic structures is presented. Based on the Feynman ratchet, a mathematical model of a thermal ratchet is constructed as the Newton equation with stochastic thermal and nonthermal forces. Conditions for directed motion of Brownian particles are discussed. An example of an exactly soluble model of stochastic transport is demonstrated. © 1999 Elsevier Science B.V. All rights reserved.

1. Introduction

Equilibrium statistical mechanics is based on the concept of Gibbs states. For a system interacting weakly with surroundings, it is described by a canonical ensemble which in some cases can be derived from the 'first principles'. Alternatively, such a thermodynamic equilibrium state is a stationary state of the system described by the Newton equation with a specific Langevin random force with imposed suitable probabilistic characteristics. This random force is modeled by Gaussian white noise, which mimics thermal fluctuations existing inherently in systems of temperature $T > 0$. The approach to statistical mechanics based on Newton–Langevin equations is very fruitful because it gives possibility of modeling of a wide class of nonequilibrium processes. Indeed, if the random force is nonthermal (or the additional random force is nonthermal) then the stationary state of the system is a nonequilibrium state. Let us consider, for example, a Brownian particle driven by thermal or/and nonthermal stochastic forces (noise, fluctuations). In the absence of a deterministic potential force, a particle diffuses and directed motion is not performed. If the Brownian particle is additionally driven by a nonzero bias force then one can observe privileged motion in a direction determined by the bias force. This is a trivial example of transport, which

---

E-mail address: luczka@us.edu.pl (J. Łuczka)

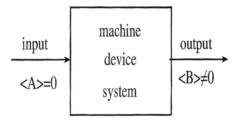

Fig. 1. A scheme of the zero mean input–nonzero mean output system.

in this case is induced by a bias force. A nontrivial case is transport when a bias force is zero and all random forces are of zero-mean values. The fundamental question arises: can transport be induced by stochastic forces, random perturbations, noise and fluctuations? The problem is schematically presented in Fig. 1 as the input–output system. As the input A we can consider: thermal and nonthermal fluctuations, noise, stochastic forces; all of zero averages over realizations, $\langle A \rangle = 0$. As the system, we consider Brownian particles in spatially periodic structures, i.e. moving in a spatially periodic potential $V(x) = V(x + L)$ of a given period L. Let us note that an average value of the potential force $f(x) = -\mathrm{d}V(x)/\mathrm{d}x$ is zero over a period L and an average bias is zero. In the output, we want to obtain a non zero mean value of a quantity B which characterizes transport and directed motion of Brownian particles. One of the simplest characteristics is stationary velocity v of particles. If an average value $\langle v \rangle = 0$ then particles are not macroscopically transported. In turn, if $\langle v \rangle \neq 0$ then particles perform macroscopic directed motion and stochastic transport occurs in the system. It is analogy to a phase transition: An average value $\langle v \rangle$ is an order parameter; in ordered state $\langle v \rangle \neq 0$ and in disordered state $\langle v \rangle = 0$.

So, we can formulate our problem in the form: How to convert diffusion motion of Brownian particles into their directed motion, without any bias forces, without any gradients (of concentration, temperature, etc.) or, jocularly, how to obtain 'something' (i.e. $\langle B \rangle \neq 0$) from 'nothing' (i.e. $\langle A \rangle = 0$). A general answer is known for physicists: the phase transition can occur due to symmetry breaking. In the ordered state $\langle v \rangle \neq 0$ symmetry should be broken! A detailed answer will be given below and is based on the analysis of the famous Feynman 'ratchet and pawl machine' [1] and its abstract generalization known in literature as the Brownian (thermal) ratchet, see, e.g. [2–11].

2. Feynman ratchet

Let us briefly consider the Feynman ratchet and pawl device, see Fig. 2. The detailed discussion is presented elsewhere [1,12]. The machine consists of the axle with vanes in one of its ends and a ratchet in the other. Two ends are inserted into two reservoirs of gas of temperatures T_1 and T_2, respectively. The pawl blocks turns of the axle in one direction and allows to turn to the other. Because the vanes as well as the ratchet are placed in boxes of gas, they undergo random motion due to collisions with molecules

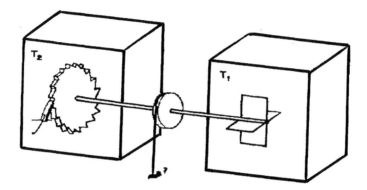

Fig. 2. The Feynmann ratchet and pawl machine.

of gas. At first sight, even at the same temperatures $T_1 = T_2$, the axle seems to turn in one direction because of the pawl mechanism. Accordingly, an average directed motion is generated due to thermal fluctuations. However, it is in contradiction with the second law of thermodynamics. The paradox is explained in [1] and critical analysis of the Feynman gadget is presented in [12]. On the other hand, if two temperatures are different, $T_1 \neq T_2$, the resulting average motion is nonzero and the efficiency of the device is the same as for the Carnot cycle. But in this case, the macroscopic gradient of temperature causes net mean motion of the axle. This is not the case we are interested in. Nevertheless, the analysis of this machine can help us to construct an abstract mathematical model of desired features. Let us note the following:

(i) The ratchet (wheel) is a spatially periodic system. It corresponds to a spatially periodic potential $V(x) = V(x + L)$.

(ii) The symmetry of the ratchet is broken, because of the pawl mechanism (teeth are asymmetrical). It corresponds to breaking of a reflection symmetry of the potential: a real number x_0 does not exist such that the relation $V(x_0 - x) = V(x_0 + x)$ is fulfilled.

(iii) The average force acting on vanes and caused by collisions of gas molecules is zero. It corresponds to zero-mean thermal fluctuations.

(iv) The directed motion is induced by temperature gradient. It is trivial and instead of this we assume that a driving nonthermal (nonequilibrium) force of zero mean acts on the system.

3. Formulation of ratchet model

An abstract mathematical model of a generalized ratchet can be based on the above observations (i)–(iv) and is formulated in the following way. We consider a Brownian particle of mass m moving in a one-dimensional spatially periodic potential $\hat{V}(\hat{x}) = \hat{V}(\hat{x} + L)$ of period L and of the barrier height $\Delta\hat{V} = \hat{V}_{\max} - \hat{V}_{\min}$ (see Fig. 3). The equation of motion for the particle is assumed to be the Newton equation with random

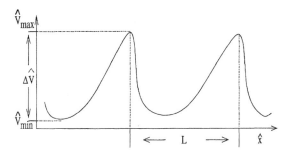

Fig. 3. An example of the spatially periodic potential $\hat{V}(\hat{x})$ of period L and barrier height $\Delta\hat{V}$.

forces, namely,

$$m\ddot{\hat{x}} + \gamma\dot{\hat{x}} = \hat{f}(\hat{x}) + \hat{\Gamma}(\hat{t}) + \hat{\xi}(\hat{t}) \,. \tag{1}$$

The inertial effects are described by the first term in the left-hand side. The dissipation process is included via the Stokes force with the friction coefficient γ which is proportional the linear size R of the particle, i.e.,

$$\gamma = c_0\eta R \tag{2}$$

and it is additionally determined by the viscosity η of the medium the particle moves in and the constant c_0 dependent on geometry of the particle [13].

The potential force

$$\hat{f}(\hat{x}) = -\frac{d\hat{V}(\hat{x})}{d\hat{x}} \tag{3}$$

is zero over the period L,

$$\langle\hat{f}(\hat{x})\rangle_L = \frac{1}{L}\int_0^L \hat{f}(\hat{x})\,d\hat{x} = \frac{1}{L}[\hat{V}(0) - \hat{V}(L)] = 0 \,. \tag{4}$$

The stochastic force $\hat{\Gamma}(\hat{t})$ describes thermal fluctuations which here are modeled by δ-correlated Gaussian white noise of the statistics

$$\langle\hat{\Gamma}(\hat{t})\rangle = 0, \qquad \langle\hat{\Gamma}(\hat{t})\hat{\Gamma}(\hat{s})\rangle = 2\hat{D}\delta(\hat{t} - \hat{s}) \,. \tag{5}$$

According to the dissipation–fluctuation theorem, the thermal-noise intensity $\hat{D}$ is related to the friction coefficient γ and temperature T of the system, i.e.,

$$\hat{D} = \gamma k_B T \,, \tag{6}$$

where k_B stands for the Boltzmann constant.

The second stochastic force $\hat{\xi}(\hat{t})$ describes nonthermal and nonequilibrium fluctuations which can be modeled by a stationary Markovian stochastic process of zero mean,

$$\langle\hat{\xi}(\hat{t})\rangle = 0 \,. \tag{7}$$

The process $\hat{x}(\hat{t})$ defined by (1) is a non-Markovian stochastic process. Its simplest probabilistic characteristics are determined by the probability density $P(v,\hat{x},\hat{t})$ in the phase space, where $v = d\hat{x}/d\hat{t}$. Unfortunately, it cannot be found in a general case. What we are interested in is rather the stationary mean velocity $\langle v \rangle = \langle \dot{\hat{x}} \rangle$ which is treated as an order parameter. Under what conditions $\langle v \rangle \neq 0$? First, let us consider the case of thermodynamic equilibrium.

4. Ratchet in equilibrium state

Under the equilibrium condition, nonthermal fluctuations are excluded, i.e.,

$$\hat{\xi}(\hat{t}) \equiv 0 \tag{8}$$

and Eq. (1) reads

$$m\ddot{\hat{x}} + \gamma\dot{\hat{x}} = \hat{f}(\hat{x}) + \hat{\Gamma}(\hat{t}). \tag{9}$$

In this case, the probability density obeys the Klein–Kramers equation (see e.g., Eq. (1.18) in [14])

$$\frac{\partial}{\partial \hat{t}} P(v, \hat{x}, \hat{t}) = \left[-\frac{\partial}{\partial \hat{x}} v + \frac{\partial}{\partial v} \left(\frac{\gamma}{m} v - \frac{1}{m} \hat{f}(\hat{x}) \right) + \frac{\gamma k_B T}{m^2} \frac{\partial^2}{\partial v^2} \right] P(v, \hat{x}, \hat{t}). \tag{10}$$

For suitable boundary conditions, Eq. (10) can be solved in the stationary state which is a Gibbs (thermodynamic equilibrium) state. The stationary probability distribution $P_{st}(v, \hat{x})$ is the Boltzmann distribution,

$$P_{st}(v, \hat{x}) = N \exp[(-mv^2/2 + \hat{V}(\hat{x}))/k_B T], \tag{11}$$

where the normalization constant N can be found from the condition

$$\int_{-\infty}^{\infty} dv \int_0^L d\hat{x} P_{st}(v, \hat{x}) = 1. \tag{12}$$

The stationary mean velocity is zero,

$$\langle v \rangle = \int_{-\infty}^{\infty} dv \int_0^L d\hat{x} \, v P_{st}(v, \hat{x}) = 0 \tag{13}$$

and there is no directed motion of the Brownian particle for any potential $\hat{V}(\hat{x})$. It is related to the principle of detailed balance. Therefore only nonthermal driving force can induce transport of particles in the case when the principle of detailed balance is violated.

5. Dimensionless variables

Let us return to the Newton–Langevin Eq. (1). It is useful to work in dimensionless variables, because from the physical point of view only relations between scales of length, time and energy play a role and not their absolute values. A characteristic length of the system is determined in a natural way by the period L of the potential

$\hat{V}(\hat{x})$. Accordingly, the position of the Brownian particle is scaled as $x = \hat{x}/L$. Time can be scaled in several ways. E.g., if the right-hand side of (1) is zero then

$$m\dot{v} + \gamma v = 0, \qquad v = \mathrm{d}\hat{x}/\mathrm{d}\hat{t}. \tag{14}$$

From this equation it follows that

$$v(t) = v(0)\exp(-\hat{t}/\tau_L), \tag{15}$$

where

$$\tau_L = m/\gamma \tag{16}$$

is called the Langevin time. It is the relaxation time of the velocity or the correlation time of velocity of the Brownian particle driven by thermal noise (cf. Eqs. (3.1) and (3.10) in [14]).

To identify the next characteristic time τ_0, let us consider a deterministic, overdamped motion of a particle in the potential $\hat{V}(\hat{x})$, namely [15]

$$\gamma \frac{\mathrm{d}\hat{x}}{\mathrm{d}\hat{t}} = -\frac{\mathrm{d}\hat{V}(\hat{x})}{\mathrm{d}\hat{x}}. \tag{17}$$

Then τ_0 can be defined by the above relation as follows:

$$\gamma \frac{L}{\tau_0} = \frac{\Delta \hat{V}}{L} \tag{18}$$

and takes the form

$$\tau_0 = \frac{\gamma L^2}{\Delta \hat{V}}. \tag{19}$$

During this time interval, an overdamped particle moves a distance of length L under the influence of the constant force $\Delta \hat{V}/L$. Other characteristic times related to deterministic dynamics can be introduced as well. The next characteristic time is the characteristic (e.g. correlation) time τ of nonthermal noise $\hat{\xi}(\hat{t})$. So, a few time scales can be distinguished. However, let us note that the Langevin time and the characteristic time of nonthermal driving noise do not depend on the system, i.e., on the potential $\hat{V}(\hat{x})$. Therefore, only τ_0 is different for different systems and we propose the scaling of time as $t = \hat{t}/\tau_0$. In this case, Eq. (1) can be transformed into its dimensionless form

$$\mu^* \ddot{x} + \dot{x} = f(x) + \Gamma(t) + \xi(t), \qquad f(x) = -\mathrm{d}V(x)/\mathrm{d}x, \tag{20}$$

where the dimensionless mass

$$\mu^* = \frac{m}{\gamma \tau_0} = \frac{\tau_L}{\tau_0} \tag{21}$$

takes into account inertial effects of Brownian particles and in fact it is a ratio of two characteristic times τ_L and τ_0. The rescaled spatially periodic potential $V(x) \equiv \hat{V}(\hat{x})/\Delta \hat{V} = V(x+1)$ has *unit period* and *unit barrier height*. The dimensionless intensity D of rescaled Gaussian white noise $\Gamma(t)$ is a *ratio of thermal energy to activation energy* over a barrier of the non-rescaled potential (i.e. the barrier height),

$$D = \frac{k_B T}{\Delta \hat{V}}. \tag{22}$$

Finally, the rescaled nonthermal noise $\xi(t) = (L/\Delta \hat{V})\hat{\xi}(\hat{t})$.

6. Overdamped equation of motion

Let us discuss the problem of overdamped dynamics [16]. As an example, we consider the kinesin which is one of biological (molecular) motors. It moves along microtubules inside the cells [17–19]. Microtubules are spatially periodic structures built of tubulin heterodimers which are arranged in rows called protofilaments which, in turn, are oriented nearly parallel to the microtubule axis. A heterodimer is about 8 nm long and is composed of two various globular subunits: α- and β-tubulin. It leads to the reflection symmetry breaking of the microtubules. In consequence, the corresponding potential $\hat{V}(\hat{x})$ with period $L = 8$ nm is asymmetric. The mass of kinesin $m = 6 \times 10^{-22}$ kg and its radius $R = 10^{-8}$ m. The friction coefficient $\gamma = 2 \times 10^{-8}$ kg/s which is calculated from the Stokes formula with $\eta = 10^{-1}$ kg/ms (it is assumed that the effective viscosity coefficient of the medium is hundred times greater than of water). In a generic Brownian domain $\Delta \hat{V} = 5k_B T$ and temperature inside of a cell is of order $T = 310$ K. In this case the Langevin time

$$\tau_L = m/\gamma = 3 \times 10^{-14} \text{ s} . \tag{23}$$

In turn

$$\tau_0 = \frac{\gamma L^2}{\Delta \hat{V}} = 6 \times 10^{-5} \text{ s} . \tag{24}$$

We can define the Einstein diffusion time τ_E by the relation

$$\tau_E = \frac{L^2}{2D_E} = 1.6 \times 10^{-4} \text{ s} , \tag{25}$$

where

$$D_E = k_B T/\gamma = 2 \times 10^{-13} \text{ m}^2/\text{s} \tag{26}$$

is the classical Einstein diffusion coefficient.

Let us note that time scales of τ_E and τ_0 are of the same order of magnitude. Indeed, from the above definitions it follows that

$$\tau_E = \frac{\Delta \hat{V}}{2k_B T} \tau_0 . \tag{27}$$

and the *ratio of two time scales depends on the ratio between activation and thermal energies*. On the other hand, usually the maximal efficiency of the Brownian ratchet is for the potential barrier height several times larger than the intensity of thermal fluctuations. The analysis of time scales explains this phenomenon as a region where the diffusive and convective effects have the same time scales.

Let us note that in Eq. (20) the dimensionless mass $\mu^* = 5 \times 10^{-10} \ll 1$. Hence, the acceleration term is 10 orders smaller than the dimensionless friction coefficient 1 at the velocity term. For that reason inertial effects can completely be neglected and the second order differential equation (20) can be approximated by the first-order differential equation

$$\dot{x} = f(x) + \Gamma(t) + \xi(t) . \tag{28}$$

This is an equation describing overdamped dynamics of Brownian particles. For the above example of kinesin it is a very good approximation to the starting equation (20). Below, we study this equation.

Let us consider now the stationary mean velocity of the particle. Starting from non-scaled variables we obtain

$$\langle v \rangle \equiv \langle d\hat{x}/d\hat{t} \rangle = \frac{L}{\tau_0} \langle dx/dt \rangle = v_0 \langle f(x) \rangle = v_0 \int_0^1 f(x) p_{st}(x) \, dx \,, \tag{29}$$

where the characteristic velocity $v_0 = L/\tau_0$ and $p_{st}(x)$ is a stationary probability distribution of the process $x(t)$ defined by the Langevin Eq. (28). Here, we have utilized (28) remembering that mean value of both sources of fluctuations is zero.

7. Overdamped ratchet in equilibrium state

If we assume that $\xi(t) \equiv 0$ then Eq. (28) reduces to the form

$$\dot{x} = f(x) + \Gamma(t) \tag{30}$$

and the master equation for the probability density $p(x,t)$ obeys the Fokker–Planck equation [14]

$$\frac{\partial p(x,t)}{\partial t} = -\frac{\partial J(x,t)}{\partial x} \,, \tag{31}$$

where the probability current

$$J(x,t) = f(x)p(x,t) - D\frac{\partial p(x,t)}{\partial x} \,. \tag{32}$$

In the stationary state, the density $p_{st}(x) = \lim_{t\to\infty} p(x,t)$, the probability current $J = \lim_{t\to\infty} J(x,t) = const.$ and (32) takes the form

$$J = f(x)p_{st}(x) - Dp_{st}'(x) \,, \tag{33}$$

where the prime denotes a derivative with respect to x. From (29) one gets

$$\langle v \rangle = v_0 J \,. \tag{34}$$

It can be shown from (33) that $J \equiv 0$ when imposing the periodic boundary condition for the stationary density $p_{st}(x)$. To explain in a simpler manner why $J = 0$, let us note that the stationary density [20]

$$p_{st}(x) \propto \exp[-\Psi(x)] \,, \tag{35}$$

where the generalized potential

$$\Psi(x) = -\int_0^x \frac{f(y)}{D} \, dy = \frac{V(x)}{D} = \frac{V(x+1)}{D} = \Psi(x+1) \,. \tag{36}$$

This generalized potential has zero slope, $\Delta\Psi = \Psi(x+1) - \Psi(x) = 0$ (see Fig. 4). It means that transition rates from a state of local minimum of the generalized potential $\Psi(x)$ are the same to the left valley (J_-) and to right valley (J_+). The stationary

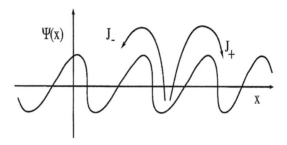

Fig. 4. The generalized potential $\Psi(x)$ in the case when its slope is zero. The probability currents J_+ and J_- to the positive and negative directions have the same values and the stationary mean velocity of particles is zero.

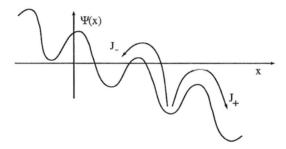

Fig. 5. The generalized potential $\Psi(x)$ exhibiting nonzero slope. Now, the probability currents J_+ and J_- to the positive and negative directions are different and particles are transported into positive direction.

mean velocity depends on the difference between transition rates in the positive and negative directions, $\langle v \rangle \propto J = J_+ - J_-$. Hence, $J = 0$. It is nothing but detailed balance. How to violate the principle of detailed balance? The hint comes from (36): It is seen that change of the constant diffusion coefficient D into a diffusion function $D(x)$ is a chance to obtain a nonzero slope $\Delta \Psi \neq 0$ of the generalized potential (Fig. 5). Then transition probabilities from the local minimum of $\Psi(x)$ to the positive and negative directions are different. An example is presented in the next section.

8. Transport induced by random telegraph signal

As an example of nonthermal driving force $\xi(t)$, we will consider the random telegraph signal, known also as dichotomic Markovian noise. This stationary stochastic process takes two values (see Fig. 6),

$$\xi(t) = \{-a, b\}, \quad a, b > 0 . \tag{37}$$

The probabilities of jump per unit time from one state to the other are

$$P(-a \to b) = \mu_a, \qquad P(b \to -a) = \mu_b \tag{38}$$

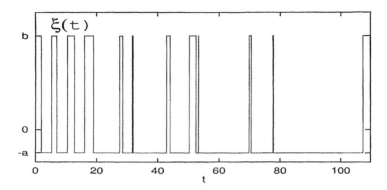

Fig. 6. An example of a realization of the random telegraph signal (dichotomic noise) of amplitudes b and $-a$.

If we assume that $a\mu_b = b\mu_a$ then the process is of zero-mean, $\langle \xi(t) \rangle = 0$. The correlation function of dichotomic noise reads

$$\langle \xi(t)\xi(s) \rangle = \frac{Q}{\tau} e^{-|t-s|/\tau} , \qquad (39)$$

where its correlation time $\tau = 1/(\mu_a + \mu_b)$ and intensity $Q = \tau ab$. We can also define times $\tau_a = 1/\mu_a$ and $\tau_b = 1/\mu_b$ which are the mean waiting times in states a and b, respectively. So, this noise is characterized by three parameters: the intensity Q, relaxation time τ and asymmetry $\theta = b - a$. If $a = b$ the process $\xi(t)$ is symmetric. Otherwise, it is asymmetric.

Let dynamics of the Brownian particle be described by the equation

$$\dot{x} = f(x) + \xi(t) \qquad (40)$$

with thermal noise being absent, $\Gamma(t) = 0$.

The stationary distribution and probability current are determined by the expression [21]

$$J = \frac{f(x) - D'_{eff}(x)}{1 + \tau f'(x)} p_{st}(x) - \frac{D_{eff}(x)}{1 + \tau f'(x)} p'_{st}(x) , \qquad (41)$$

where

$$D_{eff}(x) = \tau[a - f(x)][b + f(x)] . \qquad (42)$$

Note that Eq. (41) has a similar structure as (33) but now with changed 'force' at $p_{st}(x)$ and changed 'diffusion' function $D = D(x)$ at the derivative $p'_{st}(x)$.

We assume that the effective diffusion function $D_{eff}(x)$ in (42) is positive-valued for all values of x. It restricts a range of noise amplitudes a and b. In this case the stationary current J is of the form [21]

$$J = \frac{1 - e^{\Delta \Psi}}{\int_0^1 D_{eff}^{-1}(x) e^{-\Psi(x)} \int_x^{x+1} [1 + \tau f'(y)] e^{\Psi(y)} \, dy \, dx} , \qquad (43)$$

where

$$\Delta\Psi = \Psi(x+1) - \Psi(x) = -\int_x^{x+1} \frac{f(y)}{D_{eff}(y)}\,dy \tag{44}$$

is the slope of the generalized potential

$$\Psi(x) = -\int_0^x \frac{f(y)}{D_{eff}(y)}\,dy. \tag{45}$$

If the function $V(x)$ is symmetric and noise $\xi(t)$ is symmetric, $a=b$, then $\Delta\Psi \equiv 0$ and in consequence $J \equiv 0$. It can easily be demonstrated if, for example, $V(x) = V(-x)$. Then $f(x) = -f(x)$ and Eq. (44) can be rewritten in the form

$$\Delta\Psi = -\frac{1}{\tau}\int_{-1/2}^{1/2} \frac{f(y)}{a^2 - f^2(y)}\,dy \equiv 0. \tag{46}$$

The force $f(x)$ is an odd function and the denominator is an even function of x. Hence, the integral equals zero and the stationary velocity of the Brownian particle is zero (no transport). One can conclude from (44) that the current (43) is nonzero if the spatially periodic potential $V(x)$ is asymmetric and/or the noise $\xi(t)$ is asymmetric, i.e. when $a \neq b$. In other way, when the reflection symmetry of the potential and/or the statistical symmetry of nonthermal noise is broken then transport occurs and it is induced by the stochastic driving force $\xi(t)$.

9. Transport controlled by thermal noise

In the previous section, we presented an example of nonthermal stochastic driving force which induces directed motion of Brownian particles. In this section we briefly analyze the influence of temperature in such a system. The dynamics of the Brownian particle is described by Eq. (28) with $\xi(t)$ being the random telegraph signal (37). The master equation for the probability density $P(x,t)$ of the output process $x(t)$ satisfies the continuity equation [16]

$$\frac{\partial P(x,t)}{\partial t} = -\frac{\partial J(x,t)}{\partial x}, \tag{47}$$

where the probability current

$$J(x,t) = f(x)P(x,t) - D\frac{\partial P(x,t)}{\partial x} + W(x,t) \tag{48}$$

and the auxiliary distribution $W(x,t)$ obeys the equation

$$\frac{\partial W(x,t)}{\partial t} = -\frac{\partial}{\partial x}\left\{[f(x) + \theta]W(x,t) - D\frac{\partial}{\partial x}W(x,t)\right\}$$

$$-\frac{1}{\tau}W(x,t) - \frac{Q}{\tau}\frac{\partial}{\partial x}P(x,t). \tag{49}$$

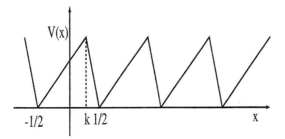

Fig. 7. The rescaled spatially periodic piecewise (sawtooth) potential $V(x)$ of period $L = 1$, barrier height $\Delta V = 1$ and asymmetry parameter $k \in (-\frac{1}{2}, \frac{1}{2})$.

The parameter $\theta = b - a$ is a measure of asymmetry of the telegraph signal $\xi(t)$. The distributions $P(x,t)$ and $W(x,t)$ are normalized in the following way:

$$\int_0^1 P(x,t)\,\mathrm{d}x = 1, \qquad \int_0^1 W(x,t)\,\mathrm{d}x = \langle \xi(t) \rangle = 0 \,. \tag{50}$$

Eqs. (47)–(49) form a closed set of partial differential equations. In the stationary state, $P(x) = P(x,t)$, $W(x) = \lim_{t\to\infty} W(x,t)$ and $J = \lim_{t\to\infty} J(x,t)$. From (47)–(49), one obtains a closed set of equations for $P(x)$, $W(x)$ and J. They read

$$-DP'(x) + f(x)P(x) + W(x) = J \,, \tag{51}$$

$$\tau DW''(x) - \tau[\theta + f(x)]W'(x) - [1 + \tau f'(x)]W(x) = QP'(x) \,. \tag{52}$$

For a general form of the periodic potential $V(x)$, Eqs. (51) and (52) cannot be solved analytically. However, exact results can be obtained for special forms of $V(x)$, e.g., for a piecewise linear potential (Fig. 7)

$$V(x) = \begin{cases} \dfrac{2x + 1}{2k + 1}, & x \in [-\frac{1}{2}, k]\,\mathrm{mod}\,1, \\[3mm] \dfrac{2x - 1}{2k - 1}, & x \in [k, \frac{1}{2}]\,\mathrm{mod}\,1, \end{cases} \tag{53}$$

where $k \in (-\frac{1}{2}, \frac{1}{2})$ determines asymmetry of the potential $V(x)$: for $k = 0$ the reflection symmetry holds; for $k \neq 0$ the reflection symmetry is broken. The method of finding the current J in this case is presented in detail in [16], where full analysis of transport properties of Brownian particle is carried out. Here only the most interesting features will be shown. The dimensionless probability current J depends on five dimensionless parameters, $J = J(D; Q, \tau, \theta; k)$. In turn, the stationary mean velocity $\langle v \rangle$ is related to J via the relation (34). For a given system, the barrier height $\Delta \hat{V}$, period L and asymmetry k of the potential are fixed. The remaining parameters can be treated as control parameters. The variation of the rescaled intensity $D = k_B T/\Delta \hat{V}$ of thermal noise denotes variation of temperature T of the system. A typical dependence of J (and $\langle v \rangle$)

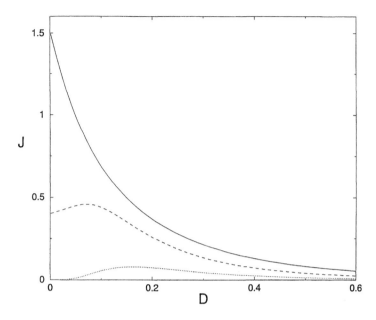

Fig. 8. Dependence of the current J on thermal-noise intensity D for positive asymmetry $k = 0.25$ of the periodic potential, fixed correlation time $\tau = 1$, symmetric dichotomic fluctuations ($\theta = 0$) and for three values of the intensity of dichotomic fluctuations, $Q = 16$ (solid line), $Q = 4.5$ (dashed line), $Q = 1.2$ (dotted line).

on temperature is shown in Fig. 8. If only one of symmetries is broken, i.e. when the potential is asymmetric ($k \neq 0$) and dichotomic fluctuations are symmetric ($\theta = 0$) or *vice versa*, one observes the following behavior of J: If dichotomic fluctuations induce both backward and forward transitions over the potential barrier (the case of large values on noise amplitudes a and b) then the velocity decreases as the temperature increases (the solid line in Fig. 8). If dichotomic fluctuations induce transitions only in one direction or do not induce transitions over the potential barrier (small values of a and b), the velocity exhibits a bell-shaped extremum (the dashed and dotted lines in Fig. 8) and then the optimal temperature exists which maximizes the stationary average velocity of particles.

An interesting phenomenon — *the current reversal* — occurs in the case of symmetry breaking of both the potential and dichotomic noise. If the signs of k and θ are opposite then in some domain of parameters the velocity changes sign as temperature varies. It is shown in Fig. 9. One can see that there exists a critical temperature $T \neq 0$, below and above which particles are transported in opposite directions. Moreover, there are two other characteristic values of temperature at which the velocity to positive and negative directions are maximal. It is worth to stress that in the critical regime, thermal energy $k_B T$ is several times smaller than the barrier height $\Delta \hat{V}$ of the potential. This is the regime of competition between jumps over a potential barrier induced by thermal fluctuations and directed movement to minima induced by local slopes of the potential.

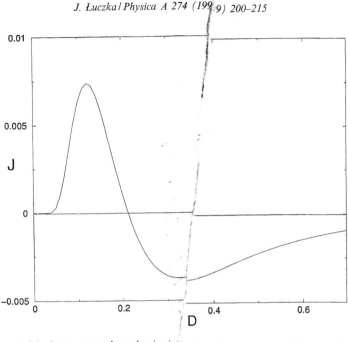

Fig. 9. The current J is shown versus thermal-noise intensities D for fixed $k = 0.25$, $\tau = 1$, $Q = 1.2$ and for asymmetric dichotomic fluctuations, $\theta = -0.9$.

If we want to analyze transport in dependence on size R of Brownian particles, then we should analyze the mean velocity $\langle v \rangle$ instead of the probability current J. Both J and $\langle v \rangle$ depend on the size of particles via the friction coefficient γ, see the Stokes formula (2), but dependence is different. Among six parameters $\{v_0, D, Q, \tau, \theta, k\}$ three depend on R, namely,

$$v_0 \equiv v_0(R) = \frac{\Delta \hat{V}}{c_0 \eta L R}, \quad Q \equiv Q(R) = \frac{\hat{Q}}{c_0 \eta \Delta \hat{V} R}, \quad \tau \equiv \tau(R) = \frac{\hat{\tau} \Delta \hat{V}}{c_0 \eta L^2 R}, \tag{54}$$

where $\hat{Q}$ and $\hat{\tau}$ are non-rescaled quantities, which do not depend on R. Inserting above to $\langle v \rangle$ we get the relative velocity

$$\frac{\langle v \rangle}{v_0(R_0)} = \frac{R_0}{R} J \left(D; \frac{R_0}{R} Q(R_0), \frac{R_0}{R} \tau(R_0), \theta; k \right) \tag{55}$$

in dependence on relative size R/R_0 of Brownian particles. It is depicted in Fig. 10 for three various temperatures. For higher temperature (the case corresponding to dotted line in Fig. 10), particles move on average to the 'left' direction. For lower temperature (the case corresponding to the dashed and solid lines in Fig. 10), particles of smaller sizes move to the 'right' while particles of larger sizes move to the 'left'. There is a region of temperature in which particles of various sizes R move in opposite directions. This phenomenon could allow to separate particles in dependence of their sizes. Maybe this is the mechanism leading to opposite movement of kinesin and dynein

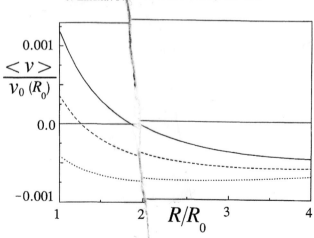

Fig. 10. The relative mean velocity $\langle v \rangle / v_0(R_0)$ of Brownian particles is depicted versus relative size R/R_0 of particles for fixed $k = 0.25$, $\tau = 0.9$, $Q = 1$, $\theta = -0.9$ and for various thermal-noise intensities (proportional to temperature of the system), $D = 0.18$ (solid line), $D = 0.19$ (dashed line), $D = 0.20$ (dotted line). The figure from [16].

along the same microtubule filaments: kinesin and dynein are of different effective sizes R.

10. Summary

In the paper we presented a mechanism of stochastic transport of particles in periodic structures. Transport can be induced by zero-mean nonthermal fluctuations acting as a driving force from which energy is pumped to the system and converted into mechanical one for directed motion of Brownian particles. The presence of non-thermal (nonequilibrium) forces is necessary but not sufficient. To generate transport, the principle of detailed balance has to be violated. It can be achieved by symmetry breaking. There are two radically different symmetries which can be broken. The first is the reflection symmetry of the spatially periodic system. The second is the statistical symmetry of temporal nonthermal fluctuations, which is characterized by multitime correlation functions and corresponding cumulants [22]. If at least one of these two symmetries is broken, stationary average velocity $\langle v \rangle$ being an 'order parameter' does not identically vanish and stochastic transport is induced by driving zero-mean non-thermal force.

Acknowledgements

The author would like to thank Prof. P. Hänggi for discussions on various aspects of ratchet problems and M. Kostur for preparing figures. The work supported by Komitet Badań Naukowych, Poland (Grant 2P03B 160 17).

References

[1] R.P. Feynman, R.B. Leighton, M. Sands, The Feynman Lectures on Physics, Addison Wesley, Reading, MA, 1963.

[2] M.O. Magnasco, Phys. Rev. Lett. 71 (1993) 1477.

[3] C.R. Doering, W. Horsthemke, J. Riordan, Phys. Rev. Lett. 72 (1994) 2984.

[4] R. Bartussek, P. Hänggi, J.G. Kissner, Europhys. Lett. 28 (1994) 459.

[5] R.D. Astumian, M. Bier, Phys. Rev. Lett. 72 (1994) 1766.

[6] J.F. Chauwin, A. Ajdari, J. Prost, Europhys. Lett. 27 (1994) 421.

[7] M.M. Millonas, M. Dykman, Phys. Lett. A 185 (1994) 65.

[8] R. Bartussek, P. Reimann, P. Hänggi, Phys. Rev. Lett. 76 (1996) 1166.

[9] J. Łuczka, R. Bartussek, P. Hänggi, Europhys. Lett. 31 (1995) 431.

[10] T. Czernik, J. Kula, J. Łuczka, P. Hänggi, Phys. Rev. E 55 (1997) 4057.

[11] J. Łuczka, T. Czernik, P. Hänggi, Phys. Rev. E 56 (1997) 3968.

[12] J.M.R. Parrondo, P. Espanol, Am. J. Phys. 64 (1996) 1125.

[13] L.D. Landau, E.M. Lifshitz, Hydrodynamics, Nauka, Moscow, 1986.

[14] H. Risken, The Fokker–Planck Equation, Springer, Berlin, 1996.

[15] J. Kula, T. Czernik, J. Łuczka, Phys. Rev. Lett. 80 (1998) 1377.

[16] J. Kula, M. Kostur, J. Łuczka, Chem. Phys. 235 (1998) 27.

[17] K. Svoboda, Ch.F. Schmidt, B.J. Schnapp, S.M. Block, Nature 365 (1993) 721.

[18] J. Łuczka, Cell. Mol. Biol. Lett. 1 (1996) 311.

[19] R.D. Astumian, Sciences 276 (1997) 917.

[20] M. Büttiker, Z. Phys. B 68 (1987) 161.

[21] J. Kula, T. Czernik, J. Łuczka, Phys. Lett. A 214 (1996) 14.

[22] P. Hänggi, R. Bartussek, P. Talkner, J. Łuczka, Europhys. Lett. 35 (1996) 315.

ELSEVIER

Physica A 274 (1999) 216–221

www.elsevier.com/locate/physa

Applications of statistical mechanics to finance

Rosario N. Mantegna[a,b,*], Zoltán Palágyi[c], H. Eugene Stanley[d,e]

[a]*Istituto Nazionale per la Fisica della Materia, Unità di Palermo, Palermo, I-90128, Italy*
[b]*Dipartimento di Energetica ed Applicazioni di Fisica, Università di Palermo, Applicazioni di Fisica,
Viale delle Scienze, Palermo, I-90128, Italy*
[c]*Department of Mathematics, Budapest University of Economic Sciences, Fővám tér 8,
1093 Budapest, Hungary*
[d]*Center for Polymer Studies, Boston, MA 02215, USA*
[e]*Department of Physics, Boston University, Boston, MA 02215, USA*

Abstract

We discuss some apparently "universal" aspects observed in the empirical analysis of stock price dynamics in financial markets. Specifically we consider (i) the empirical behavior of the return probability density function and (ii) the content of economic information in financial time series. © 1999 Elsevier Science B.V. All rights reserved.

PACS: 89.90.+n

Keywords: Econophysics

1. Introduction

The analyses and modeling of financial markets started in 1900 with the pioneering work of the French mathematician Bachelier [1]. Since the 1950s, the analysis and modeling of financial markets have become an important research area of economics and financial mathematics [2]. The researches pursued have been very successful, and nowadays a robust theoretical framework characterizes these disciplines [3–6]. In parallel to these studies, starting from the 1990s a group of physicists became interested in the analysis and modeling of financial markets by using tools and paradigms of their own discipline (for an overview, consider, for example, [7–11]). The interest of physicists in such systems is directly related to the fact that predictability has assumed a meaning in physics over the years, which is quite different from the one originally associated with the predictability of, for example, a Newtonian linear system. The

---

* Corresponding author. Fax: +39-091-236306.
E-mail address: mantegna@unipa.it (R.N. Mantegna)

0378-4371/99/$ - see front matter © 1999 Elsevier Science B.V. All rights reserved.
PII: S 0 3 7 8 - 4 3 7 1 (9 9) 0 0 3 9 5 - 7

degree of predictability of physics systems is nowadays known to be essentially limited in nonlinear and complex systems. This makes the physical prediction less strong, but on the other hand the area of research covered by physical investigations and of its application may increase [12].

The research approach of physicists to financial modeling aims to be complementary to the ones of financial mathematicians and economists. The main goals are to (i) contribute to a better understanding and modeling of financial markets and (ii) promote the use of physical concepts and expertise in the multidisciplinary approach to risk management.

In this communication, we review some results of our work on (i) the statistical properties of returns in financial markets and (ii) the characterization of the simultaneous dynamics of stock prices in a financial market. Specifically, we recall statistical properties of price changes empirically observed in different markets worldwide. In particular, we discuss studies performed on the New York stock exchange [13,14], the Milan stock exchange [15] and the Budapest stock exchange [16]. The communication is organized as follows, we first discuss the results of empirical analyses performed in these different markets and then we address the problem concerning the presence of economic information in a financial time series.

2. Statistical properties of price dynamics

The knowledge of the statistical properties of price dynamics in financial markets is fundamental. It is necessary for any theoretical modeling aiming to obtain a rational price for a derivative product issued on it [17] and it is the starting point of any valuation of the risk associated with a financial position [18]. Moreover, it is needed in any effort aiming to model the system. In spite of this importance, the modeling of such a variable is not yet conclusive. Several models exist which show partial successes and unavoidable limitations. In this research, the approach of physicists maintain the specificity of their discipline, namely to develop and modify models by taking into account the results of empirical analysis.

Several models have been proposed and we will not review them here. Here, we wish to focus only on the aspects which are *"universally"* observed in various stock price and index price dynamics.

2.1. Short- and long-range correlations

In any financial market — either well established and highly active as the New York stock exchange, *"emerging"* as the Budapest stock exchange, or *"regional"* as the Milan stock exchange — the autocorrelation function of returns is a monotonic decreasing function with a very short correlation time. High-frequency data analyses have shown that correlation times can be as short as a few minutes in highly traded stocks or indices [14,19]. A fast decaying autocorrelation function is also observed in the empirical analysis of data recorded transaction by transaction. By using as time index the number of transactions emanating from a selected origin, a time memory as short as a few transactions has been detected in the dynamics of most traded stocks of the Budapest "emerging" financial market [16].

The short-range memory between returns is directly related to the necessity of absence of continuous arbitrage opportunities in efficient financial markets. In other words, if correlation were present between returns (and then between price changes) this would allow one to devise trading strategies that would provide a net gain continuously and without risk. The continuous search for and the exploitation of arbitrage opportunities from traders focused on this kind of activity drastically diminish the redundancy in the time series of price changes. Another mechanism reducing the redundancy of stock price time series is related to the presence of the so-called "noise traders". With their action, noise traders add into the time series of stock price information, which is unrelated to the economic information decreasing the degree of redundancy of the price changes time series.

It is worth pointing out that not all the economic information present in stock price time series disappears due to these mechanisms. Indeed the redundancy that needs to be eliminated concerns only price change and not any of its nonlinear functions [20].

The absence of time correlation between returns does not mean that returns are identically distributed over time. In fact, different authors have observed that nonlinear functions of return such as the absolute value or the square are correlated over a time scale much longer than a trading day. Moreover, the functional form of this correlation seems to be power-law up to at least 20 trading days approximately [19,21–26].

A final observation concerns the degree of stationary behavior of the stock price dynamics. Empirical analysis shows that returns are not strictly sense stationary stochastic processes. Indeed the volatility (standard deviation of returns) is itself a stochastic process. Although a general proof is still lacking, empirical analyses performed on financial data of different financial markets suggest that the stochastic process is locally non-stationary but asymptotically stationary. By asymptotically stationary we mean that the probability density function (pdf) of returns measured over a wide time interval exists and it is uniquely defined. A paradigmatic example of simple stochastic processes which are locally non-stationary but asymptotically stationary is provided by ARCH [27] and GARCH [28] processes.

2.2. The distribution of returns

The pdf of returns shows some "universal" aspects. By "universal" aspects we mean that they are observed in different financial markets at different periods of time provided that a sufficiently long time period is used in the empirical analysis. The first of these "universal" or stylized facts is the leptokurtic nature of the pdf. A leptokurtic pdf characterizes a stochastic process having small changes and very large changes more frequently than in the case of Gaussian distributed changes. Leptokurtic pdfs have been observed in stocks and indices time series by analyzing both high-frequency and daily data. Thanks to the recent availability of transaction-by-transaction data, empirical analyses on a transaction time scale have also been performed. One of these studies performed by analyzing stock prices in the Budapest stock exchange shows that return pdf is leptokurtic down to a "transaction" time scale [16].

The origin of the observed leptokurtosis is still debated. There are several models trying to explain it. Just to cite (rather arbitrarily) a few of them: (i) a model of Lévy stable stochastic process [29]; (ii) a model assuming that the non-Gaussian behavior occurs as a result of the uneven activity during market hours [30]; (iii) a model where a geometric diffusive behavior is superimposed on Poissonian jumps [31]; (iv) a quasi-stable stochastic process with finite variance [32]; and (v) a stochastic process with rare events described by a power-law exponent not falling into the Lévy regime [33–35]. The above processes are characterized by finite or infinite moments. In an attempt to find the stochastic process that best describes stock price dynamics, it is important to try to preliminarily conclude about the finiteness or infiniteness of the second moment.

The above answer is not simply obtained [38] and careful empirical analyses must be performed to reach a reliable conclusion. It is our opinion that an impressive amount of empirical evidence has been recently found supporting the conclusion that the second moment of the return pdf is finite [13,33–37]. This conclusion has a deep consequence on the stability of the return pdf. The finiteness of the second moment and the independence of successive returns imply that the central limit theorem asymptotically applies. Hence, the form expected for the return pdf must be Gaussian for very long time horizons. We then have two regions — at short time horizons we observe leptokurtic distributions whereas at long time horizons we expect a Gaussian distribution. A complete characterization of the stochastic process needs an investigation performed at different time horizons. During this kind of analyses, non-Gaussian scaling and its breakdown has been detected [13,14].

3. Collective dynamics

In the previous sections we have seen that "universal" facts suggest that the stock price change dynamics in financial markets is well described by an unpredictable time series. However, this does not imply that the stochastic dynamics of stock price time series is a random walk with independent identically distributed increments. Indeed the stochastic process is much more complex than a customary random walk.

One key question in the analysis and modeling of a financial market concerns the independence of the price time series of different stocks traded simultaneously in the same market. The presence of cross-correlations between pairs of stocks has been known since a long time and it is one of the basic assumptions of the theory for the selection of the most efficient portfolio of stocks [39]. Recently, physicists have also started to investigate theoretically and empirically the presence of such cross-correlations.

It has been found that a meaningful economic taxonomy may be obtained by starting from the information stored in the time series of stock price only. This has been achieved by assuming that a metric distance can be defined between the synchronous time evolution of a set of stocks traded in a financial market and under the essential *ansatz* that the subdominant ultrametric associated with the selected metric distance is controlled by the most important economic information stored in the time evolution dynamics [40].

Another kind of study is devoted to the detection of the statistical properties of eigenvalues and eigenvectors of the covariance matrix of n stocks simultaneously traded. Also, with this approach the hypothesis that the dynamics of stock price in a portfolio of n stocks is described by independent random walks is falsified [41–43]. Moreover, information about the number of terms controlling eigenvectors can be detected.

The observation of the presence of a certain degree of statistical synchrony in the stock price dynamics suggests the following conclusion. Consideration of the time evolution of only a single stock price could be insufficient to reach a complete modeling of all essential aspects of a financial market.

4. Summary

This communication briefly discusses some of the stylized "universal" facts that are observed in financial markets and are considered robust by several researchers working in the field. Starting from these results, one can devise studies trying to enrich and expand this knowledge to provide theoreticians and computer scientists the empirical facts that need to be explained by their models progressively proposed. The ultimate goal is to contribute to the search for the best model describing a financial market, one of the most intriguing *"complex systems"*.

Acknowledgements

R.N.M. wishes to thank INFM and MURST for financial support. Z.P. thanks Cashline Broker Securities Ltd. for financial support, and H.E.S. thanks the NSF.

References

[1] L. Bachelier, Théorie de la spéculation [Ph.D. Thesis in mathematics], Ann. Sci. de l'Ecole Normale Supérieure III-17 (1900) 21–86.
[2] P.H. Cootner (Ed.), The Random Character of Stock Market Prices, MIT Press, Cambridge, MA, 1964.
[3] J.E. Ingersoll Jr., Theory of Financial Decision Making, Rowman & Littlefield, Savage, MD, 1987.
[4] R.C. Merton, Continuous-Time Finance, Blackwell, Cambridge, MA, 1990.
[5] I. Karatzas, S.E. Shreve, Brownian Motion and Stochastic Calculus, 2nd Edition, Springer, Berlin, 1991.
[6] J.Y. Campbell, A.W. Lo, A.C. MacKinlay, The Econometrics of Financial Markets, Princeton University Press, Princeton, 1997.

[7] J. Kertész, I. Kondor (Eds.), Econophysics: Proceedings of the Budapest Workshop, Kluwer Academic Press, Dordrecht (in press).

[8] R.N. Mantegna (Ed.), Proceedings of the International Workshop on Econophysics and Statistical Finance, Physica A (special issue) 269 (1999).

[9] K. Lauritsen, P. Alstrøm, J.-P. Bouchaud (Eds.), Proceedings of the International Conference on Application of Physics in Financial Analysis, Int. J. Theor. Appl. Finance (special issue) (in press).

[10] J.-P. Bouchaud, M. Potters, Théories des Risques Financiers, Aléa-Saclay, Paris, 1997.

[11] R.N. Mantegna, H.E. Stanley, An Introduction to Econophysics: Correlations and Complexity in Finance, Cambridge University Press, Cambridge, 1999.

[12] G. Parisi, Physica A 263 (1999) 557.

[13] R.N. Mantegna, H.E. Stanley, Nature 376 (1995) 46.

[14] R.N. Mantegna, H.E. Stanley, Nature 383 (1996) 587.

[15] R.N. Mantegna, Physica A 179 (1991) 232.

[16] Z. Palágyi, R.N. Mantegna, Physica A 269 (1999) 132.

[17] J.C. Hull, Options, Futures, and Other Derivatives, 3rd Edition, Prentice-Hall, Upper Saddle River, NJ, 1997.

[18] D. Duffie, J. Pan, J. Derivatives, spring issue (1997) 7.

[19] Y. Liu, P. Gopikrishnan, P. Cizeau, M. Meyer, C.-K. Peng, H.E. Stanley, Phys. Rev. E 60 (1999) 1390.

[20] R. Baviera, M. Pasquini, M. Serva, D. Vergni, A. Vulpiani, Efficiency in foreign exchange markets, cond-mat 9901225.

[21] M.M. Dacorogna, U.A. Müller, R.J. Nagler, R.B. Olsen, O.V. Pictet, J. Int. Money Finance 12 (1993) 413.

[22] R. Cont, M. Potters, J.-P. Bouchaud, in: B. Dubrulle, F. Graner, D. Sornette (Eds.), Scale Invariance and Beyond, Springer, Berlin, 1997.

[23] P. Cizeau, Y. Liu, M. Meyer, C.-K. Peng, H.E. Stanley, Physica A 245 (1997) 441.

[24] Y. Liu, P. Cizeau, M. Meyer, C.-K. Peng, H.E. Stanley, Physica A 245 (1997) 437.

[25] M. Pasquini, M. Serva, Physica A 269 (1999) 140.

[26] M. Raberto, E. Scalas, G. Cuniberti, M. Riani, Physica A 269 (1999) 148.

[27] R.F. Engle, Econometrica 50 (1982) 987.

[28] T. Bollerslev, J. Econometrics 31 (1986) 307.

[29] B.B. Mandelbrot, J. Bus. 36 (1963) 394.

[30] P.K. Clark, Econometrica 41 (1973) 135.

[31] R.C. Merton, J. Financial Econom. 3 (1976) 125.

[32] R.N. Mantegna, H.E. Stanley, Phys. Rev. Lett. 73 (1994) 2946.

[33] T. Lux, Appl. Financial Econom. 6 (1996) 463.

[34] T. Lux, J. Econ. Behav. Organ. 33 (1998) 143.

[35] P. Gopikrishnan, M. Meyer, L.A.N. Amaral, H.E. Stanley, Eur. Phys. J. B 3 (1998) 139.

[36] P. Gopikrishnan, V. Plerou, L.A.N. Amaral, M. Meyer, H.E. Stanley, Scaling of the distributions of fluctuations of financial market indices, Phys. Rev. E 60 (1999) in press, cond-mat 9905305.

[37] V. Plerou, P. Gopikrishnan, L.A.N. Amaral, M. Meyer, H.E. Stanley, Scaling of the distribution of price fluctuations of individual companies, Phys. Rev. E, cond-mat 9907161 (submitted for publication).

[38] A.L. Tucker, J. Bus. Econom. Stat. 10 (1992) 73.

[39] H. Markowitz, Portfolio Selection: Efficient Diversification of Investment, Wiley, New York, 1959.

[40] R.N. Mantegna, Eur. Phys. J. B 11 (1999) 193.

[41] S. Galluccio, J.-P. Bouchaud, M. Potters, Physica A 259 (1998) 449.

[42] L. Laloux, P. Cizeau, J.-P. Bouchaud, M. Potters, Phys. Rev. Lett. 83 (1999) 1468.

[43] V. Plerou, P. Gopikrishnan, B. Rosenow, L.A.N. Amaral, H.E. Stanley, Phys. Rev. Lett. 83 (1999) 1471.

ELSEVIER   Physica A 274 (1999) 222–228

www.elsevier.com/locate/physa

Portfolios with nonlinear constraints and spin glasses

Adrienn Gábor[a,*], I. Kondor[a,b]

[a]*Department of Physics of Complex Systems, Eötvös University, 1518 Budapest, Pf. 32, Hungary*
[b]*Raiffeisen Bank, 1052 Budapest, Váci u. 19-21, Hungary*

Abstract

In a recent paper Galluccio, Bouchaud and Potters demonstrated that a certain portfolio problem with a nonlinear constraint maps exactly onto finding the ground states of a long-range spin glass, with the concomitant nonuniqueness and instability of the optimal portfolios. Here we put forward geometric arguments that lead to qualitatively similar conclusions, without recourse to the methods of spin glass theory, and give two more examples of portfolio problems with convex nonlinear constraints. © 1999 Elsevier Science B.V. All rights reserved.

1. Introduction

Spin glasses provide a powerful paradigm for complex systems involving competition and cooperation or conflicting internal constraints, not only in physics and various other branches of the natural sciences (see [1–3] for an introduction to spin glasses), but also in the social sciences and economics [4,5]. Recently Galluccio et al. [6] added a striking new example to the growing list of "nonphysical spin glasses" by exactly mapping the problem of optimal composition of a portfolio containing short sales with obligatory deposits (margin accounts) onto that of finding the ground states of a long-range Ising spin glass. The coupling constants of this spin glass (SG) are related to the covariance matrix of the instruments making up the portfolio. Galluccio et al. argue that a sufficiently complicated covariance matrix can be regarded as a sample drawn from an ensemble of random matrices and they identify the particular matrix ensemble that offers an appropriate representation of real markets. Invoking techniques and results from SG theory, they then proceed to show that instead of the single solution of the classical Markowitz problem [7], here, as a consequence of the nonlinear

---

* Corresponding author.
E-mail address: adri@pearl.elte.hu (A. Gábor)

0378-4371/99/$ - see front matter © 1999 Elsevier Science B.V. All rights reserved.
PII: S 0378-4371(99)00387-8

constraint expressing the obligatory deposit, one has an exponentially large number of optimal portfolios which are all completely different from each other and, in addition, are extremely sensitive to any changes in the input parameters of the problem. As Galluccio et al. point out, under these conditions the very concept of rational decision making becomes questionable. It is also clear that the problem of margin accounts is just one particular example from a large set of general optimization problems with nonlinear constraints which may lead to similar "spin glass-like behaviour", therefore their work may have far reaching implications for modelling approaches to social and economic phenomena.

The purpose of the present note is to give two more examples of portfolio problems with nonlinear constraints, and to develop a geometric approach which may help to make the appearance of this SG-like instability plausible even for those unfamiliar with the heavy machinery of spin glass theory.

In Section 2 we display a geometric representation of the portfolio selection problem in its original, linear setting. In Section 3 we extend this geometric approach to the case of margin accounts considered by Galluccio et al. [6] and show that the presence of the particular nonlinear constraint (corresponding to a convex polyhedron) makes it geometrically quite obvious that one should typically find a large number of unstable solutions. It will also transpire that the relevant feature of the constraint is its convexity and we give a few other examples where qualitatively similar behaviour is to be expected. We conclude with a brief comment on the economic significance of the existence of these SG-like solutions.

2. Geometric representation of the classical portfolio problem

In order to be able to appreciate the effect of nonlinear constraints, let us first review the original, linear problem [7]. A portfolio is a package of various financial assets (bonds, shares, derivatives, etc.). Let us denote the return on asset i by $x_i = 0, 1, 2, \ldots, n$ and the expectation value of the return by $\mu_i = \langle x_i \rangle$. The precise meaning of the ensemble used to evaluate μ_i will be of no importance in the following; in practice, it is usually calculated as a time average, but can also be thought of as an expectation value taken over a virtual ensemble, which is formed on the basis of market information available at the time that the portfolio is set up and only exists in the minds of the portfolio managers. If w_i are the weights of the different assets represented in the portfolio, $w_i > 0$,

$$\sum_i w_i = 1 , \tag{1}$$

then the expected return on the portfolio is

$$\mu = \sum_i w_i \mu_i \tag{2}$$

and the variance, a customary measure of the risk, will be given by

$$\sigma^2 = \sum_{i,j} w_i C_{ij} w_j \,, \tag{3}$$

where $C_{ij} = \langle x_i x_j \rangle - \langle x_i \rangle \langle x_j \rangle$ is the covariance matrix. Normally, one of the assets, say the zeroth, is distinguished in that it is assumed riskless (a government bond e.g.). The return on this riskless asset must be the lowest: no one would include a risky asset in their portfolio if it paid less then the riskless bond. If we wish to earn more than this riskless return, we have to include risky assets in our portfolio. However, the composition of the package, i.e. the weights w_i, must be chosen in such a way that for a given level of return the risk be as low as possible. This means we have to minimize σ^2 over the weights w_i, given the linear constraints (1) and (2).

Usually this optimization problem is broken down into two stages: first the set of efficient (i.e. minimal variance) portfolios (consisting of the risky assets only) is determined, and the riskless instrument is mixed in later with a certain weight w_0 which then becomes a fundamental parameter of the problem. This two-stage procedure is well justified both economically and (thinking of the problem as one in quadratic programming) numerically, but for our present purposes it will be perfectly sufficient to perform the optimization in one go.

Without the inclusion of the riskless instrument (and without accidental degeneracy between the other instruments, see below), the covariance matrix C_{ij} would be positive definite and (3) would be an ellipsoid in the space of weights. Inclusion of the riskless asset means adding a column and row of zeros to C_{ij}, thereby converting it into a positive semidefinite matrix. The geometrical object corresponding to (3) now becomes an elliptical cylinder: its axis is along the riskless coordinate, but its projection onto any hyperplane not parallel to this axis is an $(n-1)$-dimensional ellipsoid.

Were it not for constraints (1) and (2), the solution of the optimization problem would be trivial: all the capital ought to be invested in the riskless asset, and the "risk cylinder" would shrink into a straight line along the riskless axis. Constraints (1) and (2) correspond to two (hyper)planes. Their intersection is still a hyperplane of $n-2$ dimensions. Let us consider any $(n-1)$-dimensional subspace which contains this $(n-2)$-dimensional intersection and which is not parallel to the zeroth axis. The projection of the risk cylinder onto this subspace is an ellipsoid. Our optimization problem corresponds to inflating this ellipsoid (increasing σ^2) until it touches the $(n-2)$-dimensional intersection of planes (1) and (2). As a convex body, an ellipsoid can only have a single tangent point with a hyperplane — the solution of the optimization problem is unique. It is also (relatively) stable in the following sense: if we slightly misspecify the return plane or the risk ellipsoid (as may well happen, given that their parameters are determined from finite time series) their tangent point will be displaced. The important point, however, is that this error can be kept under control: a small error in the input data results in a commensurate error in the solution. (See, however, [8,9] for a discussion of the noise involved in the smallest eigenvalues of the covariance matrix.)

The uniqueness of the solution for linear constraints can only break down in the trivial (and rather unrealistic) case when two assets, say the ith and the jth, are strictly equivalent (same return, same volatility, same covariances with the other instruments). When this happens it is clearly immaterial how much we invest in one or the other individually, the optimization should determine the sum of their weights only. How is this reflected in the geometric picture? When assets i and j are equivalent, the ith and jth columns of C_{ij} are the same and, by the symmetry of C_{ij}, the same is true for the ith and jth rows. Then the matrix C_{ij} will have another zero eigenvalue (in addition to the one corresponding to the riskless asset) and the corresponding eigenvector will be parallel to the plane (1). The object corresponding to (3) will then have another cylindrical axis along this eigenvector, so when it first touches the intersection of (1) and (2) this happens not at a single point but along a line: instead of a unique solution, we find a continuum of solutions all at once. This, however, merely reflects the fact that we can include assets i and j in our portfolio in arbitrary proportions, as long as the sum of their weights is fixed. When, in the following, we speak about multiple solutions in the case of nonlinear constraints, we *do not mean* trivial degeneracies like the substitution of one asset by an accidentally equivalent one, as in the example above.

3. Portfolios with nonlinear constraints

Following Galluccio et al. let us now consider a portfolio containing long as well as short positions with obligatory deposits proportional to the volumes of the latter. The budget constraint of such a portfolio can be written as [6]

$$\sum_i \gamma_i |w_i| = 1 , \tag{4}$$

where the margin (deposit rate) γ_i may vary from asset to asset. (It may for example have a common value $\gamma < 1$ for the shorts and unity for the longs.) Rational portfolio selection then demands minimizing the variance (3) under constraint (2) on the return and constraint (4) on the "weights" w_i (w_i can now be negative, according to the shorts). Galluccio et al. [6] show that this task is equivalent to finding the ground state of a system of "spins" $s_i = sign(w_i)$ interacting via the "couplings" $J_{ij} = \lambda C_{ij}^{-1}$ and subject to the "local external fields" $H_i = \nu \sum_j C_{ij}^{-1} \mu_j$, where λ and ν are Lagrange multipliers associated with constraints (4) and (2), respectively.

Geometrically, constraint (4) corresponds to a convex n-dimensional polyhedron. Its intersection with the hyperplane (2) is a convex polyhedron of one less dimension. In order to solve the minimization problem we have to find the first tangent point of the risk ellipsoid with this $(n - 1)$-dimensional polyhedron as we gradually inflate the ellipsoid inside the cage formed by the polyhedron. In contrast to the previous case where the risk ellipsoid and the budget constraint could have at most one tangent point, here the maximal number of possible tangent points is the number of faces of the polyhedron, i.e. 2^{n-1}. Of course, finding 2^{n-1} solutions all at once is singularly unlikely.

Consider, however, that under appropriate conditions the eigenvalues of a sufficiently large random matrix tend to be jammed into a finite interval of length of the order of the variance of the matrix elements [10] and that, according to [8,9,11] the spectra of empirical covariance matrices measured on major markets, with well over 90% of the eigenvalues falling into this theoretical interval, come very close to this situation. The density of eigenvalues is particularly high around the lower band edge. Since the lengths of the principal axes of the ellipsoid are inversely proportional to the square root of the eigenvalues, the first tangent point will lie in the direction of the eigenvector corresponding to the smallest eigenvalue. As we increase σ^2 a little further, another tangent point will soon appear in the direction of the eigenvector corresponding to the second smallest eigenvalue, that is, on a different face of the polyhedron. Increasing σ^2 still further, we find a third, a fourth, etc. solution, all occurring within a very small interval of σ^2 and scattered about the various faces of the polyhedron in a random fashion. In addition, the smallest eigenvalues are precisely those that are the most affected by the noise inevitably present in any empirical covariance matrix [8,9,11]. This means that the slightest change in the input parameters of the problem (such as the estimated returns or covariances) will completely reshuffle the order in which all these nearly degenerate solutions appear and will make the "best" solution jump about in the space of weights.

This scenario is exactly the same as the one that describes the behaviour of the low-lying metastable states of a spin glass. Full-fledged spin glass effects can be expected to show up only in the limit of large systems, i.e. with portfolios consisting of a large number of assets. Real life portfolios do not contain a macroscopic number of instruments, but a portfolio of the size of the S & P 500 considered by Galluccio et al. [6] is more than sufficient to display all the SG instabilities in their full glory. As a matter of fact, our own numerical experiments on small portfolios formed from subsets of the assets making up the DJIA have shown that spin glass effects are already visible for n's as low as 5 or 7.

It should be obvious from the foregoing that the approximate degeneracy and extreme sensitivity of the solutions do not depend on the particular form of the nonlinear constraint. Replacing (4) by any other convex surface would allow us to repeat essentially the same argument. What other kind of nonlinear constraints can be envisaged in the context of portfolio selection? It is commonplace now that the statistical behaviour of real-life financial instruments can be substantially different from Gaussian. In order to contain the fluctuations of a portfolio composed of such instruments, we may wish to limit some higher cumulants, say the kurtosis, of the portfolio. This would introduce a quartic surface into the optimization problem and would lead to the same sort of SG effects as the polyhedron considered above.

Another interesting situation, already hinted at in [12], is, when the constraint is simply a sphere. This corresponds to what may be termed forced diversification. It may happen, especially on small markets, that a few liquid assets tend to dominate every portfolio. In such a situation we may wish to favour portfolios that contain a somewhat larger number of instruments than dictated by a blind application of the

Markowitz theory alone. This can be achieved by introducing a constraint on the sum of the squares of the weights

$$\sum_i w_i^2 = \frac{1}{K} . \tag{5}$$

K can be regarded as the effective number of assets included in the portfolio. (K is of the order of n if the weights are distributed relatively uniformly, and is 1 if the portfolio is concentrated on a single asset.) The peculiarity of this constraint is that it is a quadratic form as is the objective function σ^2. As a result, whenever the objective function and the spherical constraint have more than a given number (depending on the dimensionality of the problem) of tangent points, a continuous degeneracy of the solutions suddenly appears (e.g. in three dimensions, when a sphere and an ellipsoid have more than two tangent points, they are necessarily tangent along a circle.) In our numerical experiments we never observed more than two discrete solutions, but found evidence for a continuum.

Another question that may arise concerning the robustness of these SG effect is the characterisation of risk, i.e. the choice of the objective function. As long as our assets obey normal statistics, any plausible measure of risk will be proportional to the variance. If, however, our instruments deviate from Gaussian statistics then the usefulness of variance will be less obvious. A widespread substitute is value at risk (VaR). Do we expect to find the same kind of instabilities if we use VaR as a measure of risk and combine it with nonlinear constraints? The answer is yes. For a wide class of joint probability distribution functions of n assets, it can be shown [13] that the surfaces defined in the space of the weights by fixing the VaR *and* the safety level are closed and convex, in some cases, actually ellipsoids. The geometry of the problem of a portfolio subject to a nonlinear constraint and to a given VaR will then be quite similar to the one discussed before.

4. Concluding remarks

We have argued that under nonlinear, convex constraints the appearance of many nearly degenerate, unstable solutions is a generic feature of the portfolio selection problem. The question may arise, however, whether this SG-like phenomenon has any economic relevance — after all, the resulting portfolios have the same risk and return, so it might seem that the investor will be equally satisfied, whichever solution he happens to choose.

In a wider context, however, the problem is far from irrelevant. If the efficient portfolio is unique and if the available information is essentially the same for all market participants, then they will have a single degree of freedom when constructing their portfolios, namely choosing the weight of the riskless asset. The risky components of their portfolios will all have the same internal proportions and will then necessarily reflect the distribution of capital over the market as a whole. It is clear that such a uniform investment strategy on the part of all market participants would impose very strong conditions on the economy and would determine its future development to a large extent. Of course, such complete uniformity never occurs in real life, nevertheless the idea of a unique optimal investment strategy plays an important role in modern theoretical finance [14]. If, on the other hand, there exist a large number of equivalent solutions to the portfolio selection problem, then investors will choose between them according to factors that were deemed to be of lesser importance and not included in the original objective function. (As Galluccio et al. [6] put it, the degeneracy will be lifted by small, "irrational" effects.) The resulting investment strategies will not be uniform and will represent a whole population of competing strategies. In the long run, the existence of this "ecodiversity" may even prove more advantageous than the dominance of a single rigid investment pattern.

Looking at the problem from a practical point of view, we can also see a fundamental numerical difference between a situation with a single solution and one with a large number of nearly degenerate, unstable solutions. Traditional methods of quadratic programming cannot cope with the exponentially many SG-like solutions, and in such a situation the special algorithms developed in the statistical physics of random systems (see [15] for a review) may offer a remedy.

Acknowledgements

This work has been partially supported by the Hungarian Science Fund OTKA, contract No. T019422, and by a Széchenyi Scholarship received by I.K. We are grateful to R. Mantegna for providing the data for our numerical experiments, to P. Medvegyev for a useful comment, and to A. Czirók for continuous helpful interaction.

References

[1] M. Mézard, G. Parisi, M.A. Virasoro, Spin Glass Theory and Beyond, World Scientific, Singapore, 1997.
[2] K.H. Fischer, J.A. Hertz, Spin Glasses, Cambridge University Press, Cambridge, 1991.
[3] A.P. Young, Spin Glasses and Random Fields, World Scientific, Singapore, 1998.
[4] R. Axelrod, D.S. Bennett, British J. Political Sci. 23 (1993) 211.
[5] S. Galam, in: J. Kertész, I. Kondor (Eds.), Econophysics — an Emergent Science, Kluwer, Amsterdam, to appear.
[6] S. Galluccio, J.P. Bouchaud, M. Potters, Physica A 259 (1998) 449.
[7] H. Markowitz, Portfolio Selection: Efficient Diversification of Investment, Wiley, New York, 1959.
[8] L. Laloux, P. Cizeau, J.P. Bouchaud, M. Potters, Phys. Rev. Lett. 83 (1999) 1467.
[9] L. Laloux, P. Cizeau, J.P. Bouchaud, M. Potters, Risk Mag. 12 (1999) 69.
[10] M.L. Mehta, Random Matrices, Academic Press, New York, 1991.
[11] V. Plerou, P. Gopikrishnan, B. Rosenow, L.A.N. Amaral, H.E. Stanley, Phys. Rev. Lett. 83 (1999) 1471.
[12] J.P. Bouchaud, M. Potters, Théorie des Risques Financiers, Aléa-Saclay, Eyrolles, Paris, 1997.
[13] I. Kondor, unpublished.
[14] E.J. Elton, M.J. Gruber, Modern Portfolio Theory and Investment Analysis, Wiley, New York, 1995.
[15] J. Kertész, I. Kondor (Eds.), Advances in Computer Simulation, Springer, Berlin, 1998.

ELSEVIER

Physica A 274 (1999) 229–240

PHYSICA A

www.elsevier.com/locate/physa

Applications of statistical physics to economic and financial topics

M. Ausloos[a,*], N. Vandewalle[a], Ph. Boveroux[b],
A. Minguet[b], K. Ivanova[c]

[a] GRASP[1], Institue of Physics B5, Univeristy of Liége, B-4000 Liège, Belgium
[b] Théorie monétaire et finances, Faculté d'Economie, Gestion et Sciences Sociales, B31,
Université de Liège, B-4000 Liège, Belgium
[c] Institute of Electronics, Bulgarian Academy of Sciences, Tzarigrazdsko Chaussee 72,
1784 Sofia, Bulgaria

Abstract

Problems in economy and finance have started to attract the interest of statistical physicists. Fundamental problems pertain to the existence or not of long-, medium-, short-range power-law correlations in economic systems as well as to the presence of financial cycles. Methods like the extended detrended fluctuation analysis, and the multi-affine analysis are recalled emphasizing their value in sorting out correlation ranges and predictability. Among spectacular results, the possibility of crash predictions is indicated. The well known financial analyst technique, the so-called moving average, is shown to raise questions about fractional Brownian motion properties. Finally, the (m, k)-Zipf method and the i-variability diagram technique are presented for sorting out short range correlations. Analogies with other fields of modern applied statistical physics are also presented in view of some universal openess. © 1999 Elsevier Science B.V. All rights reserved.

1. Introduction

Problems in economy and finance have previously interested mathematicians [1], but have only started to attract the interest of statistical physicists recently [2]. A fundamental problem is the existence or not of long-, medium-, short-range power-law correlations in economic systems as well as the presence of financial cycles. Indeed, traditional methods (Fourier transforms, wavelet analysis, etc.) have corroborated that

* Corresponding author. Fax: +32-4-366-29-90.
E-mail address: ausloos@gw.unipc.ulg.ac.be (M. Ausloos).
[1] GRASP = Group for Research in Applied Statistical Physics.

there is evidence that the Brownian motion idea, models like ARCH, GARCH, etc., and the "effficient market hypothesis" are only approximately right [3–6].

The application of statistical physics ideas to the behavior of firms and forecasting stock market fluctuations have been proposed earlier following the pioneer work of physicists interested by economy laws [7–11]. Econophysics [12–17] aims to fill the huge gap separating empirical finance and econometric theories. Various subjects have been approached like the commodity prices [18], option pricing [19], foreign exchange currency markets [20], etc. from a microscopic or specific company life features on a sort of fundamental phase diagram [21], thus macroscopic view point. Statistical physicists have developed new methods, new models, and presented new ideas. A few of which are discussed here, selected from our own work [20–30], emphasizing results, defects, and fundamental questions for physicists.

Methods like (i) the extended detrended fluctuation analysis, and (ii) the multi-affine analysis will be recalled emphasizing their value in sorting out correlation ranges and predictability. Among spectacular results, we will point out (iii) the crash predictions. An interesting financial analyst technique, (iv) the moving average, will be shown to raise apparently elementary questions about fractional Brownian motion (fBm). Finally, (v) the (m, k)-Zipf method and (vi) the i-variability diagram technique will be presented for sorting out short range correlations.

In the following we cannot touch upon all such questions of interest with much detail. Other contributions of very high value are found in these Proceedings. In each case the reference lists should be consulted. For more information, see also web sites, like the Liège[2] site and the Genoa[3] site.

2. Detrended fluctuation analysis technique

The Detrended Fluctuation Analysis (DFA) technique consists in dividing a time series or random one-variable sequence $y(t)$ of length N into N/τ nonoverlapping boxes, each containing τ points [31]. The linear local trend in each box is defined by

$$z(t) = at + b. \tag{1}$$

A cubic trend, like

$$z(t) = ct^3 + dt^2 + et + f \tag{2}$$

can be also considered [23]. Other combinations should be investigated by statistical physicists. The parameters a to f are estimated through a best least-squares fit of the data points in each box. The detrended fluctuation function $F(\tau)$ is then calculated as follows:

$$F^2(\tau) = \frac{1}{\tau} \sum_{t=k\tau+1}^{(k+1)\tau} |y(t) - z(t)|^2, \quad k = 0, 1, 2, \ldots, \left(\frac{N}{\tau} - 1\right). \tag{3}$$

---

[2] http://www.supras.phys.ulg.ac.be/statphys/statphys.html.
[3] http://www.ge.infm.it/econophysics/content3.html.

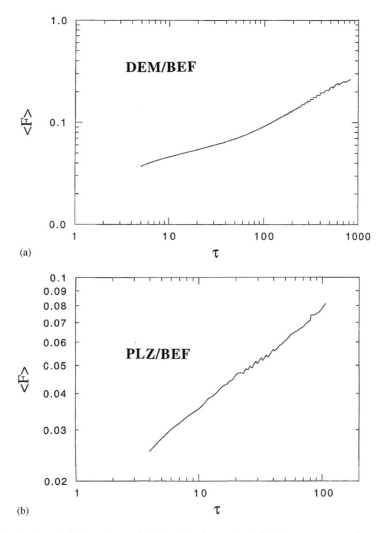

Fig. 1. Linearly detrended fluctuation analysis function for two typical foreign currency exchange rates, i.e. (a) DEM/BEF and (b) PLZ/BEF between Jan. 01, 1980 and Dec. 31, 1995. The Brownian motion behavior corresponds to a slope $\frac{1}{2}$ on this log–log plot; see the crossover on the DEM/BEF plot. Time τ is expressed in days. The notation $\langle F \rangle$ is used for $\langle F^2(\tau) \rangle^{1/2}$ for conciseness in labeling the y-axis.

Averaging $F^2(\tau)$ over all N/τ box sizes centered on time τ gives the fluctuations $\langle F^2(\tau) \rangle$ as a function of τ. The behavior is expected to be a power law

$$\langle F^2(\tau) \rangle^{1/2} \sim \tau^\alpha . \tag{4}$$

An exponent $\alpha \neq \frac{1}{2}$ in a certain range of τ values implies the existence of long-range correlations in that time interval as in the fractional Brownian motion [32]. Correlations and anticorrelations correspond to $\alpha > \frac{1}{2}$ and $\alpha < \frac{1}{2}$ respectively. Two cases are shown in Fig. 1 for DEM/BEF and PLZ/BEF. Most of the time, for the foreign exchange currency (FEXC) rates that we have examined, the scaling range is well defined (Table 1),

Table 1
Scales for different FEXC rates over the 1980–86 years. Time is expressed in weeks (w)

FEXC	Lower scale range (w)	Mid crossover (w)	Upper scale range (w)
USD/DEM	1	—	50
GBP/DEM	1	—	62
JPY/USD	1	—	101
PLN/BEF	1	—	20
BLG/DEM	1	—	4
NLG/BEF	1	10	120
DEM/BEF	1	6	120
DKK/BEF	1	7	120
FRF/BEF	1	—	246

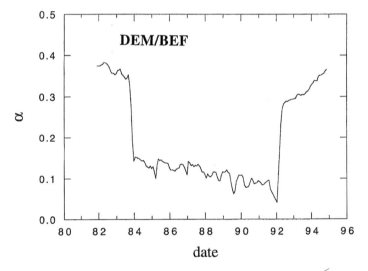

Fig. 2. The moving α exponent for DEM/BEF.

and crossovers are well observed. These crossovers suggest that correlated sequences have characteristic durations with well-defined lower and upper scales [20].

Furthermore, in order to probe the existence of *correlated and decorrelated sequences*, a so-called observation box of a given size, e.g. 2 year can be constructed. The exponent α for the data contained in that finite size box can be calculated at each step. The box can be moved along the historical time axis by a finite number of points (say, corresponding to 4 weeks) along the financial sequence. The *moving α exponent* can be displayed, e.g. for DEM/BEF (Fig. 2). This procedure leads to a local measurement of the degree of long-range correlations. This has also been done for *DNA* sequences and for liquid water content and brightness temperature of stratus clouds [33]. In the former case, the α exponent drops below $\frac{1}{2}$ in the so-called non-coding regions, while the exponent α jumps from much below $\frac{1}{2}$ to about $\frac{1}{2}$ and drops back to a low value when the clouds are breaking apart.

Analogies between *DNA*, clouds and FEXC rate behaviors, indicate that α is like a measure of information, an entropy variation. It indicates how information is managed by the system, how fractional Brownian motion fluctuation processes stabilize or not a system. In so doing, correlations can be sorted out and a strategy for profit making can be developed [20].

3. Multiaffine analysis

Since α has locally varying values, a multifractal process can be thought of. A multi-affine analysis of several financial series has been reported in Refs. [24,25,30,34]. The roughness (Hurst) exponent $H_1(=H)$ and the intermittency exponent C_1 are calculated for the so-called qth order height–height correlation function [35] $c_q(\tau)$ of the time-dependent signal $y(t)$:

$$c_q(\tau) = \langle |y(t) - y(t')|^q \rangle_\tau \,, \tag{5}$$

where only non-zero terms are considered in the average $\langle \ldots \rangle_\tau$ taken over all couples (t, t') such that $\tau = |t - t'|$. The roughness exponent H_1 describes the excursion of the signal [32]. The generalized Hurst exponent H_q is defined through the relation

$$c_q(\tau) \sim \tau^{qH_q} \,. \tag{6}$$

The C_1 exponent [35–37] is a measure of the intermittency in the signal $y(t)$

$$C_1 = -\left.\frac{\mathrm{d}H_q}{\mathrm{d}q}\right|_{q=1} \tag{7}$$

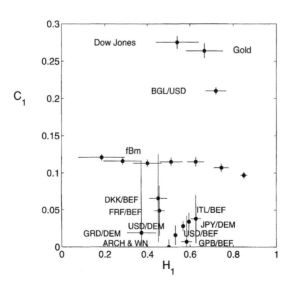

Fig. 3. Roughness (H_1), intermittency (C_1) parameter phase diagram of a few typical financial data and mathematical, i.e. fractional Brownian motion (*fBm*) and white noise (WN), signals.

which can be numerically estimated by measuring H_q around $q=1$. A (H_1, C_1) diagram can be displayed for various signals (Fig. 3), indicating the likeliness of a variety of *universality* classes.

4. Rupture phenomena and market crashes

Another investigation of the relationship between the trend and local structure of financial index signals, like DJIA, S&P500, etc. has led us into examining regions where huge variations were taking place. These are usually associated to rupture phenomena and crashes. In order to analyse crash precursors and economic behavior of stock exchanges, data from the main most complete source (DATASTREAM) was used. Moreover, due to recent past history and actual knowledge, two periods were interestingly compared, i.e. 80–87 and 90–97.

The application of statistical physics ideas to the forecasting of stock market behavior and crashes had been proposed in two independents works [9,10]. They proposed that the economic index $y(t)$ follows a complex power law-like

$$y(t) = A + B(t_c - t)^m [1 + C \cos(\omega ln(t_c - t) + \phi)] \qquad (8)$$

for $t < t_c$, where t_c is the crash-time or rupture point, A, B, m, C, ω, ϕ are parameters. This index evolution is finite at $t = t_c$ if $m > 0$, but $y(t)$ is a power law divergence if $m < 0$, laws on which log-periodic oscillations are superposed. The most singular term is similar to that of critical points at the so-called second-order phase transitions [38,39], but generalizes the scaleless situation for cases in which discrete scale invariance is presupposed [40]. This relationship[4] was already proposed in order to fit experimental measurements of sound wave rate emissions prior to the rupture of heterogeneous composite stressed up to failure [46]. The same type of complex power law behavior has been observed as a precursor of the Kobe earthquake in Japan [8] and in non-equilibrium measurements of singular quantities at critical points [47].

Fits using Eq. (1) were performed in Refs. [9,10], but it should be stressed that these fits are not robust against small perturbations. Physical arguments stipulating that m could be *universal* by analogy with phase transitions suggest to take for granted the limiting case of a power law behavior, i.e. the logarithmic behavior, corresponding to $m = 0$. Thus, the *divergence* of $y(t)$ for t below t_c could be

$$y(t) = A + B ln(t_c - t)[1 + C \cos(\omega ln(t_c - t) + \phi)] \qquad (9)$$

---

[4] This law looks exactly like the first term of a Fourier series for the density of states associated with a Hamiltonian having a singular continuous spectrum, e.g. with infinitely many holes near its edge. It is governed by the nearest complex singularities of the Mellin transform of the spectral density. The essential underlying structure is a Cantor set [41–45].

though no real divergence is expected for any stock market indices, just like there is never any infinite divergence at the critical temperature. The value of t_c has to be taken as an upper bound on t. The divergence of the correlation length, specific heat, etc. is only a virtual (mathematical) image of some physical reality. This logarithmic behavior is known in physics as e.g. characterizing the specific heat (a four point correlation function) of the 2D-Ising model, and the 2D-Kosterlitz–Thouless phase transition [48]. They are thus specific to systems with a low-order dimension of the so-called order parameter. It is nevertheless a smooth transition. Physically it represents a transformation from a disordered vortex fluid state with equal number of vortices with opposite vorticity to an ordered molecular-like state with molecules composed from a pair of vortices with diffent polarities. Such a behavior appears in systems governed by fluid-like contagions [48], in biased diffusion on random lattices [49], and in sandpile avalanche amplitudes when the underlying base is quasi-fractal [50].

The results of the fit as well as the correlation fitting factor have been previously reported [26]. In light of the theory of Ref. [44], it is appropriate to recall that these authors predicted $\omega = 2\pi/\ln(\mu) \sim 9.06$ for $\mu = 2$, for the density of states singularity, while the best fits [26] give an ω value of the order of $[6.71, 7.54]$, corresponding to $\mu \sim 2.5$, for the DJIA and S&P500. This value together with the spectral dimension $\sim 2(m+1)$ allows to determine the scaling factor of the number of degrees of freedom as well as the fractal dimension of the underlying space [44].

5. Moving averages

Consider a time series $y(t)$ given at discrete times t. At time t, the moving average $\bar{y}$ is defined as

$$\bar{y}(t) = \frac{1}{N} \sum_{i=0}^{N-1} y(t-i),$$ (10)

i.e. the average of y for the last N data points. One can easily show that if y increases (resp. decreases) with time, $\bar{y} < y$ (resp. $\bar{y} > y$). Thus, the moving average captures the trend of the signal over a time interval N.

Let two different moving averages $\bar{y}_1$ and $\bar{y}_2$ be calculated, respectively, over e.g. T_1 and T_2 intervals such that $T_2 > T_1$. If $y(t)$ increases for a long period before decreasing rapidly, $\bar{y}_1$ will cross $\bar{y}_2$ from above. This event is called a *death* cross in empirical finance [51]. On the contrary, if $\bar{y}_1$ crosses $\bar{y}_2$ from below, the crossing point coincides with an upsurge of the signal $y(t)$, i.e. a *gold* cross.

The set of crossing points between a signal $y(t)$ and the $y = 0$-level is a Cantor set with a fractal dimension $1 - H$ [52]. The related physics pertains to the so-called studies about first return time problems [49]. We have checked that the density ρ of crossing points between $\bar{y}_1$ and $\bar{y}_2$ curves is homogeneous along a signal and is thus *not* a Cantor set. In fact, the fractal dimension D of the set of crossing points is one. Thus arguments on forecasting from expected *gold* and *death* crosses, even for self-affine signals $y(t)$, and for implementing an investing strategy seem rather dubious.

However, consider the relative difference $0 < \Delta T < 1$ defined as $\Delta T = (T_2 - T_1)/T_2$. It has been found [27] that the density of crossing points $\rho(\Delta T)$ curve is fully symmetric, has a minimum and diverges for $\Delta T = 0$ and for $\Delta T = 1$, *with an exponent which is the Hurst exponent*. This remarkable and puzzling result does not seem to have been mentioned previously. It certainly raises fundamental questions on the properties of (fractional or not) Brownian motion processes. The behavior of ρ is analogous to the age distribution of domains after coarsening in spin-like models [53] and to the density of electronic states on a fractal lattice in a tight binding approximation [42,43]. Roughness or Hurst exponents are commonly measured in surface science [54] and also in time series analysis [55,56]. Notice that the moving average method can serve *to measure the Hurst exponent* in a very fast, elegant and continuous way.

6. (m, k)-Zipf

The Zipf analysis [57] consists in counting the number of words of a certain type appearing in a text, calculating the frequency of occurence f_o of each word in a given text, and sorting out the words according to their frequency, i.e. a rank R is assigned to each word, with $R = 1$ for the most frequent one, and rank R_M for the word appearing the less. A power law

$$f_o \sim R^{-\zeta} \tag{11}$$

with an exponent ζ is next searched for on a log–log plot. The appearance of this power law is due to the presence of a so-called hierarchical structure in writing a text [58]. This search for ζ value has been made on various complex signals or *texts* [57–59], economy (size of sales and firms) data [60], financial data [28,30], meteorological [61], sociological [62] or even random walk [63] and percolation [64] after translating whatever signal into a text based on an alphabet of k characters, and looking for ζ on a signal made of constant m-size words.

Since conjectures in the literature occur [28,30,65,66] on the relationship between ζ and α, some data available is shown in Fig. 4 for different cases.

7. i-Variability diagram

One disadvantage of the Zipf-method is that it is not possible to distinguish between persistent and antipersistent sequences. Only the departure from randomness is easily observed, – which e.g. leads to a description of the *domino effect* [67]. Another way to sort out short range correlations is the i-Variability Diagram technique, used for example in heart beat [68] and cloud water content [69] studies. Recall

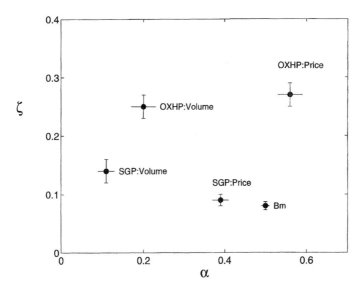

Fig. 4. Relationship between α and ζ exponents from various "samples": OXHP and SGP are a health insurance company and a pharmaceutical company quoted on the NASDAQ and NYSE, respectively; volume is the daily transaction volume; price is the daily closing price; Bm means Brownian motion.

that the first return map (r_i, r_{i-1}) or the τ-return map $(r_i, r_{i-\tau})$ of a signal are often used for revealing *strange attractors* and the embedding dimension of a signal [70–72].

The return map of the *first derivative* of the signal, i.e. the so-called first-order variability diagram (1-VD) [68] correlates every three consecutive points of the series,

$$s_{i+1} = r_{i+1} - r_i ,$$
$$s_i = r_i - r_{i-1} . \tag{12}$$

The curvature of the signal, relating every four consecutive events, i.e.

$$u_{i+1} = r_{i+1} - 2r_i + r_{i-1} ,$$
$$u_i = r_i - 2r_{i-1} + r_{i-2} \tag{13}$$

is called the second-order variability diagram (2-VD). A non-trivial shape of the point distribution on such diagrams indicates an asymmetry between the different consecutive slopes and curvatures. It has been found [29] that for free market financial series (e.g., *DJIA* and Gold price) the short-range correlations behave like $\sqrt{t}$, but for FEXC rates other power laws are observed. The differences can be conjectured to depend on economic policy. It has been also noticed that the slopes of the polygons embedding the data points have non-trivial values depending on the examined processes, e.g. financial or meteorological time series are not characterized by the same exponents.

A combination of Zipf and *i-VD* has been recently attempted for the local curvature correlations in financial signals [29]. This method suggests tests based on microscopic models.

8. Conclusion

As a conclusion, let us quote (1990) Nobel laureate Markowitz [73]: *I believe that microscopic market simulations have an important role to play in economics and finance. If it takes people from outside economics and finance – perhaps physicists – to demonstrate this role, it won't be for the first time that outsiders have made substantial contributions to these fields.*

Acknowledgements

MA and NV thank the ARC 94-99/174 for financial support. MA thanks the organizers of the NATO ARW for inviting him to present the above results. NV is financially supported by FNRS (Brussels).

References

[1] S.D. Howison, F.P. Kelly, P. Wilmott, in: Mathematical models in Finance, The Royal Society, London, 1994, pp. 449–598.
[2] N. Vandewalle, M. Ausloos, Ph. Boveroux, in: Proceedings of Econophysics Workshop, Budapest, Hungary, 1997.
[3] E.F. Fama, J. Finance 45 (1990) 1089.
[4] B.B. Mandelbrot, J. Business 36 (1963) 349.
[5] E.E. Peters, Fractal Market Analysis: Applying Chaos Theory to Investment and Economics, Wiley Finance Editions, New York, 1994.
[6] E.E. Peters, Chaos and Order in the Capital Markets: A New View of Cycles, Prices, and Market Volatility, Wiley Finance Editions, New York, 1996.
[7] R.N. Mantegna, H.E. Stanley, Nature 376 (1995) 46–49.
[8] A. Johansen, D. Sornette, H. Wakita, U. Tsunogai, W.I. Newman, H. Saleur, J. Phys. I France 6 (1996) 1391–1402.
[9] D. Sornette, A. Johansen, J.-P. Bouchaud, J. Phys. I France 6 (1996) 167–175.
[10] J.A. Feigenbaum, P.G.O. Freund, Int. J. Mod. Phys. B 10 (1996) 3737–3745.
[11] M. Levy, S. Solomon, Int. J. Mod. Phys. C 7 (1996) 65; S. Solomon, M. Levy, Int. J. Mod. Phys. C 7 (1996) 745.
[12] J.-Ph. Bouchaud, M. Potters, Théorie des risques financiers, Alea-Saclay/Eyrolles, Paris, 1997.
[13] Y.-C. Zhang, Europhys. News 29 (1998) 51–54.
[14] M. Ausloos, Europhys. News 29 (1998) 70–72.
[15] J.-Ph. Bouchaud, P. Cizeau, L. Laloux, M. Potters, Phys. World 12 (1999) 25–29.
[16] S. Moss de Oliveira, P.M.C. de Oliveira, D. Stauffer, Evolution, Money, War and Computers, Teubner, Stuttgart, 1999.
[17] R.N. Mantegna, H.E. Stanley, Econophysics: An Emerging Science, Cambridge Univ. Press, Cambridge, in press.
[18] B.M. Roehner, D. Sornette, Eur. Phys. J. B 4 (1994) 863–881.
[19] J.-Ph. Bouchaud, D. Sornette, J. Phys. I (France) 4 (1998) 387–399.

[20] N. Vandewalle, M. Ausloos, Physica A 246 (1997) 454–459.
[21] K. Ivanova, Physica A 270 (1999) 567–577.
[22] N. Vandewalle, Ph. Boveroux, A. Minguet, M. Ausloos, Physica A 255 (1998) 201–210.
[23] N. Vandewalle, M. Ausloos, Int. J. Comput. Anticipat. Syst. 1 (1998) 342–349.
[24] N. Vandewalle, M. Ausloos, Int. J. Phys. C 9 (1998) 711–720.
[25] N. Vandewalle, M. Ausloos, Eur. J. Phys. B 4 (1998) 257–261.
[26] N. Vandewalle, M. Ausloos, Eur. J. Phys. B 4 (1998) 139–141; N. Vandewalle, M. Ausloos, Ph. Boveroux, A. Minouet, Eur. J. Phys. B 9 (1999) 355–359.
[27] N. Vandewalle, M. Ausloos, Phys. Rev. E 58 (1998) 6832–6834; N. Vandewalle, M. Ausloos, Ph. Boveroux, Physica A 269 (1999) 170–176.
[28] N. Vandewalle, M. Ausloos, in: M.M. Novak (Ed.), Fractals and Beyond. Complexity in the Sciences, World Scient., Singapore, 1999, pp. 355–356.
[29] K. Ivanova, M. Ausloos, Physica A 265 (1999) 279–286.
[30] K. Ivanova, M. Ausloos, Eur. J. Phys. B 8 (1999) 665–669; M. Ausloos, K. Ivanova, Physica A 270 (1999) 526–542.
[31] H.E. Stanley, S.V. Buldyrev, A.L. Goldberger, S. Havlin, C.-K. Peng, M. Simmons, Physica A 200 (1996) 4–24.
[32] B.J. West, B. Deering, The Lure of Modern Science: Fractal Thinking, World Scient., Singapore, 1995.
[33] K. Ivanova, M. Ausloos, E.E. Clothiaux, T.P. Ackerman, unpublished.
[34] F. Schmitt, D. Schertzer, S. Lovejoy, in: Chaos, Fractals, Models, Italian U. Press, in press.
[35] A.L. Barabási, T. Vicsek, Phys. Rev. A 44 (1991) 2730–2733.
[36] A. Davis, A. Marshak, W. Wiscombe, in: E. Foufoula-Georgiou, P. Kumar (Eds.), Wavelets in Geophysics, Academic Press, New York, 1994, pp. 249–298.
[37] A. Marshak, A. Davis, R. Cahalan, W. Wiscombe, Phys. Rev. E 49 (1994) 55–69.
[38] H.E. Stanley, Phase Transitions and Critical Phenomena, Oxford Univ. Press, Oxford, 1971.
[39] M.E. Fisher, Rev. Mod. Phys. 46 (1974) 597–616.
[40] D. Sornette, Phys. Rep. 297 (1998) 239–270.
[41] M. Nauenberg, J. Phys. A: Math. Gen. 8 (1975) 925–928.
[42] R. Rammal, G. Toulouse, Phys. Rev. Lett. 49 (1982) 1194–1197.
[43] E. Domany, S. Alexander, D. Bensimon, L.P. Kadanoff, Phys. Rev. B 28 (1983) 3110–3123.
[44] D. Bessis, J.S. Geronimo, P. Moussa, J. Physique (France) Lett. 44 (1983) L977–982.
[45] D. Stauffer, D. Sornette, Physica A 252 (1998) 271–277.
[46] J.C. Anifrani, C. Le Floc'h, D. Sornette, B. Souillard, J. Phys. I (France) 5 (1995) 631–638.
[47] M. Ausloos, unpublished.
[48] J.M. Kosterlitz, D.J. Thouless, J. Phys. C 6 (1973) 1181–1203.
[49] J.-P. Bouchaud, A. Georges, Phys. Rep. 195 (1990) 127–294.
[50] N. Vandewalle, R. D'hulst, M. Ausloos, Phys. Rev. E 59 (1999) 631–635.
[51] A.G. Ellinger, The Art of Investment, Bowers & Bowers, London, 1971.
[52] J. Feder, Fractals, Plenum, New York, 1988.
[53] L. Frachebourg, P.L. Krapivsky, S. Redner, Phys. Rev. E 55 (1997) 6684–6689.
[54] A.-L. Barabási, H.E. Stanley, Fractal Concepts in Surface Growth, Cambridge Univ. Press, Cambridge, 1995.
[55] P.E. Rapp, Integrat. Physiol. Behav. Sci. 29 (1994) 311.
[56] P.G. Drazin, G.P. King (Eds.), Physica D 58 (1992) 1.
[57] G.K. Zipf, Human Behavior and the Principle of Least Effort, Addisson-Wesley, Cambridge, MA, 1949
[58] W. Ebeling, A. Neiman, Physica A 215 (1995) 233–241.
[59] B. Vilensky, Physica A 231 (1996) 705–711.
[60] M.H.R. Stanley, S.V. Buldyrev, S. Havlin, R.N. Mantegna, M.A. Salinger, H.E. Stanley, Econom. Lett. 49 (1995) 453–457.
[61] K. Ivanova, unpublished.
[62] M. Marsili, Y.-C. Zhang, Phys. Rev. Lett. 80 (1998) 2741–2744.
[63] N. Vandewalle, M. Ausloos, unpublished.
[64] M.S. Watanabe, Phys. Rev. E 53 (1996) 4187–4190.
[65] A. Czirok, R.N. Mantegna, S. Havlin, H.E. Stanley, Phys. Rev. E 52 (1995) 446–452.
[66] G. Troll, P.B. Graben, Phys. Rev. E 57 (1998) 1347–1355.
[67] N. Vandewalle, M. Ausloos, Physica A 268 (1999) 240–249.

[68] A. Babloyantz, P. Maurer, Phys. Lett. A 221 (1996) 43–55.
[69] K. Ivanova, M. Ausloos, A. Davis, T.P. Ackerman, Physica A, in press.
[70] E. Ott, Chaos in Dynamical Systems, Cambridge Univ. Press, Cambridge, 1993.
[71] K.J. Falconer, The Geometry of Fractal Sets, Cambridge Univ. Press, Cambridge, 1985.
[72] P.S. Addison, Fractals and Chaos, Inst. of Phys., Bristol, 1997.
[73] S. Moss de Oliveira, P.M.C. de Oliveira, D. Stauffer, Evolution, Money, War and Computers, Teubner, Stuttgart, 1999, p. 120

ELSEVIER

Physica A 274 (1999) 241–266

www.elsevier.com/locate/physa

Molecular motors and the forces they exert [☆]

Michael E. Fisher, Anatoly B. Kolomeisky

Institute for Physical Science and Technology, University of Maryland, College Park, MD 20742, USA

Abstract

The stochastic driving force that is exerted by a single molecular motor (e.g., a kinesin, or myosin protein molecule) moving on a periodic molecular track (such as a microtubule, actin filament, etc.) is discussed from a general theoretical viewpoint open to experimental test. An elementary but fundamental "barometric" relation for the driving force is introduced that (i) applies to a range of kinetic and stochastic models of catalytic motor proteins, (ii) is consistent with more elaborate expressions that entail further, explicit assumptions for the representation of externally applied loads and, (iii) sufficiently close to thermal equilibrium, satisfies an Einstein-type relation in terms of the observable velocity and dispersion, or diffusion coefficient, of the (load-free) motor protein on its track. Even in the simplest two-state kinetic models, the predicted velocity-vs.-load plots (that are observationally accessible) exhibit a variety of contrasting shapes that can include *nonmonotonic* behavior. Previously suggested bounds on the driving force are shown to be inapplicable in general by considering discrete jump models which feature waiting-time distributions. Some comparisons with experiment are sketched. ©1999 Elsevier Science B.V. All rights reserved.

1. Introduction

Molecular motors are individual protein molecules that are ultimately responsible for essentially all "active" biological motion including internal material transport. Important examples are myosin, kinesin, dynein, and RNA polymerase [2–6]. These molecules will move along appropriate, periodically structured, linearly polarized molecular tracks, such as actin filaments, microtubules, and DNA strands. They perform tasks vital to the life of the organism — muscle contraction, bacterial motion, cell division, intracellular transport, and genomic transcription [2–6]. Understanding how the various molecular motors operate represents a significant scientific challenge.

[☆] This article represents an expanded and significantly extended version of a more concise paper by the same authors — Ref. [1: PNAS 96 (1999) 6597–6602] — that, however, is closely followed in places. Some notational simplifications are also introduced here.

0378-4371/99/$ - see front matter ©1999 Elsevier Science B.V. All rights reserved.
PII: S 0 3 7 8 - 4 3 7 1 (9 9) 0 0 3 8 9 - 1

The hydrolysis of adenosine triphosphate (ATP), with the release of adenosine diphosphate (ADP) and inorganic phosphate (P_i), is known to be the power source for many motor proteins. In the simplest picture — which, nonetheless, seems rather accurate in many cases — a single molecule of ATP diffuses in solution, encounters a motor protein attached to a molecular track, and lodges in an active site on the motor. The motor protein then catalyzes the decomposition of the bound ATP molecule into $ADP + P_i$, releasing, in the process, a significant quantity of energy: that engenders a major conformational change in the motor protein resulting, after final discharge of the reaction products, in the net movement of the motor along the molecular track in a "forward" direction by one discrete step, say of size d. Under a sufficiently high ambient concentration of ATP in the solution the catalytic process then repeats with another ATP molecule, and the motor protein takes a further step forward. Of course, thermal fluctuations must introduce statistical features. Clearly, then, an activated motor may well be in a dynamical or, better, a stochastic steady state but it *cannot* be in full thermal equilibrium.

In recent years striking in vitro experiments have actually observed individual motor protein molecules moving along fixed tracks under controlled external loads [7–13][1] in accordance with this scenario. Such motility studies, employing optical traps, force clamps, etc., have stimulated enhanced theoretical work aimed at understanding the mechanisms by which a biological motor functions.

Now, from a broad theoretical perspective one may regard a molecular motor simply as a microscopic object that moves (predominantly in one direction) along a directed or "polar" one-dimensional periodic lattice, i.e., the molecular track [1–13] in accordance with some "laws of motion" — presumably of basically stochastic character. Furthermore, the motor may be subject to an external force F, which might vary spatially (or, even with time, t). The questions of interest are, then: "What *driving force*, say f, can the motor exert? What mean velocity, V, will it display? How will the velocity achieved depend on the load, F?" And: "Since the displacement, say x, along the track is stochastic, what *dispersion*,

$$D \approx [\langle x^2(t) \rangle - \langle x(t) \rangle^2]/2t \tag{1}$$

(or effective diffusion constant), will the motor display with respect to its mean position, $\langle x(t) \rangle \approx Vt$, at time t (under steady-state conditions)?"

To provide a more concrete conception, let us mention that for the much studied motor protein kinesin, which moves on microtubules [3–11,13], the single-step size is $d \simeq 8.2$ nm, while velocities up to $V \lesssim 800$ nm/s, and forces as large as, roughly, $f \simeq 6$ piconewtons (pN) are observed. The detailed data display dispersions of magnitude $D \simeq 1500$ nm^2/s [8,13]. Concentrations of ATP in the range $1\mu M \lesssim$ [ATP] $\lesssim 10$ mM have been studied [13]. But note, by contrast, that RNA polymerase (which

---

[1] The recent article by Visscher et al. [13], appearing after [1], provides extensive data, especially regarding the dependence on ATP concentration, which, however, is not reflected in our discussion in Section 7 here. We plan to present an analysis in the future [28].

Table 1
Forces related to a molecular motor

Force	Notation and relations	Eq.
Maximum driving force	$f_{max} = \Delta G_0/d > f$	(13)
Einstein scale	$f_E = k_B TV/D < f$	(15)
Gravitational force	$f_G = mg$	(20)
Barometric force	$f_B = k_B T\kappa \cong f$	(17, 21)
Stalling force	$f_S = k_B T\varepsilon/\theta d \overset{?}{=} f$	(31)
Load and stalling load	F, F_S, $\eta = F/F_S$	(32)

is powered by nucleoside triphosphates that release pyrophosphate, PP_i) displays velocities of 30–40 nucleotides/s and can generate forces up to $f \simeq 25$ pN [12].

The aim of the work reported here [1] is *not* to treat detailed (or realistic) models of various motor proteins that embody the quantitative features just sketched. Rather, it is to construct a general theoretical framework[2] in which to address, in particular, *how* the driving force, f, should be calculated for a broad class of molecular motor models; to relate the driving force to the velocity and to the dispersion, inasfar as that is, in fact, appropriate; and to discuss the way in which the external load force, F, should be incorporated in a model. A sense of where the reader will be taken may be gained by perusing Table 1: this lists the various forces (and force scales) we will be led to consider and, for reference, includes the corresponding defining equations presented below.

To prepare the ground, we outline, in the following section, a general class of discrete-state stochastic/kinetic models which embody basic features of a multitude of more specific treatments found in the literature. Following a previous lead [15], we show how a striking analysis by Derrida [16] for arbitrary periodic one-dimensional random walks (an extension of which we plan to publish), provides an exact and *explicit* analytical tool for the task in hand. (see also the appendix here and, for a more restricted scheme, the appendix of Ref. [8].) In the subsequent section we consider the maximum driving force and an "Einstein force scale", that is related to the dispersion D (see Table 1). Our main result, presented in Section 4, is the proposal of a general "barometric" expression for the force, f, which, we believe, is the most appropriate candidate for predicting the driving force from a specific model. Then, in Section 5, we consider the introduction of external loads, demonstrating how it is essential to allow for *load distribution factors* which determine the response of the internal transitions in the motor protein to imposed stresses. With this formulation in hand, we investigate the velocity vs. load plots that may be derived: the allowed shapes, even for the simplest ($N=2$)-state models, present a surprising diversity: see Figs. 3–4 below. To check the general theory against reality, a brief discussion of experiments on kinesin is presented on the basis of the ($N=2$)-state models in Section 7: significant features of the theory are confirmed, although further detailed experimental tests are

---

[2] Aspects of our treatment appear, although in a considerably less general setting, in work by Hong Qian [14].

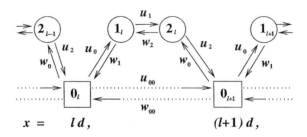

$$x = \quad ld, \qquad\qquad (l+1)d,$$

Fig. 1. Schematic representation of the sequential kinetic scheme (2) for describing a motor protein in the case $N=3$. The squares denote resting states free of any power supplying molecule, the circles correspond to "active" internal states. The initial forward rate, u_0, may be expected to be proportional to the concentration of ATP or other power source. The dotted lines represent the possibility of "spontaneous" forward and backward processes *not* dependent on an explicit molecular power source (such as ATP): see text.

most certainly desirable and simulations could be useful. Finally, Section 8 exposes an artificial limitation of the simplest kinetic descriptions by introducing discrete jump models which have waiting-time distributions [17]. (It is for such models, in particular, that we have extended Derrida's analysis [16].) Our main conclusions are summarized briefly in Section 9.

2. Molecular motor models

Traditionally, and in simplest terms one studies catalytic reactions, as exemplified by motor proteins, via kinetic chemical descriptions: see, for example [14,18] and references therein. Recently, in addition to various more detailed schemes [12,13,19–21], so-called "isothermal ratchet" models (that postulate pairs of periodic 'saw-tooth' potentials) have been proposed to account for the mechanics: see the reviews [22,23] and, e.g. [14,24,25].

Now, a common feature of most approaches is that a motor protein molecule is associated with a labeled site l ($=0, \pm 1, \pm 2, \ldots$) on the track (or linear lattice) and can be pictured as residing in one of N essentially discrete states j, which may be *free of* or *bound to* a power-source molecule, say, ATP and its various hydrolysis products. We will take $j = 0$ to label the free state and $j = 1, 2, \ldots, N-1$ to label the various bound states. Consider, as an illustration, a kinesin molecule, K, on a microtubule, M: the ($N = 4$) states identified might be: $M \cdot K$, $M \cdot K \cdot ATP$, $M \cdot K \cdot ADP \cdot P_i$, and $M \cdot K \cdot ADP$ [9,18]. Transition rates between these states can be introduced via the sequential kinetic scheme (see Fig. 1).

$$
\begin{array}{ccccc}
u_0 & u_1 & u_{N-2} & & u_{N-1} \\
0_l \rightleftharpoons 1_l & \rightleftharpoons \cdots & \rightleftharpoons (N-1)_l & \rightleftharpoons 0_{l+1}, \\
w_1 & w_2 & w_{N-1} & & w_0
\end{array}
\qquad (2)
$$

where the subscripts indicate that the states j are associated with successive sites, l and $l+1$, on the track spaced at distances $\Delta x = x_{l+1} - x_l = d$: this defines the *step size d*, as introduced above. Of course, states j_l, $j_{l+1}, \ldots, j_{l+n}$ differ physically only in their spatial displacements d, $2d$, $\ldots$, nd, along the track. By the same token, the rates u_j and w_j are independent of l (or $x = ld$);[3] however, in the subsequent developments it will prove useful to allow for spatially varying rates $u_j(l)$ and $w_j(l)$. The "laws of motion" are now given by the standard rate equations

$$\frac{\partial}{\partial t} P_j(l;t) = u_{j-1}(l) P_{j-1}(l;t) + w_{j+1}(l) P_{j+1}(l;t) - [u_j(l) + w_j(l)] P_j(l;t), \quad (3)$$

for $j = 0, 1, \ldots, N-1$, where $P_j(l;t)$ is the probability of finding the motor in state j at site l at time t, and, in order to maintain the underlying periodicity we make the identifications

$$P_{-1}(l;t) \equiv P_{N-1}(l-1;t), \quad P_N(l;t) \equiv P_0(l+1;t),$$

$$u_{-1}(l) = u_{N-1}(l-1) \quad \text{and} \quad w_N(l) = w_0(l). \quad (4)$$

One may, conveniently for many purposes, suppose that the motor starts from the origin $l = x = 0$ in a free state so that $P_j(l;0) = \delta_{j0}\delta_{l0}$.

To properly represent physicochemical reality (that is, microscopic reversibility) *none* of the forward rates, u_j, or backward rates w_j may strictly vanish even though in reality some, such as the last reverse rate, $w_N \equiv w_0$ might be extremely small [12,15,18]. On the other hand, if, as indeed observed in the presence of free ATP (or other power source), the motor moves under no external load to the right (increasing x), the transition rates *cannot* (all) satisfy the usual conditions of detailed balance that would characterize thermal equilibrium if the scheme (2) were regarded as a set of chemical reactions (near equilibrium) between effective species j_l [22].

It is of practical significance to notice that the first forward transition in (2), in which, say, a free ATP molecule initially binds to the motor protein, may be envisaged chemically as a second-order rate process, e.g., $M \cdot K + ATP \rightleftharpoons M \cdot K \cdot ATP$. Then, for sufficiently low concentrations of ATP one can conclude that $u_0 = k_0[ATP]$, where $k_0(T)$ is a concentration-independent rate constant. This, in turn, can lead to Michaelis–Menten-type rate-vs.-concentration relations [7,13] of the overall form

$$\mathcal{R} \simeq \mathcal{R}_{max}[ATP]/([ATP] + K_M), \quad (5)$$

where $\mathcal{R}$ is some rate of interest (that might be the motor velocity, V). The Michaelis–Menten constant, K_M, in fact sets the concentration at which $\mathcal{R} = \frac{1}{2}\mathcal{R}_{max}$.

However, as illustrated in Fig. 1 for an ($N=3$)-state system, one might also contemplate a small "spontaneous" or first-order background rate, $u_{00} > 0$, that exists even in the absence of ATP. A corresponding backward rate w_{00}, should then also be included. Except in the trivial $N=1$ case (where u_{00} can simply be included in u_0 and w_{00} in w_0) this changes the linear nature of the kinetic scheme (2): however, while many of our explicit algebraic expressions (for $V(F)$, D, etc.) will then hold only if $u_{00} = w_{00} = 0$

---

[3] Note that in [1] the forward rates denoted here by $u_0, \ldots, u_{N-1}$, were called $u_1, \ldots, u_N$, respectively, while the backward rate w_0 here ("*out* of the state $j = 0$") was called w_N. Beyond these changes which, in particular, simplify Eq. (3) and other expressions, we follow the notations of [1].

(as we shall assume henceforth), all the general conceptual principles of our discussion will still apply.

Now, within statistical physics, the kinetic scheme in Eqs. (2)–(4) represents a one-dimensional hopping process of a particle on a periodic but, in general, *asymmetric* lattice. After initial transients, the particle will move with steady (mean) velocity V, and diffuse [with respect to the mean position, $\langle x \rangle = Vt$, at time t: see (1)] with a diffusion constant D [16,17,26,27]. Complicated, but exact and explicit equations for V and D in terms of the rate constants u_j and w_j have been obtained for all N by Derrida [16]: for reference, these are exhibited in our notation in the appendix. To describe the transient behavior more labor is required: however, for the case $N = 2$ explicit integral expressions can be found for the probabilities $P_j(l; t)$ introduced in (3): see [15] and, more generally, the appendix of [1].

One immediately observes from Eq. (A.1) that a dimensionless, overall rate factor, that appears rather naturally, is given by the product

$$\Gamma = \prod_{j=0}^{N-1} \left(\frac{u_j}{w_j} \right) \equiv e^\varepsilon . \tag{6}$$

This will play a central role in our discussion. Note, indeed, that viewing (2) as a standard set of chemical reactions and requiring detailed balance would impose $\Gamma \equiv 1$ (or $\varepsilon = 0$) whereas $\Gamma > 1$ (or $\varepsilon > 0$) is evidently needed for a positive velocity V. (One might recall, however, [15] that as regards the *full* chemistry, the complex of motor protein plus track may be regarded simply as catalyzing the hydrolysis of ATP (or other power source): the reaction rates for the corresponding *overall* isothermal process may then be expected to satisfy detailed balance.)

The simplest or "minimal" physical models for a motor protein must clearly have $N = 2$. As mentioned, one can then calculate analytically not only the steady-state behavior but also the full transient responses. In Ref. [15] only the special (limiting) case with $w_0 = 0$ was treated; but as will be seen below, this limit can be misleading and so the general $N = 2$ results were presented in [1]. Here we will use only the velocity and diffusion constant for $N = 2$: these can conveniently be written in the forms

$$V = \frac{(u_0 u_1 - w_0 w_1)d}{u_0 + u_1 + w_0 + w_1} \equiv (\Gamma - 1)\omega d , \tag{7}$$

$$D = \tfrac{1}{2}[\Gamma + 1 - 2(\Gamma - 1)^2 \omega/\sigma]\omega d^2 , \tag{8}$$

where $\Gamma = u_0 u_1 / w_0 w_1$, as in (6), and we have introduced the associated overall rates

$$\sigma = u_0 + u_1 + w_0 + w_1, \qquad \omega = w_0 w_1/\sigma. \tag{9}$$

As already illustrated in Fig. 1, one can, of course, envisage more complicated schemes than (2) that include various branches, internal loops, parallel pathways, detachment, etc. Thus, for example, a backwards reaction directly from, say, state $j_l^\dagger$ to the original unbound or free state 0_l could account for "futile" ATP hydrolysis, i.e., "consumption" of an ATP molecule but without any forward motion of the motor [19]. Note, however, that within the $N = 2$ models (which enforce $j^\dagger = 1$) this phenomenon may be described simply by including the futile-hydrolysis parallel reaction rate in the backward rate w_1. In all cases, however, there will be a well-defined (zero-load) steady-state velocity V and a diffusion constant D (both independent of the particular states, j); and these are susceptible to estimation by simulation even should their explicit mathematical expressions be intractable (although for general u_{00}, w_{00} (see Fig. 1) and for branches and detachments, etc., closed-form results can also be obtained [28]). Furthermore, in real systems both V, as often demonstrated [7,9–13], and D [8,13] are susceptible to experimental measurement as explained.

Now there arises an obvious but crucial question, namely: "What (mean) driving force, f, will such a general motor protein model exert as it moves along its track?" That is the principal issue we have to address.

3. Maximum and Einstein force scales

For concreteness, let us suppose the molecular motor is kinesin and, thus, is powered by ATP. If the hydrolysis

$$\text{ATP} + (\text{M} \cdot \text{K})_l \rightarrow \text{ADP} + \text{P}_i + (\text{M} \cdot \text{K})_{l+1} \tag{10}$$

was, hypothetically, performed in vacuo at low temperatures, so that all reactants and products were in their (quantum mechanical) ground states, a definite energy ΔE_0 would be released by the reaction. This would then be available as mechanical energy and, so, could move the motor through the step size $\Delta x = d$ from state 0_l to 0_{l+1} exerting an effective force $f_0 = \Delta E_0 / d$. In practise, however, the catalytic hydrolysis process, *in vitro* or *in vivo*, takes place in solution and essentially isobarically and isothermally, at ambient pressures and temperatures, say (p_0, T_0). It is then more reasonable to regard the reaction as releasing a free energy, say ΔG_f, which, physically, accounts for the presence of solvent, of thermal fluctuations, etc.

If one were considering an overall bulk chemical reaction

$$A + B \rightleftharpoons C + D + E \tag{11}$$

with (relatively low) concentrations $[A], [B], \ldots$, one would compute the free energy released via

$$\Delta G = \Delta G_0 + RT \ln([C][D][E]/[A][B]) , \tag{12}$$

where ΔG_0 is referred to standard conditions, which, to sufficient accuracy, we may take as (p_0, T_0): see, e.g. [29,30]. Then, comparing (10) and (11), it is tempting to make identifications $A = \text{ATP}$, $B = (\text{M} \cdot \text{K})_l$, etc.: accepting those and the value ΔG_0 of about 0.50×10^{-19} J, corresponding to 7.3 kcal/M or 12 $k_B T$ at typical *in vitro* temperatures, T [4], one obtains $\Delta G \simeq 20 k_B T$ by using reasonable estimates for typical cellular concentrations [ATP],..., etc.[4] However, that is not, we believe, the appropriate way to identify ΔG_f. Rather, in understanding the operation of a molecular motor, one should be concerned with the microscopically local release of free energy by ATP adsorbed on the motor-protein–track complex. The appropriate concentrations to use in (12) are then to be determined essentially only by the stoichiometry of the reaction. Thus, in addition to the obvious concentration ratio $[E]/[B] = [(M \cdot K)_{l+1}]/[(M \cdot K)_l] = 1$, we should, in (12), also take $[C][D]/[A] = [ADP][P_i]/[ATP] = 1$. This leads directly to $\Delta G = \Delta G_0$; consequently, we accept ΔG_0 as a most reasonable estimate of ΔG_f, the locally released free energy.

If all the free energy $\Delta G_f = \Delta G_0$ could be converted into mechanical energy and move the motor protein through the step size, d, the force exerted would be

$$f_{max} = \Delta G_0/d .$$ (13)

Accepting that one molecule of ATP (or other power source) is sufficient to translocate the motor protein by one step [8], this expression clearly represents the maximal driving force that can be exerted. For a kinesin moving on a microtubule [7–13] with $d \simeq 8.2$ nm [10,13] it yields $f_{max} \simeq 6.2$ pN. Then, if f is the driving force actually realized, the *efficiency* of a molecular motor protein may sensibly be defined by $\mathscr{E} = f/f_{max}$.

To gain further insight, consider a small ("mesoscopic") particle with "instantaneous" position $x(t)$ and velocity $v(t)$ that undergoes one-dimensional Brownian motion in a fixed, slowly varying external potential, $\Phi(x)$. Under a constant external force, $F = -(d\Phi/dx)$, the particle will diffuse with a diffusion constant which, for long times, t, satisfies the relation (1) in which, now $\langle\cdot\rangle$ denotes an equilibrium, statistical mechanical average [26,27,31,32]. In addition, the particle experiences an (effective) *frictional force*, $f_E = \zeta v(t)$, where ζ is a friction coefficient determined by the environment [26,27,31,32]. In a steady state, the friction balances the external force, F, leading to a drift motion, $\langle x(t)\rangle \approx Vt$, with mean velocity given by $V = F/\zeta = f_E/\zeta$. Now, by definition, Brownian motion takes place within *full thermal equilibrium*: that fact dictates [26,27,31,32] the Einstein relation

$$\zeta = k_B T/D$$ (14)

which, in turn, implies the result

$$f_E = k_B TV/D$$ (15)

for what we will call the *Einstein force scale* — the second entry in Table 1.[5]

In the present context this is an appealing formula since it determines a force in terms *only* of the velocity, V, and the dispersion, D. As discussed, these are directly predictable by a motor-protein model — see, e.g. (7)–(9) above; likewise, V and

---

[4] We acknowledge stimulating remarks conveyed in correspondence with Hong Qian.
[5] Note that Svoboda et al. [8] choose to characterize fluctuations in the movement of the motor in terms of a "randomness parameter" r which, in our notation, is given simply by $r = 2D/Vd$. Thus the Einstein scale can also be expressed as $f_E = 2k_B T/dr$.

D are observable in an experiment or a simulation. However, because an activated molecular motor is *not* a Brownian particle and *cannot*, as explained, be described by thermal equilibrium alone, there are really no grounds for expecting f_E to be related to the proper driving force, f. Nevertheless, we will show that in a certain limit such a Brownian motion "mimic" of an activated motor protein does provide an adequate prediction for f. Indeed, Ref. [15] accepted the identification $f = f_E$ without discussion and used relation (15) to estimate driving forces for the restricted ($w_1 = 0$) $N = 2$ models. The values of f so obtained were not unreasonable in comparison with experimental data [15]: see further discussion below.

It is also worth pointing out that Ref. [33] (see also [19]) invokes an Einstein relation in an analysis of observations of "protein friction". However, this is a rather different context in which many "blocked" motor proteins (that cannot hydrolyze ATP) are attached to a substrate and a rigid microtubule diffuses, apparently freely, close-by in the medium above. Quantitative arguments [19,33] explain the large frictional slow-down seen — relative to an appropriate Einstein-relation estimate using the solvent viscosity — as due to weak protein binding on to and unbinding off the microtubule.

4. Barometric formulation for the driving force

Although the identification of the motor driving force f with the Einstein scale, f_E, is unjustified, it is certainly desirable to have a soundly based, general expression for f which, like f_E, does *not* entail any intrinsic modifications or extensions of the motor model or of the associated physicochemical picture beyond the specified rate constants. To that end, let us consider the placement of an "impassable block" or *barrier* on the molecular track, say, between sites L and $L+1$ ($\gg 1$) or at distance $x = X = Ld$ from the origin $x = 0$ (fixed, as we have already supposed, by where the motor starts): see Fig. 2.

Such a barrier may be realized theoretically by decreeing that all states j_l for $l \geqslant L+1$ are inaccessible. This may be achieved simply by setting one of the local forward rate constants, say, $u_J(l = L)$, equal to zero so the motor can never pass beyond the state J_l. No other rate constants need be modified: thus essentially no change of the basic molecular model is entailed. Nevertheless, if further nearby rate constants are changed, it will have no consequences for our main conclusions. One might actually want to do this to take cognizance of some aspects of a real barrier that might be attached to a molecular track in an experimental set-up.

It is intuitively clear that running a (real or model) molecular motor up to such a barrier will lead — provided it does not detach from the track or "freeze" irreversibly, as might happen in practice [7,11] — to some stationary probability distribution, as sketched in Fig. 2. It is convenient to write this distribution as

$$P_j(l; t \to \infty) = P_j^\infty(L - l) \tag{16}$$

with $z = (L - l)d = X - x$ so that z measures the distance back from the barrier: see Fig. 2. On very general theoretical grounds one should expect this distribution to

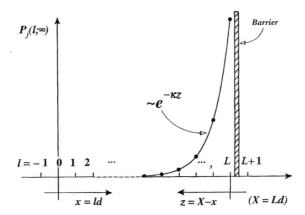

Fig. 2. Depiction of a barrier placed on a molecular motor track between sites L and $L + 1$, and of the exponentially decaying stationary probability distribution that builds up when a motor is run up against the barrier: see relation (17) and the accompanying text.

decay exponentially with increasing z except for possible deviations close to the barrier. Consequently, we can write

$$P_j^\infty(z/d) \approx A_j e^{-\kappa z}, \quad (z \gg d), \tag{17}$$

where the (positive) decay constant κ should, in principle, be experimentally measurable (although this may be difficult if κd is large). The amplitude ratios A_j/A_0 ($j = 0, 1, \ldots, N - 1$) must depend on the various rate ratios, u_i/w_i, while A_0 may be determined simply by normalization of the overall probability distribution.

To justify this surmise for the kinetic equations (3) (although it is of much more general validity), note that the mean flow between adjacent states $(N - 1)_{l-1}$ and 0_l and between $(j - 1)_l$ and j_l [for $j = 1, 2, \ldots, (N - 1)$] must vanish for a stationary distribution (with no net current flow). Balancing local forward and backward rates thus yields

$$u_{N-1}(l - 1)P_{N-1}^\infty(L - l + 1) = w_0(l)P_0^\infty(L - l),$$
$$u_{j-1}(l)P_{j-1}^\infty(L - l) = w_j(l)P_j^\infty(L - l), \tag{18}$$

for $j = 1, 2, \ldots, (N - 1)$. Starting from an initial nonzero value $P_j^\infty(0)$, one can then recursively determine $P_{J-1}^\infty(0),\ P_{J-2}^\infty(0), \ldots, P_0^\infty(0),\ P_{N-1}^\infty(1), {}_{,N-2}(1), \ldots$. By induction, this leads directly to the exponential decay (17) [since the $u_j(l)$, and $w_j(l)$ become independent of l for, say, $l < L - l_0$ where l_0 is some small fixed integer representing the extent of influence of the barrier on the transition rates]. Most crucially one finds, recalling the definition (6), that the decay constant is simply given by

$$\kappa = (\ln \Gamma)/d = \varepsilon/d . \tag{19}$$

Now, to interpret these results in terms of some effective driving force, consider a dilute gas of molecules of mass m moving in a gravitational field that acts "downwards" along the vertical or z-axis. Each molecule then has a weight $f_G = mg$; in addition, the equilibrium density distribution is given by [34]

$$\rho(z) = \rho(0)e^{-mgz/k_B T} = \rho(0)\exp[-(f_G/k_B T)z] , \tag{20}$$

where $\rho(0)$ is the density at the level $z=0$. (Any deviations arising close to the "lower" wall (at $z \simeq 0$) due to molecular size, structure, etc., have, of course, been neglected.) Comparing this well-known barometric formula with the analogous barrier distribution (17) leads us to identify the driving force f of the molecular motor with f_G and, thence, with

$$f_B = k_B T (\ln \Gamma)/d = k_B T \varepsilon/d . \tag{21}$$

This is one of our principal results: the subscript B serves merely to indicate the barometric analogy underlying our identification. It is significant to note that, by comparison with (13) for $f_{\max}$, we may expect

$$\varepsilon \lesssim \Delta G_0/k_B T \tag{22}$$

for any real molecular motor.

4.1. Barometric vs. Einstein scale

Before studying this result in relation to extensions of the simple kinetic scheme (2) that are needed to describe a motor functioning under external loads, let us compare f_B with f_E. To start, let us suppose the molecular motor operates close to equilibrium in the sense that $\varepsilon = \ln \Gamma$ is small. (Recall that detailed balance, in equilibrium, would require $\Gamma = 1$ and $\varepsilon = 0$.) Then, on expanding in $\varepsilon = 0$ at fixed ω/σ, Eqs. (6)–(9) and (21) yield

$$f_B/f_E = 1 + [\tfrac{1}{12} - (\omega/\sigma)]\varepsilon^2 - \tfrac{1}{2}(\omega/\sigma)\varepsilon^3 + \cdots \tag{23}$$

for $N = 2$. Evidently, the coefficient of ε vanishes identically! Furthermore, one finds $0 < \omega/\sigma \leqslant \tfrac{1}{16}$ so that the coefficient of ε^2 is small, lying between $\tfrac{1}{48}$ and $\tfrac{1}{12}$. Consequently, and as might well have been anticipated, the Einstein scale approximates the barometric result very well when the motor operates sufficiently close to equilibrium. Indeed, for $\Gamma < 10$, calculations show that f_B can exceed f_E by no more than 44%. Furthermore, the series in (23) truncated at $O(\varepsilon^2)$ proves reasonably accurate up to $\varepsilon \simeq 5$ ($\Gamma \simeq 150$) where one has $1.473 < f_B/f_E < 2.535$; beyond that one can establish the effective bounds,

$$\tfrac{1}{4}\varepsilon < f_B/f_E \lesssim \tfrac{1}{2}\varepsilon . \tag{24}$$

These specific results are limited to $N = 2$; however, the vanishing of the $O(\varepsilon)$ term in (23) is independent of N. Indeed, the basic symmetry of the kinetic scheme (2) under the forward–backward transition rate interchanges: $w_0 \Leftrightarrow u_{N-1}$ and $w_j \Leftrightarrow u_{j-1}$ ($j = 1, \ldots, N-1$), and $\varepsilon \Leftrightarrow -\varepsilon$, leads to $\Gamma \Leftrightarrow 1/\Gamma$, $V \Leftrightarrow -V$, while D remains invariant: consequently,[6] the ratio f_B/f_E is essentially an even function of ε. (It is

---

[6] We are indebted to B. Widom for a remark on this point.

only because we chose, in (23), to expand at fixed ω, which is not an invariant under the rate exchange, that an $O(\varepsilon^3)$ term appears).

By the same token, we expect f_B always to rise steadily above f_E when ε increases. Indeed, on recalling (16) for Γ, one observes from (21) that f_B is unbounded above and so, with an injudicious assignment of rate constants, it may even exceed $f_{\max}$, as given in (13)! Conversely, one may show from (7), (8) and (15), that f_E, the Einstein force scale, is bounded above by $4k_BT/d$ for $N=2$ [15]. However, we will demonstrate in Section 8 that this bound on f_E is rather artificial and does *not* apply for models that account in a more direct fashion for the discreteness of the hydrolysis of ATP (or other power source molecules).

5. The effects of a load

In a typical experiment on motor proteins [7–11], optical tweezers are used to carry a silica bead coated with a few molecules of the motor protein up to a molecular-track filament secured on a glass surface. Then a single motor binds to the track and, in the presence of a power source, spontaneously starts to move, exerting a force against the opposing load, F, as it pulls the bead away from the center of the optical trap. In leading approximation, the external force F is a linear function of the displacement of the motor from the trap center, and the constant of proportionality can be measured. Thus the trap and bead work like a calibrated spring acting on the molecular motor. (Alternatively [13], with the aid of appropriate feedback controls, a constant force can be applied.) To represent such experiments, the load-free scheme embodied in (2) must, clearly, be extended.

To this end, suppose the motor moves on the track in a slowly varying external potential, $\Phi(x)$, so that in translocating from site l to $l+1$ (say, in the free state $j=0$), additional mechanical work

$$W_l \equiv \Delta\Phi(x=ld) = \Phi(x+d) - \Phi(x), \tag{25}$$

must be done (relative to the load-free situation). Of course, this corresponds to imposition of a local external force, $F(x)=\Delta\Phi(x)/d$ (or $F_l=W_l/d$), directed negatively. For an (ideal) optical trap of spring constant K we may take

$$\Phi(x)=\tfrac{1}{2}Kx^2, \qquad F(x)=K(x+\tfrac{1}{2}d). \tag{26}$$

Our analysis will not, however, depend on any specific form for $\Phi(x)$ although, for conceptual simplicity, we will suppose $F(x)$ increases with x.

In such a situation the motor should, in effect, compress the spring and, as t increases, attain a stationary distribution, say $P_0^S(l)$, where, for simplicity, we focus only on the (free) states 0_l. This distribution should peak at some l_S, corresponding to a mean (or most probable) compression of the spring by a displacement $x_S = l_S d$. Then the measured "stalling force" [in the harmonic situation (26)] would be $f_S = Kx_S$.

Now it is evident physically that under any local load, $F(x)$, the transition rates, $u_j(l)$ and $w_j(l)$, must change. If, as traditional, one views the chemical transitions between successive states, j and $j + 1$, as proceeding in quasiequilibrium over various free energy barriers [19], one expects (in leading approximation) the rates to change exponentially with $F(x)d/k_BT$. But a crucial question now arises, namely, "How should the exponential loading factors be distributed among the various reaction processes, $j \rightleftharpoons (j + 1)$, occurring "inside" the motor protein ?" This is far from obvious: indeed, the way in which the load is shared should, clearly, be of considerable interest in understanding the motor mechanism at a more detailed microscopic level.

To avoid prejudice, therefore, we advance the *quasiequilibrium hypothesis* that under a local load, F, acting "at" site l, the individual transition rates change in accord with

$$u_j \Rightarrow u_j^{(F)} = u_j^{(0)} e^{-\theta_j^+ Fd/k_BT}, \qquad w_j \Rightarrow w_j^{(F)} = w_j^{(0)} e^{+\theta_j^- Fd/k_BT} . \qquad (27)$$

The *load distribution factors*, θ_j^+ and θ_j^-, introduced here need not be of uniform sign: but we certainly expect the overall factor

$$\theta = \sum_{j=0}^{N-1} (\theta_j^+ + \theta_j^-) , \qquad (28)$$

to be positive, since that simply implies that the load force acts to oppose motion. Indeed, should the motor undergo diffusion in thermal equilibrium when *not* activated by ATP [as suggested in Fig. 1 and in the discussion following (5)], detailed-balance considerations would dictate $\theta = 1$. As a *supplement* to our quasiequilibrium hypothesis this value of θ is also plausible for an activated motor that operates not too far from equilibrium.

Notice, however, that an individual *negative* θ_j^+ or θ_j^- simply means that the corresponding forward rate, u_J, is *enhanced*, or the reverse rate, w_J, is *diminished* by the internal molecular strain induced in the motor by the load. There are no good reasons for excluding such possibilities. Indeed, it is not difficult to imagine concrete mechanisms that would lead to such effects: for example, suppose an adsorption site on the protein were covered by a "lid" that was pulled open by imposing a load against a spring that otherwise held it closed.

If we accept the hypothesis (27), we can find the stationary "spring-compression" distribution $P_0^S(l)$ with the aid of the rate-balance the relations (18), simply by replacing $P_j^\infty(L - l)$ by $P_j^S(l)$, and the rates u_j and w_j in accord with (27). By iterating on j the relations (18) lead to

$$P_0^S(l+1) = \frac{u_{N-1}(l)}{w_0(l)} P_{N-1}^S(l) = \frac{u_{N-1}(l)}{w_0(l)} \frac{u_{N-2}(l)}{w_{N-1}(l)} P_{N-2}^S(l) = \cdots . \qquad (29)$$

The most probable motor location, l_S, then follows by equating $P_0^S(l)$ and $P_0^S(l + 1)$, which yields the condition

$$\Gamma^{(F)}(l) \equiv \prod_{j=0}^{N-1} [u_j^{(F)}(l)/w_j^{(F)}(l)] = \Gamma^{(0)} e^{-\theta F(x)d/k_BT} = 1 . \qquad (30)$$

Solving this determines $x_S = l_S d$ and hence, by identifying $F(x_S)$ with f_S, the measured spring or *stalling force*, yields our second principal result, namely,

$$f_S = k_B T (\ln \Gamma)/\theta d = k_B T \varepsilon / \theta d , \qquad (31)$$

see Table 1. Here, of course, we have identified the zero-load rate factor, $\Gamma^{(0)}$, with the original rate factor Γ defined in (6) in terms of the *unmodified* transition rates u_j and w_j $(j = 0, 1, \ldots, N - 1)$.

It is striking that this expression for the stalling force — which rests on the quasiequilibrium hypothesis (27) that is needed to extend the original kinetic model — agrees *precisely* with the barometric expression (21) for f_B, provided one accepts the natural, near-equilibrium evaluation $\theta = 1$. We regard this overall consistency as strengthening both approaches.

6. Velocity versus load

The extended rate constants $u_j^{(F)}$ and $w_j^{(F)}$ introduced in hypothesis (27) also serve to provide a relation for $V(F)$, the motor velocity, as a function of a steady load force, F, and, equally, for the load-dependent diffusion constant $D(F)$: see [13] for recent experimental results. For arbitrary N one may appeal to (A1) which shows, as expected, that the stalling load, F_S, which brings $V(F)$ to zero, agrees with (31), i.e., $F_S = f_S$. To write an explicit result for $N = 2$ in an illuminating form, we introduce the reduced force and modified load distribution factors

$$\eta = F/F_S \quad \text{and} \quad \Delta_j^\pm = \tfrac{1}{2} - (\theta_j^\pm/\theta) . \qquad (32)$$

Then by combining (7), (27) and (31) we can construct the expression

$$\frac{V(F)}{V(0)} = \frac{\sigma \sinh[\tfrac{1}{2}\varepsilon(1 - \eta)]/\sinh(\tfrac{1}{2}\varepsilon)}{u_0 e^{-\Delta_1^+ \varepsilon \eta} + u_1 e^{-\Delta_0^+ \varepsilon \eta} + w_0 e^{\Delta_1^- \varepsilon \eta} + w_1 e^{\Delta_0^- \varepsilon \eta}} , \qquad (33)$$

where, naturally, $V(0)$ is simply the no-load result stated in (7), so that the right-hand side must reduce to unity when $\eta = 0$ (while it vanishes when $\eta \to 1$). For convenience, we recall that $\varepsilon = \ln(u_0 u_1 / w_0 w_1)$.

Now for ε small (say, $\lesssim 2$), so that the motor is operating not too far from equilibrium, one has

$$V(F) \approx V(0)(1 - \eta)/(1 + c\varepsilon\eta) . \qquad (34)$$

This represents a *hyperbolic* force law which will be *concave* or *convex* depending on the sign, $+$ or $-$, of c: see the illustrative examples in Fig. 3.[7] Concave plots, like (b) in Fig. 3, are characteristic of experiments on animal muscles: see, e.g. [2, Fig. 2.19].

---

[7] Note, that in the caption for Fig. 1 of [1] the data for the plot (c) — the same plot as reproduced here — contains a misprint: the value of ε should read 9.2 (as specified in the caption here).

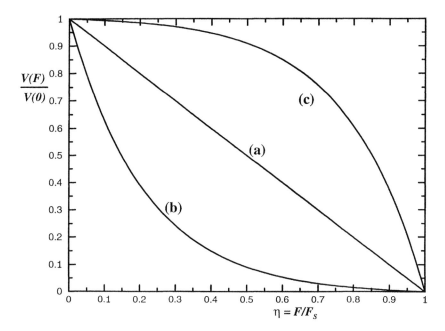

Fig. 3. Examples of nearly linear and of hyperbolic velocity–load plots for $N = 2$ models with rate exponent ε, and reduced transition rate ratios $\bar{w}_j \equiv w_j/u_0$, and load distribution factors $\bar{\theta}_j^{\pm} \equiv \theta_j^{\pm}/\theta$, given by the parameter sets $\{\varepsilon;\ \bar{w}_0 = \bar{w}_1;\ \bar{\theta}_1^+, \bar{\theta}_0^- = \bar{\theta}_1^-\}$: (a) $\{0.01;\ 0.99;\ \frac{1}{2}, 0\}$, (b) $\{9.2;\ 0.01;\ \frac{1}{2}, 0\}$, and (c) $\{9.2;\ 0.01;\ 0,\ \frac{1}{2}\}$. Note that $V(0)$ is the velocity at zero load [see (7)] while F_S denotes the stalling load.

For small c the law is close to *linear* [see (a) in Fig. 3] and, in fact, c vanishes whenever

$$u_0 \Delta_1^+ + u_1 \Delta_0^+ = w_0 \Delta_1^- + w_1 \Delta_0^- .\tag{35}$$

This condition has many solutions; for example, if the backward rates are small, so that $\delta \equiv (w_0 + w_1)/(u_0 + u_1) \lesssim 0.1$, say, the load-distribution scheme $\theta_0^+ \simeq \theta_1^+ \approx \frac{1}{2}\theta/(1 + \delta)$ yields a near-vanishing c. Velocity–load plots that are fairly linear have frequently been observed in experiments, particularly on kinesin over quite wide ranges of ATP concentration: see, e.g. [7]. Indeed, if u_0 greatly exceeds u_1, w_0, and w_1, the reduced (V, F) plots become insensitive to u_0. Then if, as discussed above in Section 2, one has $u_0 \simeq k_0[\text{ATP}]$, the plots will become independent of the ATP concentration [7]. Furthermore, if ε is large but $(\theta_1^+/\theta)\varepsilon \simeq 1$, the (V, F) plots again become close to linear. On the other hand, recent experiments on kinesin [13] have found convex velocity–load plots — resembling (c) in Fig. 3 — at high concentrations: $[\text{ATP}] = 2$ mM.

Although straight, convex, and concave velocity–load plots are readily generated within the $N = 2$ models, other reasonable values of the six parameters:

$$\varepsilon, \quad \bar{w}_0 \equiv w_0/u_0, \quad \bar{w}_1 \equiv w_1/u_0,$$

$$\bar{\theta}_1^+ \equiv \theta_1^+/\theta, \quad \bar{\theta}_0^- \equiv \theta_0^-/\theta, \quad \bar{\theta}_1^- \equiv \theta_1^-/\theta ,\tag{36}$$

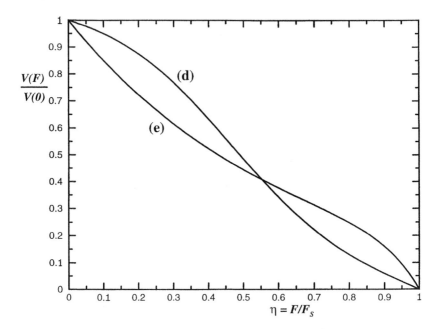

Fig. 4. Velocity–load plots for $N = 2$ models displaying points of inflection of opposite sense: the parameter sets are $\{\varepsilon;\ \bar{w}_0, \bar{w}_1;\ \bar{\theta}_1^+, \bar{\theta}_0^- = \bar{\theta}_1^-\}$: (d) $\{11.1;\ 0.15, 10^{-4};\ 0, \frac{1}{2}\}$, (e) $\{23.0;\ 10^{-5}, 10^{-5};\ 0.07, 0.43\}$.

yield plots exhibiting points of inflection of either sense, as illustrated in Fig. 4. Plots with a *positive* inflection point, such as (d), have been observed in experiments on RNA-polymerase [12]. However, plots with *negative* inflection points, such as (e), appear to be realized in relatively small regions of the $(N = 2)$ parameter space. If negative θ_1^+ or θ_0^+ are admitted — as discussed after Eq. (28) — plots with two inflection points are also allowed as example (f) in Fig. 5 demonstrates. Furthermore, in such cases the velocity may even *rise* when a load is initially imposed! See plot (g) in Fig. 5. Thus if one could determine plausible values for the no-load transition-rate ratios, experimental (V,F) plots might, at least within the scope of $N = 2$ models, throw some light on the load distribution parameters, $\theta_j^\pm$; these, we repeat, must be of significance in understanding a motor protein's operation at a molecular level.

7. Relation to Kinesin data

Let us recapitulate briefly: in order to understand the driving force, f, exerted by a molecular motor that takes steps of size d on a molecular track, we have analyzed a broad class of stochastic models: in particular, expressions (2) and (6), embody a general, "linear" motor reaction sequence. In the presence of a constant free-energy source, the motor will achieve a steady velocity V (> 0) but with fluctuations about the mean position, Vt, described by a dispersion or diffusion constant, D, as introduced

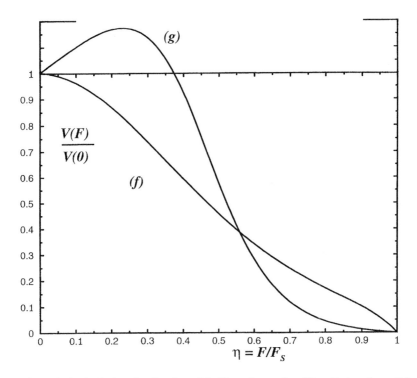

Fig. 5. Further velocity–load plots for $N = 2$ models illustrating a plot, (f), with two (opposite) points of inflection, and one, (g), which is nonmonotonic as well as having a point of inflection: both examples entail one negative load distribution factor, namely, $\theta_1^+ < 0$. The specific parameter values are: $\{\varepsilon;\ \bar{w}_0, \bar{w}_1;\ \bar{\theta}_1^+,\ \bar{\theta}_0^-,\ \bar{\theta}_1^-\} = \{23.0;\ 10^{-5}, 10^{-5};\ -0.07, 0.48, 0.48\}, \{10;\ 3.4\times10^{-4}, 2.5\times10^{-3};\ -0.1, 0.1, 0.2\}$, respectively.

in (1). Table 1 lists various force scales that have arisen in our analysis and summarizes their relation to the driving force, f. It is useful at this point to attempt to check, at least semiquantitatively, the degree to which our theory for the driving force, the velocity, the dispersion, and their interrelations, satisfactorily corresponds with available experimental data. Because kinesin moving on a microtubule is well studied [1–5, 7–11,13,18], it can provide some concrete numerical evidence. [8]

Recall, first, that the microtubule-kinesin step-size, d, is close to 8.2 (or 8.3) nm [10,13]. In 1994 Svoboda et al. [8] observed a zero-load velocity $V \simeq 670$ nm/s for a concentration [ATP] = 2 mM; in addition, they measured the variance of the motor's position, $x(t)$, from which we have derived the estimated dispersion $D \simeq 1400$ nm²/s (see also [13]). At $T = 300$ K these results yield the Einstein scale $f_E \simeq 2.0$ pN. On the other hand, the observed stalling force was $f_S \simeq 5$–6 pN [7,8]. This is significantly larger than f_E, as we have argued it should be: see Section 4. Note also, by comparing with the maximal force estimate, $f_{max} \simeq 6.2$ pN [see (13)], that

---

[8] Recall footnote 1 regarding Visscher et al. [13].

the observed efficiency $\mathscr{E}$, is in the range 80–95%. (It may be remarked, however, that this observational estimate of the efficiency does *not* allow for the possible "wastage" of ATP by futile hydrolysis [19] that occurs without translocation of the motor: recall the discussion following (9) in Section 2.)

The "barometric" force scale, f_B, was derived by considering an obstacle that blocks the motor's motion on the track as illustrated in Fig. 2. The resulting statistically stationary distribution should, quite generally, decay with the distance z from the obstacle like $e^{-\kappa z}$, as asserted in (17). It would, indeed, be interesting and valuable to measure κ and to compare f_B, so derived, with the observed stalling force f_S. It seems likely, however, that such a measurement for kinesin is not currently practicable: indeed, our analysis would suggest a decay length, $\xi \equiv \kappa^{-1}$, of only 0.7 or 0.8 nm.

For the $(N = 2)$-state models on which we have focused (see Section 2), with transition rates u_0, u_1, w_0 and w_1, one has $f_B = (k_B T/d)\ln(u_0 u_1/w_0 w_1)$. For kinesin (from *Drosophila*) Gilbert and Johnson [18] have studied the kinetics using chemical quench-flow methods. Assuming [ATP] $= 2$ mM their data show that the values $u_0 = 3800$ s^{-1}, $u_1 = 15$ s^{-1}, and $w_1 = 200$ s^{-1} represent a sensible map on to an $N = 2$ model; however, the backwards rate from the free state, w_0, proved unobservably small. Merely for illustration, therefore, let us consider the guess $w_0 = u_1/100 = 0.15$ s^{-1}. Via (7) and (8), these rates lead to $V \simeq 116$ nm/s and $D \simeq 474$ nm^2/s, which values yield the Einstein scale $f_E \simeq 1.0$ pN (at $T = 300$ K), while the barometric approach gives $f_B \simeq 3.8$ pN. The agreement with the estimates based on the results of Svoboda et al. is not so impressive. Nevertheless, the orders of magnitude, the inequality $f_B > f_E$, and the rough equality $f_B \simeq f_S$, are in full accord with our theoretical predictions.

More recently, Higuchi et al. [10] obtained data (for bovine brain kinesin) from which we estimate $u_0 \simeq 1400$ s^{-1} and $u_1 \simeq 45$ s^{-1}, in only rough agreement with the values derived from Gilbert and Johnson [18]. The further ad hoc assumption $w_1/u_0 \simeq w_0/u_1 \simeq 1/100$ then yields the values $V \simeq 355$ nm/s and $D \simeq 1370$ nm^2/s which are rather closer to the observations of Svoboda et al. [7,8]. Likewise, the corresponding values $f_E \simeq 1.1$ pN and $f_B \simeq 4.7$ pN, now accord better with the direct experiments (although, of course, depending logarithmically, that is, weakly, on our guesses for w_0 and w_1). We may conclude that, while our general theoretical picture is supported, further experiments (such as [13]) and the use of standardized kinesin samples will certainly be valuable and could provide more stringent tests.

As explained in Section 5, to discuss the velocity $V(F)$ of a motor under a load F, the transition rates in any model must be modified: thus, in our quasiequilibrium hypothesis, (27), the load-distribution factors, $\theta_j^{\pm}$, recognize that the various transitions in a real motor protein molecule almost certainly accept quite different fractions of the total stress: see also [13]. Indeed, the *acceleration* of some forward rates (corresponding to negative θ_j^{+}) could provide a mechanism to conserve, e.g., ATP, under "no-load" conditions: recall Fig. 5(g) where $V(F)$, and hence the rate of free-energy consumption, reaches a maximum only under imposition of a load. [9]

---

[9] A question conveyed to us from Jonathan Widom stimulated these remarks.

It is natural, as explained, to take the overall load-distribution factor θ in (28) as unity since this leads to the equality of f_S and f_B: compare (21) and (31). However, except for operation close to equilibrium, the hypothesis $\theta = 1$ can be doubted for real motors or more realistic models. It might well be tested by experiment or, probably more feasibly, by simulation.

As seen in (33), the dependence of $V(F)$ on the transition rates and load factors is quite complex even for the simplest two-state models. Indeed, Figs. 3–5 demonstrate that the six independent parameters (36) permit velocity–load plots of extremely varied shapes (including further forms not shown). Although certain types, such as (e), seem to characterize fairly small regions of the parameter space, it seems that, in general, the variation of V with F may reveal comparatively little about the motor mechanism or specific parameter values. Currently, therefore, beyond noting the variety of shapes seen experimentally (as mentioned in Section 6 and note especially [13]), we may say that the observed motility data provide mild support for the concrete aspects of the theory.

Nevertheless, it is worth noting that *negative*, i.e., assisting loads ($F < 0$) are predicted to speed up the motor and this has been observed [11]. Conversely, under super-stalling loads ($F > F_S$), *backwards velocities* are predicted: single reverse steps of kinesin have then been seen [11] which are, thus, *consistent* with our concept of blocked distribution as in (17). However, no steady reverse velocities have been reported. These facts probably reflect the very small terminal reverse rates, w_0, of kinesin [18] already commented upon. Indeed, we may note, complementing the discussion leading to the Michaelis–Menten relation (5), that these transitions presumably describe second (or higher)-order chemical reactions controlled by the low concentrations of hydrolysis products. The frequently observed process in which a kinesin molecule detaches itself from or "falls off" the track [7,11] should also be included in a fuller account. (Indeed, the exact analysis of Derrida [16] we have used in our discussions, can be generalized to allow for detachment or "death" processes [28].)

The adequacy of the stochastic models encompassed in the kinetic schemes (2)–(4), might be challenged by the existence of *lower bounds* on the dispersion, D, which yield the *upper* bounds on the Einstein scale, f_E, that were mentioned briefly at the end of Section 4. For kinesin at $T = 300$ K the upper bound on f_E is 2.03 pN for any ($N = 2$)-state model. The data of Svoboda et al. [7,8] essentially satisfy this bound; but were the bound violated, one might conclude that a kinetic model with $N = 3$ or more states was needed since the bound increases with N. However, as we demonstrate in the next section, models in which the transitions are described by *discrete jumps* occurring after certain waiting times, are *not* susceptible to these constraints. Such models might well prove more realistic — especially in more complicated molecular motors like RNA polymerase [12] — although, at present, the simpler kinetic representations may suffice. Nevertheless, in assessing our (essentially tentative) comparisons with experiment, it should be borne strongly in mind that the main principles we have enunciated are *not* restricted to the $N = 2$ sequential kinetic models specifically analyzed. Consequently, the observation of significant violations would indicate serious deficiencies in the general understanding of molecular motor mechanisms embodied in the analysis presented.

8. Jump-and-wait models

As mentioned above, the Einstein force scale, f_E, obtained from the kinetic scheme (2) is subject to a fairly stringent bound. From a more fundamental point of view this can be regarded as arising from a lower bound on the dispersion D for a given velocity V. Since, by (15), $f_E \propto V/D$ this yields an upper bound on the Einstein scale. Specifically, analysis of the explicit expressions (7)–(9) with (6), enables us to establish the bound

$$D \geqslant Vd/2N \quad \text{(all } u_j, w_j > 0) \tag{37}$$

for $N = 2$ kinetic models. The bound is achieved (when $N = 2$) by *uniform* rates, that is $u_j = u_0 \gg w_j = w_0$ (all j). This observation can be understood heuristically since in a uniform situation there are no distinguished "rate-limiting steps" in the reaction cycle. For general N, the same uniformity condition yields (37) as an *equality*. Examining numerical examples for small $N \geqslant 3$, convinces one that, in accord with the heuristic argument, any departure from uniform rates increases D. Thus we believe that inequality (37) is valid for *all* N. [10] (In passing, we may mention that the particular $N = 2$ model studied in [15] also respects an *upper* bound on D and, hence, obeys the *lower* bound $f_E/k_B T > 2/d$. However, this is directly attributable to the special limiting situation, $w_0 = 0$, studied there which, as mentioned initially, cannot be literally true in reality. [11])

Now any general lower bound on the dispersion of a molecular motor is open to suspicion since, if the motor were "purely mechanical", it would move forward, under any fixed load (including $F = 0$), at a strictly constant rate exhibiting no variance at all. For such an ideal or 'clockwork motor' one would, thus, have $D \equiv 0$. Since highly accurate clocks exist in the animal world — albeit made by humans — one should prefer models that allow the dispersionless, purely mechanical limit to be attained. How closely real molecular motors can or do approach the limit is certainly a matter of interest.

In light of these remarks, our purpose in this section is to demonstrate that the bounds on D and f_E are directly related to the *continuous-time* picture of the rate process that is embodied in the kinetic master equations (3) and (4). In essence, these enforce a minimum value of the dispersion D given a value of V. To see this most directly, consider an $(N = 1)$-state model with master equation

$$\frac{\partial P_0}{\partial t}(l,t) = u P_0(l - 1, t) + w P_0(l + 1, t) - (u + w) P_0(l, t) , \tag{38}$$

---

[10] In terms of the randomness parameter of Svoboda et al. [8] this amounts to the lower bound r ($\equiv 2D/Vd$) $\geqslant 1/N$.

[11] A similar comment applies to the scheme discussed by Svoboda et al. [8] in their appendix where randomness r is subject to the upper bound $r \leqslant 1$, which again implies $f_E/k_B T \geqslant 2/d$.

where, for brevity, we have put $u_0 = u \geqslant w_0 = w > 0$. Then one finds (say, from the expressions in the appendix) the simple results

$$V = (u - w)d, \quad D = \tfrac{1}{2}(u + w)d^2 .$$ (39)

Note, now, the lower bound $D > \tfrac{1}{2}ud^2$, which is approached when $w/u \to 0$; in this limit V approaches ud so that (37) is recaptured (for the case $N = 1$). Likewise, the upper bound $f_E/k_B T < 2/d$ (for $N = 1$) follows.

By contrast, consider instead a *discrete* event sequence in which a forward or backward jump is attempted at (mean) time intervals $\Delta t = \tau$ (triggered, one might picture for a molecular motor, by the arrival of individual ATP molecules). If $\check{P}_0(l; n)$ is the probability that the (motor) particle is at site l after n jump attempts, one now has [19,21,31]

$$\check{P}_0(l; n + 1) = p_+ \check{P}_0(l - 1; n) + p_0 \check{P}_0(l; n) + p_- \check{P}_0(l + 1; n) ,$$ (40)

where p_+ and p_- are the probabilities of completing a positive or negative step while $p_0 = 1 - p_+ - p_-$ is the probability of remaining at the same site. If one sets $p_+ = u\tau$ and $p_- = w\tau$, and identifies the time as $t \approx n\tau$, this discrete master equation, reduces to the continuous form (38) in the limit $\tau \to 0$ [17].

Now the mean displacement $\langle x \rangle_1$ after just one attempt is clearly $(p_+ - p_-)d$. Since, by the assumptions of the model, successive jumps are uncorrelated, one has $\langle x \rangle_n = n\langle x \rangle_1$ so that the mean velocity is

$$V = (p_+ - p_-)d/\tau = (u - w)d .$$ (41)

Note that the identifications appropriate for reaching the continuous-time limit yield agreement with the corresponding result (39) for V. Because successive jump attempts are uncorrelated, we can compute D using definition (1) with only a *short* time interval: specifically, we may take $t = \tau$. Thus, from $\langle x^2 \rangle_1 = (p_+ + p_-)d^2$ we obtain

$$D = \tfrac{1}{2}(d^2/\tau)[p_+ + p_- - (p_+ - p_-)^2]$$

$$= \tfrac{1}{2}[u + w - (u - w)^2\tau]d^2 .$$ (42)

To see that D now has no positive lower bound for fixed V, we may either specialize to the case $p_0 = 0$ or consider the limit p_- $(= w\tau) \ll p_+$: then one finds

$$D \propto (d^2/\tau)p_+(1 - p_+) \approx Vd(1 - p_+) ,$$ (43)

which becomes indefinitely small when p_+ approaches unity (while $V \to d/\tau$). Hence there is no lower bound on D or upper bound on f_E in such a discrete jump model. Indeed, it is intuitively clear that in the limit $p_+ = 1$ (so that $p_0 = p_- = 0$) the particle moves in clockwork manner at speed d/τ with no dispersion.

It is important notice, however, that the barometric formulation can be applied directly to the jump model by introducing a barrier, as before, such that site $L + 1$ (> 0) cannot be reached. Clearly, this can be accomplished by changing only the master equation (40) for $\check{P}_0(L; n+1)$ by setting $p_- = 0$ so that $p_0 = 1 - p_+$. With the initial condition $\check{P}_0(l; 0) = \delta_{l0}$, this leads precisely to the previous form, (21), but with $\Gamma = p_+/p_-$.

Furthermore, this jump model result agrees exactly with the continuous-time ($N = 1$) expression $\Gamma = u/w$ when, as above, one puts $p_+ = u\tau$ and $p_- = w\tau$. Beyond that, the ratio $R(\varepsilon) = f_B/f_E$ still obeys (23) but with, in leading order, (ω/σ) replaced by $\frac{1}{2}p_- = (1 - p_0)/2(\Gamma + 1)$. For $\varepsilon > 2$ one must have $\frac{1}{2}p_- < 0.06$ and one finds that when ε increases the ratio $R(\varepsilon)$ varies much as above for the continuous case.

The master equation (40) can be readily extended to periodic N-state jump models, in analogy to the kinetic equations (3) and (4), by introducing forward and backward jump probabilities, p_j^+ and p_j^- with $p_j^0 = 1 - p_j^+ - p_j^-$ (for $j = 0, 1, \ldots, N - 1$). The absence of a lower bound on D follows merely by considering the uniform situation, $p_j^+ = p_+$, $p_j^- = p_-$ (all j), which obviously reduces the model to the $N = 1$ case just discussed. The barometric formulation can, equally, be implemented and, again, leads to (21) but now with

$$\Gamma = \prod_{j=0}^{N-1} (p_j^+/p_j^-) . \tag{44}$$

Likewise, in order to account for the effects of a load on the motor particle, the quasiequilibrium hypothesis (27) can be adopted for the spatially dependent jump probabilities $p_j^+(l)$ and $p_j^-(l)$.

8.1. Waiting-time distributions

At a deeper level, however, it is reasonable to object that our arguments for the one-state jump models have more or less tacitly assumed that the jump attempts occur with clockwork regularity at times $n\tau$ whereas, more realistically, there should be some *distribution*, say $\psi(t)$, of *waiting times* between one event and the next. More specifically, after arriving at a site one may suppose that the probability of attempting a jump between subsequent times t and $t + dt$ is $\psi(t)\,dt$. In that case τ should be identified with the mean time between attempts, defined via

$$\tau = \bar{t} \quad \text{with} \quad \overline{t^n} = \int_0^\infty t^n \psi(t)\,dt \quad \text{and} \quad \overline{t^0} = 1 . \tag{45}$$

Such a waiting-time model may be studied along the lines of Montroll and Scher [17].[12] Provided $\psi(t)$ decreases sufficiently fast when $t \to \infty$ that the second moment $\overline{t^2}$ remains finite, the analysis for V and D can be carried through: it shows again that D is, in general, unbounded below while $f_E \propto V/D$ is unbounded above. Indeed, expression (41) for V remains valid. On the other hand, expression (42) for D is no longer accurate: it must be replaced by the, albeit, similar form

$$D = \tfrac{1}{2}(d^2/\tau)[p_+ + p_- - (1 - \Theta)(p_+ - p_-)^2]$$

$$= \tfrac{1}{2}[u + w - (1 - \Theta)(u - w)^2\tau]d^2 , \tag{46}$$

---

[12] A restricted N-state version, with no reverse-reaction transitions (which simplifies the analysis appreciably), was discussed in the appendix of Ref. [8].

the new parameter here, namely,

$$\Theta = (\overline{t^2} - \overline{t}^2)/\overline{t}^2 \geqslant 0 , \tag{47}$$

is the reduced variance of the waiting time and so measures the relative width or "spread" of the distribution $\psi(t)$. As a conveniently general example, suppose $\psi(t) \propto t^{\nu-1}e^{-\gamma t}$ with $\nu, \gamma > 0$. Then one readily finds [13] $\tau = \nu/\gamma$ and $\Theta = 1/\nu$. The sharp distribution originally assumed evidently corresponds to the limit $\nu \to \infty$ and then result (46) reproduces (42). Conversely, for $\nu = 1$, when $\psi(t)$ reduces to the simple exponential or Poisson process form $e^{-t/\tau}$, one has $\Theta = 1$ and the ($N = 1$) kinetic model result (39) is recaptured. Indeed, the full kinetic description becomes valid for exponential waiting distributions.

Finally, once one introduces a waiting-time distribution one would clearly prefer to have *distinct* distributions, say $\psi_j^+(t)$ and $\psi_j^-(t)$, for forward and backward jumps out of states j in a multi-state model (in place of $p_+\psi(t)$ and $p_-\psi(t)$ for the ($N=1$)-state model described). By extending Derrida's analysis with the aid of the theorem on generalized master equations due to Landman et al. [35], one can, in fact, handle such a general jump-and-wait model precisely. Thereby we obtain [28] closed-form expressions for V and D that depend only on the low-order moments of $\psi_j^+(t)$ and $\psi_j^-(t)$. Indeed the explicit calculations can be carried further by allowing for finite branching processes, say with waiting-time distributions $\psi_j^\beta(t)$, that lead off the main linear reaction sequence (2). Such branching models may be useful, in particular, for describing RNA-polymerase where individual motors exhibit lengthy "pauses" in their motion on a DNA strand [12]. In addition, at some further cost in calculation, we can [28] include "death processes" with waiting-time distributions, $\psi_j^\delta(t)$: the overall probability of finding a motor particle anywhere on the track then decays in time (ultimately at an exponential rate) but those particles that remain on the track should still be characterized by a drift velocity V and dispersion D. As mentioned, this enables one to include irreversible detachment of a motor from its track as seen, e.g., in kinesin experiments [7,11].

9. Conclusions

We have presented a general theoretical framework for addressing the questions of the driving force, f that a molecular motor protein can exert and the relations of f to the velocity under a load and to the positional dispersion of the motor as it moves along its molecular track. While many of the general concepts advanced should be widely applicable, the detailed analysis has focused on a fairly broad and basic class of discrete-state stochastic models — kinetic descriptions in the simplest instance but extended to jump-and-wait models at a somewhat more elaborate level of description. These models prove amenable to a surprising degree of exact analysis.

---

[13] The specific results quoted in Ref. [17], Eqs. (75) for $\nu = \frac{1}{2}$ and 2 are in error. In addition the factor 4 in Eq. (76) should read 2. Dr. Harvey Scher has kindly acknowledged that these corrections are needed.

It proves physically essential to recognize that, although in some sense isothermal and isobaric, the crucial physicochemical operations in a motor protein are highly local and take place intrinsically far from equilibrium. Thus close-to-equilibrium, Einstein-type relations between friction, diffusion, and velocity yield only lower bounds on f; and, in examples like kinesin on a microtubule, these are too small to use as guides to the value of f by factors of three (or more).

By contrast, a "barometric formulation" in terms of a limiting spatial distribution of a model motor faced with a rigid barrier, provides a simple relation for f, in terms of the intrinsic rate constants, that appears quite consistent with available experimental data (although, that is relatively limited).

The barometric approach agrees, in turn (subject to a rather natural, small proviso), with a more elaborate quasiequilibrium hypothesis for the dependence of the rate constants on imposed loads. It is essential in this connection to recognize that the induced stresses in the motor protein molecule will, in general, cause quite different changes in the various "internal" forward and backward rate processes. As a result, a full specification of even the simplest two-state kinetic model requires six independent parameters. The resulting plots of velocity *vs.* force can be quite varied in shape: see Figs. 3–5. However, even precise experimental knowledge of such motility plots and of the dispersion may not suffice to pin down the model parameters. And, of course, the parameters of even a very successful model may, by their nature, provide comparatively little insight into the detailed molecular mechanisms employed by a real motor protein. Nevertheless, we believe that a systematic and general theoretical approach, such as we have expounded, should play a useful role in analyzing and classifying data, simulations, and more elaborate models.

Acknowledgements

We are indebted to Daniel S. Fisher, David A. Huse, Stanislas Leibler, Michelle D. Wang and Benjamin Widom for valuable comments on our work. Correspondence with Hong Qian and Steven M. Block has been appreciated. The support of the National Science Foundation (under Grant CHE 96-14495) is gratefully acknowledged.

Appendix : General expressions for velocity and dispersion

For a one-dimensional hopping model with N states and arbitrary transition rates u_j and w_j, as introduced in Eqs. (2)–(4), Derrida [16] obtained the exact steady-state behavior. For the drift velocity he found

$$V = V[\{u_j, w_j\}_N] = \frac{d}{R_N}\left(1 - \prod_{j=0}^{N-1}\frac{w_j}{u_j}\right) , \qquad (A.1)$$

where d is the lattice spacing (or step size) while

$$R_N = \sum_{j=0}^{N-1} r_j, \qquad r_j = \frac{1}{u_j}\left(1 + \sum_{k=1}^{N-1}\prod_{i=1}^{k}\frac{w_{j+i}}{u_{j+i}}\right) . \tag{A.2}$$

The expression for the dispersion (or diffusion constant) — defined as in (1) — is more elaborate: it may be written as

$$D = D[\{u_j, w_j\}_N] = \{(VS_N + dU_N)/R_N^2 - \tfrac{1}{2}(N+2)V\}\frac{d}{N} , \tag{A.3}$$

where the further sums are given by

$$S_N = \sum_{j=0}^{N-1} s_j \sum_{k=0}^{N-1}(k+1)r_{k+j+1}, \qquad U_N = \sum_{j=0}^{N-1} u_j r_j s_j , \tag{A.4}$$

while the supplementary coefficients are

$$s_j = \frac{1}{u_j}\left(1 + \sum_{k=1}^{N-1}\prod_{i=1}^{k}\frac{w_{j+1-i}}{u_{j-i}}\right) . \tag{A.5}$$

Derrida's methods will also yield further moments of the steady-state walk distribution; but the expressions become increasingly cumbersome and have not been published to our knowledge.

References

[1] M.E. Fisher, A.B. Kolomeisky, Proc. Natl. Acad. Sci. USA 96 (1999) 6597–6602.
[2] R.C. Woledge, N.A. Curtin, E. Homsher, Energetic Aspects of Muscle Contraction, Academic Press, London, 1985
[3] L. Stryer, Biochemistry, 3rd Edition, W.H. Freeman, San Francisco, 1988, pp. 927–944.
[4] J. Darnell, H. Lodish, D. Baltimore, Molecular Cell Biology, 2nd Edition, Scientific American Books, New York, 1990, pp. 832–835.
[5] R.H. Abeles, P.A. Frey, W.P. Jencks, Biochemistry, Jones and Bartlett, New York, 1992, Chapter 30.
[6] S. Leibler, Nature (London) 370 (1994) 412–413.
[7] K. Svoboda, S.M. Block, Cell 77 (1994) 773–784.
[8] K. Svoboda, P.P. Mitra, S.M. Block, Proc. Natl. Acad. Sci. USA 91 (1994) 11 782–11 786.
[9] H. Kojima, E. Muto, H. Higuchi, T. Yanagida, Biophys. J. 73 (1997) 2012–2022.
[10] H. Higuchi, E. Muto, Y. Inoue, T. Yanagida, Proc. Natl. Acad. Sci. USA 94 (1997) 4395–4400.
[11] C.M. Coppin, D.W. Pierce, L. Hsu, R.D. Vale, Proc. Natl. Acad. Sci. USA 94 (1997) 8539–8544.
[12] M.D. Wang, M.J. Schnitzer, H. Yin, R. Landick, J. Gelles, S.M. Block, Science 282 (1998) 902–907.
[13] K. Visscher, M.J. Schnitzer, S.M. Block, Nature (London) 400 (1999) 184–189.
[14] H. Qian, Biophys. Chem. 67 (1997) 263–267.
[15] A.B. Kolomeisky, B. Widom, J. Stat. Phys. 93 (1998) 633–645.
[16] B. Derrida, J. Stat. Phys. 31 (1983) 433–450.
[17] E.W. Montroll, H. Scher, J. Stat. Phys. 9 (1973) 101–135.
[18] S.P. Gilbert, K.A. Johnson, Biochemistry 33 (1994) 1951–1960.
[19] S. Leibler, D.A. Huse, J. Cell Biol. 121 (1993) 1357–1368.
[20] T. Duke, S. Leibler, Biophys. J. 71 (1996) 1235–1247.
[21] I. Derényi, T. Vicsek, Proc. Natl. Acad. Sci. USA 93 (1996) 6775–6779.
[22] F. Jülicher, A. Ajdari, J. Prost, Rev. Mod. Phys. 69 (1997) 1269–1281.
[23] F. Jülicher, in: Transport and Structure in Biophysical and Chemical Phenomena, eds. S.C. Müller, J. Parisi, W. Zimmerman, Lecture Notes in Physics, Springer, Berlin, 1999.

[24] H. Qian, Phys. Rev. Lett. 81 (1998) 3063–3066.

[25] A. Parmeggiani, F. Jülicher, A. Ajdari, J. Prost, Phys. Rev. E 60 (1999) 2127.

[26] G.E. Uhlenbeck, L.S. Ornstein, Phys. Rev. 36 (1930) 823–841.

[27] S. Chandrasekhar, Rev. Mod. Phys. 15 (1943) 1–89, Chapter II.

[28] A.B. Kolomeisky, M.E. Fisher, to be published.

[29] W.F. Sheehan, Physical Chemistry, 2nd Edition Allyn and Bacon, Inc., Boston, 1970, Chapter 9.

[30] P. Atkins, Physical Chemistry, 6th Edition, W.H. Freeman and Company, New York, 1999, Chapters 4 and 5.

[31] N.G. van Kampen, Stochastic Processes in Physics and Chemistry, North-Holland, Amsterdam, 1981, Chapters 8, 5, 6.

[32] M. Doi, S.F. Edwards, The Theory of Polymer Dynamics, Clarendon Press, Oxford, 1986, Chapter 3.

[33] K. Tawada, K. Sekimoto, J. Theor. Biol. 150 (1991) 193–200.

[34] L.D. Landau, E.M. Lifshitz, Statistical Physics, 3rd Edition Pergamon Press, London, Part 1, 1980, p. 114.

[35] U. Landman, E.W. Montroll, M.F. Shlesinger, Proc. Natl. Acad. Sci. USA 74 (1997) 430–433.

ELSEVIER

Physica A 274 (1999) 267–280

www.elsevier.com/locate/physa

Application of braid statistics to particle dynamics

Arne T. Skjeltorp[a,b,*], Sigmund Clausen[a,b,1], Geir Helgesen[a]

[a]*Institute for Energy Technology, POB 40, 2007 Kjeller, Norway*
[b]*Physics Department, University of Oslo, Oslo, Norway*

Abstract

How, in a simple and forceful way, do we characterize the dynamics of systems with several moving components? The methods based on the theory of braids may provide the answer. Knot and braid theory is a subfield of mathematics known as topology. It involves classifying different ways of tracing curves in space. Knot theory originated more than a century ago and is today a very active area of mathematics. Recently, we have been able to use notions from braid theory to map the complicated trajectories of tiny magnetic beads confined between two plates and subjected to complex magnetic fields. The essentially two-dimensional motion of a bead can be represented as a curve in a three-dimensional space–time diagram, and so several beads in motion produce a set of braided curves. The topological description of these braids thus provides a simple and concise language for describing the dynamics of the system, as if the beads perform a complicated dance as they move about one another, and the braid encodes the choreography of this dance. © 1999 Elsevier Science B.V. All rights reserved.

1. Introduction

The objective of this review is to describe the dynamical behaviour of a few-body system using notions from knot and braid theory as discusssed in recent publications [1–3]. For a popular science article about magnetic holes and braids written in French, see [4]. The idea is that one particle moving in a plane can be represented as a curve in a three-dimensional space–time diagram, and so several particles in motion produce a set of braided curves. This is a way of "freezing" the dynamics and the braid becomes the time history of the moving particles. Braids as time histories are useful in the study of dynamical systems. There has been some work using braids to describe the intertwining of a set of phase curves $[x_i(t), \dot{x}_i(t)]$ in one-dimensional dynamical systems [5]. In this article we present another approach which allows for

---

* Corresponding author.
E-mail address: arne.skjeltorp@ife.no (A.T. Skjeltorp)
[1] Present address: SINTEF, Blindern N-0314, Oslo, Norway.

applying symbolic dynamics to a wider range of dynamical systems. Once a sequence of motions of n particles has been converted into a braid, it can be further examined to obtain numbers characterising the structure and complexity of the braid. The spheres carry magnetic moments and they are moved around in the plane as the result of external ac magnetic fields. For certain values of the parameters of the driving field the dynamics is intermittent and a hierarchical ordering takes place in both space and time. The rotational part of the motion of the spheres is then well modelled by a one-dimensional *Lévy walk* where large fluctuations lead to a so-called *super diffusive* behaviour, i.e., an enhanced diffusion [6]. The potential for such *anomalous* diffusive behaviour exists in any physical situation where there is some hierarchical ordering of the processes. This ordering can take place in both space and time and anomalous diffusion has been shown to be intimately connected to the notion of *fractal space and time* [7,8]. Recently, several authors have investigated the connection between anomalous transport and Lévy statistics. The Lévy walk describes particularly well Hamiltonian chaos such as diffusion of tracer particles in a two-dimensional flow [9] and phase diffusion in Josephson junctions [10].

2. Experimental and model

Our experimental system consisted of the so-called magnetic holes [11–13] confined to a planar geometry and placed under an optical microscope as shown in Fig. 1. The magnetic holes were nonmagnetic uniformly sized plastic microspheres [14] of typical sizes $d \approx 100$ μm dispersed in a ferrofluid. The ferrofluid [15] with microspheres dispersed in it was confined between two parallel glass plates, where the distance between the plates was typically twice the diameter d of the microspheres. We placed the cell within a system of two pairs of coils producing a magnetic field rotating within the plane of the cell. The motion of the magnetic holes was also the same xy-plane.

When a ferrofluid sample containing monodisperse microspheres is placed in a uniform magnetic field H, the voids created by the microspheres acquire an apparent magnetic dipole moments μ antiparallel to the external field:

$$\mu = -V\chi_{eff} H .$$

(1)

Here, V is the volume of a microsphere and χ_{eff} is the effective volume susceptibility of the ferrofluid [15]. We used circularly or elliptically polarised magnetic fields rotating within the sample (x, y)-plane with angular velocity ω_H:

$$H(t) = [H_x \cos(\omega_H t), \varepsilon H_x \sin(\omega_H t)]$$

(2)

with $\varepsilon = H_y/H_x$ as a measure of the field anisotropy.

The dipolar interaction energy of n magnetic holes of diameter d is given by

$$U(r_1, r_2, \ldots, r_n, t) = \begin{cases} \displaystyle\sum_{i>j}^{n} \left\{ \frac{\mu^2(t)}{r_{ij}^3} - \frac{3[\mu(t) \cdot r_{ij}]^2}{r_{ij}^5} \right\} & \text{if all } r_{ij} > d , \\ \infty & \text{if any } r_{ij} \leqslant d . \end{cases}$$

(3)

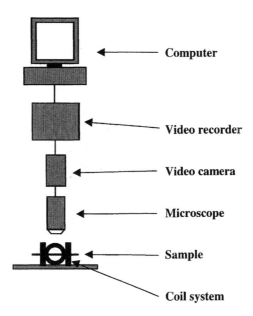

Fig. 1. Schematic view of the experimental set-up. The sample contains plastic microspheres dispersed in a magnetised ferrofluid confined between two plane-parallel glass plates. Two pairs of coils are used to produce a magnetic field rotating in the (x, y)-plane.

Here $r_{ij} = r_j - r_i$ is the vector joining the centres of the interacting microspheres and $r_i = (x_i, y_i)$. The components of the magnetic force acting on the ith magnetic hole are given by

$$F_{i,\xi} = -\sum_{i>j}^{n} \frac{\partial}{\partial \xi_j} U(r_1, r_2, \ldots, r_n, t),$$ (4)

where ξ denotes either x or y. The system is overdamped due to the large viscosity of the ferrofluid, and we may therefore neglect inertial forces. Thus, we assume that at any time the velocity of the ith magnetic hole is proportional to the force given by Eq. (4):

$$\frac{d\xi_i}{dt} = \beta F_{i,\xi},$$ (5)

where $\beta = (3\pi\eta d)^{-1}$ and η is the viscosity of the ferrofluid. Eqs. (3)–(5) can be transformed into a dimensionless form suitable for numerical integration by letting $H_x = 1$ and $\beta = \frac{1}{6}$. By this choice of parameters, the threshold angular velocity for the stable rotation of a single pair of magnetic holes is equal to 1. The equations of motion were simulated using a fourth-order Runge–Kutta algorithm and compared with our experimental results. With the existing experimental set-up it is possible to grab and digitise up to 25 images per second which gives us a continuous motion-picture of the particle dynamics. A two-dimensional projection of the (x, y, t) space–time braid traced by the motion of five magnetic holes is shown in Fig. 2(a). Due to the relatively

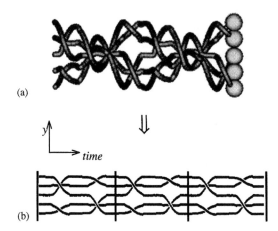

Fig. 2. (a) A space–time plot of five magnetic holes moving in the (x, y)-plane. The x-axis goes into the paper. (b) The resulting braid structure after removal of the total twist from the braid in (a). The periodicity is easily visible and the repeating braid structure is shown within the vertical bars.

high viscosity of the ferrofluid the chain of microspheres has to divide into two smaller pieces containing two and three microspheres. These smaller pieces are able to rotate with the field. One magnetic hole is interchanged between the two chain pieces making the overall motion complicated. In addition, there is an overall twisting of the space–time trajectories. The total twist in the braid can be extracted by running the braid word through Garside's word algorithm [16], which can separate the external twist of an arbitrary braid from the local intertwining of the strands. Garside's solution may be given by means of *normal forms* [17]. Here, we shall follow a refinement of this solution due to Elrifai and Morton [18].

Fig. 2(b) shows the resulting braid structure after removal of the total twist from the braid in Fig. 2(a). The motion proves to be periodic and it is possible to extract phase portraits of the periodic modes by this type of analysis [1]. In the next section we describe the fluctuation analysis of the rotational motion after a short introduction to braid theory.

3. Braids

A thorough introduction to braid theory may be found in the book by Birman [17]. Here we will only give a brief description of the notion used in our analysis. Once a sequence of motions has been converted into braid notation, it can be further examined to obtain numbers characterising the structure and complexity of the braid and thereby the dynamics. In order to describe a braid without having to draw it, one may decompose it into a product of elementary braids; see Fig. 3a. For an n-strand system there exist $n - 1$ elementary braids called generators and denoted by $\sigma_1, \sigma_2, \ldots, \sigma_{n-1}$. In an

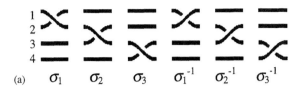

Fig. 3. The generators of the 4-strand Artin braid group (a) and an example of a braid composed of them (b).

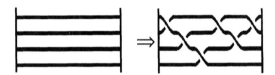

Fig. 4. A positive half twist in the 4-strand Artin braid group.

elementary braid σ_i the strand located at the ith position exchanges its place with the strand located at the $(i + 1)$th position.

In order to characterise the structure and complexity of a braid different numbers or *topological invariants* can be calculated. One such number is the *writhe* of the braid, Wr, which is simply the sum of the exponents of the braid word, a positive crossing adds $+1$ and a negative crossing adds -1. The writhe is therefore equal to the number of positive crossings minus the number of negative crossings. As an example consider the braid in Fig. 3(b), where $Wr=2$. A second useful number to calculate is the number of *twists* in the braid. A geometric picture of a *positive half-twist* of n strands, Δ_n, is obtained by imagining the n strings attached to a rod which is given a $180°$ twist as shown in Fig. 4. A twist commutes with all other braid structures and therefore opposite senses of twist in a braid may be moved nearby to each other and cancel out.

The recipe for the fluctuation analysis is now as follows: we make a video recording of the motion of n magnetic holes where the output is a braid word describing the space–time braid of the motion. This braid is then divided into what may be denoted *half-period braids*. One half-period braid is simply the space–time braid describing the motion of the microspheres during one half of the period T of the rotating field, see Fig. 5. As the total braid grows with time t the value of Wr is extracted every time a new half-period braid is plaited, i.e., whenever $t = mT/2$, where m is the number of completed half-periods. We set $t = m$ so that the unit of time is half a period of the rotating field. This approach results in a time series, $Wr(t)$. However, this time

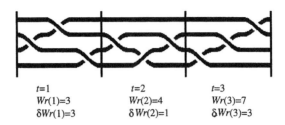

$t=1$ $t=2$ $t=3$
$Wr(1)=3$ $Wr(2)=4$ $Wr(3)=7$
$\delta Wr(1)=3$ $\delta Wr(2)=1$ $\delta Wr(3)=3$

Fig. 5. Schematic illustration of the division of a braid into $t=3$ half-period braids. The accumulated value of the signed crossing number $Wr(t)$ and the half-period differences $\delta Wr(t)$ are written below the braid.

series does not give a complete topological description of the dynamics as there are several combinations of crossings of strands which results in the same value of $Wr(t)$. Nevertheless, our primary goal now is to study the aperiodic motion of the magnetic holes and the fluctuations in the patterns of the strands. In addition to the total writhe as a function of time, we are also interested in the successive half-period variations defined by

$$\delta Wr(t) = Wr(t) - Wr(t-1) \tag{6}$$

which might be thought of as an average *writhe velocity*. $\delta Wr(t)$ equals the writhe Wr of half-period braid number t, see Fig. 5.

4. Selected results

In the following analysis we limit ourselves to study the dynamics of $n=7$ magnetic holes. The number $n=7$ was chosen more or less arbitrarily – it is not too small ($n=2,3,4$) to make the motion relatively simple, and it is not too large ($n>10$) making a full analysis very time consuming. However, we observe qualitatively similar statistical behaviour for all n up to $n=20$ magnetic holes, which are the maximum number analysed in this study. We fix the driving frequency to $f_H = \omega_H/2\pi = 0.25$ Hz and vary the anisotropy parameter ε. Different types of dynamical behaviour are observed ranging from periodic to intermittent and random. For circularly polarised magnetic fields the motion is periodic and the results has been described in detail elsewhere [1]. Here, we want to describe more complex aperiodic dynamics. Only experimental results are presented unless stated otherwise.

4.1. Random fluctuations

In the following we shall analyse the fluctuations in the writhe number, $Wr(t)$. This increases steadily with time t due to the large majority of positive crossings of the space–time strands. In order to observe the fluctuations more easily the average

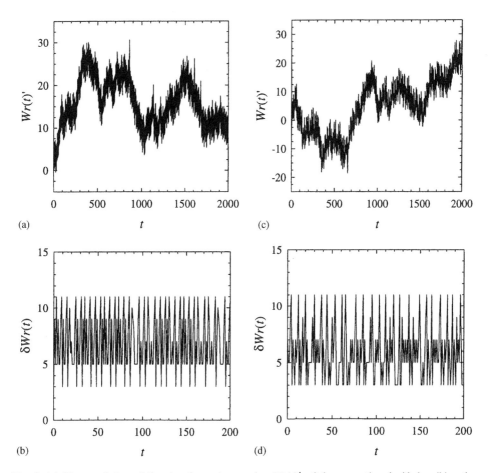

Fig. 6. (a) Time evolution of the signed crossing number $Wr(t)'$ of the space–time braid describing the motion of seven magnetic holes when $\varepsilon = 0.80$. Time t is measured in units of half a period of the driving field. (b) The associated half-period differences $\delta Wr(t)$. (c) and (d) shows the same quantities for $\varepsilon = 0.70$.

increasing trend is subtracted from the original value of $Wr(t)$. The difference is denoted by $Wr(t)'$:

$$Wr(t)' = Wr(t) - t \cdot \overline{\delta Wr}, \tag{7}$$

where $\overline{\delta Wr}$ is the average value of $\delta Wr(t)$ averaged over the total number of half-periods N:

$$\overline{\delta Wr} = \frac{1}{N} \sum_{t=1}^{N} \delta Wr(t). \tag{8}$$

For $\varepsilon < 0.85$ the motion becomes aperiodic, Fig. 6(a) shows $Wr(t)'$ for $\varepsilon = 0.80$. The associated half-period variations are shown Fig. 6(b). Fig. 6(c) and (d) show the same quantities for $\varepsilon = 0.70$. Using $Wr(t)'$ instead of $Wr(t)$ in this definition will only shift $\delta Wr(t)$ by $\overline{\delta Wr}$ and the fluctuations will be centred around zero instead of around $\overline{\delta Wr}$.

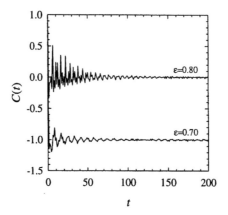

Fig. 7. Plots of the auto-correlation function $C(t)$ for $\varepsilon = 0.80$ (upper curve) and $\varepsilon = 0.70$ (lower curve). Time t is measured in units of half a period of the driving field. In both cases $C(t)$ fluctuates around zero, but for $\varepsilon = 0.70$ the data are shifted for clarity.

For both $\varepsilon = 0.80$ and 0.70, the motion of the magnetic holes was recorded for a total of 10 000 half-periods of the rotating field and the braids can then be divided into a total of 10 000 half-period braids. Only short sequences of the whole time series are shown in Fig. 6 in order to resolve the fluctuations.

Any periodic behaviour is quantified by calculating the auto-correlation function defined by

$$C(t) = \frac{\langle [\delta Wr(\tau) - \langle \delta Wr(\tau)\rangle] \cdot [\delta Wr(\tau + t) - \langle \delta Wr(\tau)\rangle]\rangle}{\langle [\delta Wr(\tau) - \langle \delta Wr(\tau)\rangle]^2\rangle}, \qquad (9)$$

where brackets indicate averaging over all times τ and t is a time interval. $C(t)$ is then a measure of the spatiotemporal correlations in the system, that is, it measures the correlations between half-period braids separated by t half-periods in time.

Fig. 7 shows the auto-correlation function $C(t)$ for the two cases discussed above. It decays exponentially with a time correlation length of about 30 half-periods. This is a clear indication of random behaviour with only short time correlations. In addition to calculating the auto-correlation function, we focus attention on the dynamics of the variations

$$\delta Wr(t, \tau) = Wr(\tau)' - Wr(\tau - t)', \qquad (10)$$

where both τ and t is measured in units of half a period of the rotating field. By setting $t = 1$ in this equation one gets back the definition of the half-period variations in Eq. (6). It is important to notice that we select the complete set of *non-overlapping* records separated by t half-periods. Varying the time interval t enables us to study the fluctuations of $Wr(t)'$ on different time scales. A standard method of extracting information about the fluctuations in a time series is to calculate the variance of the successive variations in that time series. The variance of the variations defined in

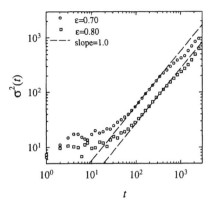

Fig. 8. Variance $\sigma^2(t)$ characterising the increments $\delta Wr(t,\tau)$ versus time t for $\varepsilon = 0.80$ and 0.70. The time t is measured in units of half a period of the driving field. The two dashed lines are the best fits to the experimental data with slope $\eta = 1.0$.

Eq. (10) as a function of the time interval t is given by the following expression:

$$\sigma^2(t) = \langle [\delta Wr(t,\tau) - \langle \delta Wr(t,\tau) \rangle]^2 \rangle, \qquad (11)$$

where the brackets indicate averaging over τ. Fig. 8 shows the variance for both $\varepsilon = 0.80$ and 0.70 calculated from Eq. (11). The figure indicates random fluctuations over a relatively large time span. After an initial time span of about 30 half-periods the data approaches the behaviour

$$\sigma^2(t) \propto t^{1.0} \qquad (12)$$

characteristic of random processes with independent increments. The deviations from the straight line in the figure for $t > 1000$ are a finite size effect due to the limited time range in the data set.

Both the auto-correlation function and the variance show a crossover from periodic to random behaviour for relatively short times. The flat region in the plot of the variance for $t < 30$ half-periods is consistent with the short time behaviour of $C(t)$. For short times the motion *appears* periodic and the magnetic holes apparently try to find a periodic solution. However, these solutions are only stable for short times and the motion appears random for longer times.

4.2. Critical behaviour, intermittence and Lévy motion

The motion of the seven magnetic holes seems to be characterised by random fluctuations down to $\varepsilon \approx 0.62$ where a new type of behaviour is observed. At this value of the anisotropy parameter there is a finite probability for smaller chain pieces to stay separated from each other for long times. A continuous-time one-dimensional Lévy walk with a power law distribution of step lengths can model the overall motion. This model of a Lévy motion was proposed by Klafter et al. [6]. Each time the chain of microspheres separates into smaller pieces a new step of the random walk is started.

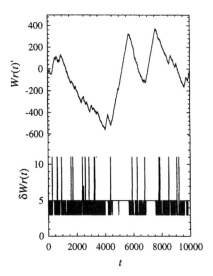

Fig. 9. Time evolution of the signed crossing number $Wr(t)'$ of the space–time braid describing the motion of seven magnetic holes in an elliptically polarised field with $\varepsilon = 0.55$ (upper curve). Time t is measured in units of half a period of the driving field. The lowermost curve shows the associated half-period differences $\delta Wr(t)$.

The separation times are defined as the times the chain pieces stay separated from each other and they are power law distributed with an exponent equal to the exponent of the distribution of the step lengths. This is always the case for a one-dimensional Lévy walk where the velocity is independent of the step length [19].

Fig. 9 shows $Wr(t)'$ for $\varepsilon = 0.55$ with the associated half period variations $\delta Wr(t)$ plotted below. The driving frequency is still $f_H = 0.25$ Hz. The motion of the magnetic holes was recorded for 10 000 half-periods of the rotating field, and the braids can thus be divided into a total of 10 000 half-period braids. Clearly, the dynamical behaviour differs from the one observed in the previous case. The fluctuations are much larger and the figure also shows long steps in $Wr(t)'$ with constant velocity $\delta Wr(t)$. There is a *distribution* of step lengths, or equivalently a distribution of time intervals where long steps in $Wr(t)'$ are observed. During these long time intervals the microspheres move in a regular manner with the chain of magnetic holes divided into three smaller pieces which rotate with the field. Two of these chain pieces contain two microspheres whereas the third one contains three. The three chain pieces stay separated from each other for long times and the half-period variations $\delta Wr(t) = 5$ during the steps, as seen in Fig. 9. This value equals the number of crossings of a half-period braid describing $2 + 2 + 3$ magnetic holes rotating with the field. When the chain pieces are forced together again, the magnetic holes move in an aperiodic way for some time before they separate once more. The dynamical evolution is intermittent, and thus consists of both quiescent and more chaotic phases, which alternate temporally in an interspersed way. During the separation times the signed crossing number $Wr(t)'$ increases at a constant rate causing the large-scale fluctuations seen in Fig. 9. We will show that the

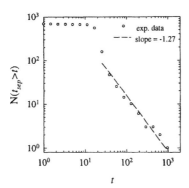

Fig. 10. Plots of the distribution $N(t_{sep} \geqslant t)$ of separation times t for $\varepsilon = 0.55$. The dashed line is the best fit to the experimental data in the region $t \in (20, 1000)$.

motion is well modelled by a one-dimensional Lévy walk with a distribution $\varphi(t)$ of quiescent time intervals or separation times t. The tail of this distribution follows a power law

$$\varphi(t) \propto t^{-(\alpha+1)} \tag{13}$$

with an exponent α. The number N of separation times t_{sep} larger than t goes as

$$N = N(t_{sep} \geqslant t) = \int_t^\infty \varphi(t_{sep}) \, dt_{sep} \propto t^{-\alpha} . \tag{14}$$

Fig. 10 shows the distribution N of separation times t extracted from the time series shown in Fig. 9. It is possible to fit the data to a power law with an exponent $\alpha = 1.27 \pm 0.13$ for $t > 20$.

We want to find out how the long tails in the distribution influence the behaviour of the auto-correlation function. The experimental data of $C(t)$ in Fig. 11 is fitted to a power law decay over almost three decades in time indicating long-range correlations in the half-period variations $\delta Wr(t)$:

$$C(t) \propto t^{-\gamma} \tag{15}$$

with the exponent $\gamma = 0.34 \pm 0.12$. According to theory [19] one expects $\gamma = \alpha - 1$ for a one-dimensional Lévy walk with constant velocity when $\alpha > 1.0$. This is in good agreement with our experimental observations, $\alpha - 1 = 0.27 \pm 0.13$.

Further confirmation of the proposed Lévy motion is obtained by calculating the variance $\sigma^2(t)$ of the fluctuations of the experimental data using Eq. (11) as shown in Fig. 12. In the same figure we also display the results from a numerical simulation of the motion of the magnetic holes. In that case the dynamics was recorded for 500 000 half-periods and the parameters were the same as in the experiment. There is a good fit of the data for long times to

$$\sigma^2(t) \propto t^\eta \tag{16}$$

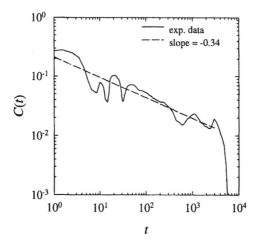

Fig. 11. Plots of the auto-correlation function $C(t)$ characterising the half-period differences $\delta Wr(t)$. Time t is measured in units of half a period of the driving field. The dashed line is the best fit to the experimental data in the region $t \in (1, 3000)$.

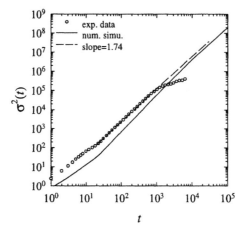

Fig. 12. Variance $\sigma^2(t)$ characterising the increments $\delta Wr(t, \tau)$ versus time t when $\varepsilon = 0.55$. The time t is measured in units of half a period of the driving field. The dashed line is the best fit to the experimental data in the region $t \in (20, 1000)$. The fitted line is extended beyond $t = 1000$ for clarity and for comparison with the numerical data (full line).

with an exponent $\eta = 1.74 \pm 0.03$ for the experimental case. The numerical data approaches the same behaviour for longer times. In both cases a super diffusive behaviour is observed with $\eta > 1$. As Fig. 12 shows, there is a finite size effect in the experimental data above approximately 1000 half-periods which is due to the limited time range in the data set. This observation is consistent with the results of the numerical simulation, where the super diffusive behaviour extends to longer times due to the finite probability of larger separation times in the numerical distribution. The exponent

η is related to the separation time exponent α by $\eta = 3 - \alpha$ when $\alpha > 1.0$ [19]. For $\alpha = 1.27 \pm 0.13$ one gets $\eta = 1.73 \pm 0.13$, which is consistent with this exponent relation.

A Lévy walk is a fractal generalisation of Brownian motion and it is possible to relate the diffusion exponent η to the fractal dimension of space and time. In this simple model of diffusion, space and time are coupled and a step in space is associated with a certain time span. The ensemble of time instants where jump events occur, form a fractal set with fractal dimension:

$$d_t = \begin{cases} \alpha, & \alpha \leqslant 1, \\ 1, & \alpha > 1. \end{cases} \tag{17}$$

When do the stopover points form a fractal in space? An ordinary Brownian trajectory wiggles so much that it is actually two-dimensional independent of the dimension of the embedding space. For a Lévy walk, where $1 \leqslant \alpha < 2$, the ensemble of stopover points form a set with fractal dimension [19]

$$d_r = \frac{2}{3 - \alpha}, \tag{18}$$

Thus, for the case studied here the ensemble of stopover points form a fractal set with $d_r = 1.15 \pm 0.11$. The coupling of space and time through d_r and d_t is explicitly given by an expression for the diffusion exponent

$$\eta = \frac{2d_t}{d_r} = 3 - \alpha, \tag{19}$$

from the considerations discussed above.

Recently, Wang and Tokuyama [20] has used a generalised Langevin equation to show theoretically that for the superdiffusive case, $\alpha > 1.0$, one expects $\eta = 1.66 \pm 0.12$, which is also in good agreement with our experimental result. They have also found that $d_r = 1.21 \pm 0.09$, again consistent with our experimental result.

5. Conclusions

The random motion of microspheres in a plane has been studied. The two-dimensional trajectories of n spheres generate a braid in a three-dimensional space–time. By studying the fluctuations of the signed crossing number a wide range of different dynamical behaviour is observed, ranging from periodic to random motion. For certain parameter values the fluctuations were shown to be highly intermittent and a hierarchical ordering takes place in both space and time. In that case the motion is well modelled by a one-dimensional Lévy walk with a power law distribution of step lengths, which determines the fluctuation behaviour. The dynamical evolution consists of both quiescent (regular) and more chaotic phases which alternate temporally in an interspersed way.

In conclusion, our experimental model system is simple and well defined with precision control of all the parameters. This coupled with computer simulations in good agreement with experiments allows us to look for general features of non-equilibrium phenomena.

Acknowledgements

This work has been supported in part by the Research Council of Norway (S.C.). We want to thank A. Berge at NTNU in Trondheim, and Dyno Particles A.S. for kindly providing the microspheres used in these experiments.

References

[1] S. Clausen, G. Helgesen, A.T. Skjeltorp, Int. J. Bifurcation Chaos 8 (7) (1998) 1383–1397.
[2] S. Clausen, G. Helgesen, A.T. Skjeltorp, Phys. Rev. E 58 (4) (1998) 4229–4237.
[3] S. Clausen, Braid description of particle dynamics, Thesis, Physics Dept., Univ. of Oslo, 1998.
[4] P. Pieranski, A.T. Skjeltorp, Les tresses tissées par des trous magnétiques, Pour la Science, Avril 1997.
[5] F.A. Mc. Robbie, J.M.T. Thompson, Int. J. Bifurcation Chaos 3 (6) (1993) 1343.
[6] J. Klafter, A. Blumen, M.F. Shlesinger, Phys. Rev. A 35 (1987) 3081.
[7] M.F. Shlesinger, G. Zaslavsky, J. Klafter, Nature (London) 363 (1993) 31.
[8] J. Klafter, M.F. Shlesinger, G. Zumofen, Phys. Today 49 (2) (1996) 33.
[9] T.H. Solomon, E.R. Weeks, H.L. Swinney, Phys. Rev. Lett. 71 (1993) 3975.
[10] T. Geisel, J. Nierwetberg, A. Zacherl, Phys. Rev. Lett. 54 (1985) 616.
[11] A.T. Skjeltorp, Phys. Rev Lett. 51 (1983) 2306.
[12] G. Helgesen, A.T. Skjeltorp, J. Appl. Phys. 69 (1991) 8277.
[13] G. Helgesen, P. Pieranski, A.T. Skjeltorp, Phys. Rev. A 42 (1990) 7271.
[14] J. Ugelstad et al., Adv. Colloid Int. Sci. 13 (1980) 101. Produced under the trade name Dynospheres by Dyno Particles A.S., P.O.B. 160, N-2001 Lillestrøm, Norway.
[15] Type EMG 909, produced by Ferrofluidics Corporation, 40 Simon St., Nashua, NH 03061.
[16] F.A. Garside, Quart. J. Math. 20 (2) (1969) 235–254.
[17] J.S. Birman, Braids, links and mapping class group, Annals of Math. Study, Vol. 82, Princeton University Press, Princeton, NJ, 1974.
[18] E.A. Elrifai, H.R. Morton, Quart. J. Math. Oxford 2 (45) (1994) 479.
[19] X.-J. Wang, Phys. Rev. A 45 (1992) 8407.
[20] K.G. Wang, M. Tokuyama, Physica A 265 (1999) 341.

ELSEVIER

Physica A 274 (1999) 281–293

www.elsevier.com/locate/physa

Applications of statistical mechanics in subcontinuum fluid dynamics

Marek Cieplak[a,*], Joel Koplik[b], Jayanth R. Banavar[c]

[a]*Institute of Physics, Polish Academy of Sciences, 02-668 Warsaw, Poland*
[b]*Benjamin Levich Institute and Department of Physics, City College of New York, NY 10031, USA*
[c]*Department of Physics and Center for Materials Physics, 104 Davey Laboratory,
The Pennsylvania State University, University Park, PA 16802, USA*

Abstract

We show results of molecular dynamics studies of fluid flows in the Knudsen regime in which the mean free path is comparable to the system size. We elucidate the boundary conditions at the wall–fluid interface in such flows and find scenarios envisioned by Maxwell. We also find scenarios which do not agree with Maxwell's hypothesis. We focus primarily on the case of repulsive walls and discuss similarities and differences in behavior when the wall–fluid interaction is attractive or repulsive. Striking many body effects are found on increasing the fluid density as one interpolates between the dilute gas and the dense fluid regimes. © 1999 Elsevier Science B.V. All rights reserved.

PACS: 51.10.+y; 34.10.+x; 92.20.Bk

Novel technological applications often involve flows of rarefied fluids. The density of such fluids may become so low that the description of the flow within the scope of continuum hydrodynamics either breaks down or needs to be supplemented by new parameters. These parameters depend on an atomic level understanding of the phenomena taking place near a confining wall. The interest in such subcontinuum flows goes back to the 19th century studies by Fresnel (1985) and Crookes (1874–1878) of the physics underlying the workings of a radiometer and the phenomena of photophoresis and thermal creep. It also provided motivation for Maxwell to interpret the boundary conditions for fluid in the vicinity of a solid surface [1] from the point of view of his kinetic theory of gases [2]. The 1938 monograph by Kennard [3] gives an exhaustive account of the three basic new phenomena that arise at high dilution: thermal creep, thermal slip, and viscous slip. The latter has been studied by Kundt and Warburg [4],

---

* Corresponding author.
E-mail address: ciepla@ifpan.edu.pl (M. Cieplak)

Timiriazeff [5], and especially by Knudsen [6,7] after whose name the high dilution regime is called. The Knudsen number, Kn, is defined as a ratio of the mean free path, λ, to the characteristic linear system size, L. At high values of Kn, the flows consist of essentially ballistic movements of molecules between the walls combined with a usually more complex behavior at the walls.

The current return of interest to this classic problem is due to the fact that it is an essential aspect of aerosol dynamics [8,9], high altitude aerodynamics [10], isotope separation schemes [11], and flow cooling techniques [10]. It is also at the core of functioning of micro- and nanoscale mechanical and electromechanical systems [12,13] that are currently being tested and produced as a result of the advances made in silicon-based technology.

Molecular dynamics based simulations offer a unique tool to study subcontinuum aspects of flows in the vicinity of a solid surface [4–19]. At present, there are no other techniques (experimental or theoretical) for deducing the boundary conditions from first principles. Recently [20], we have adopted a fully atomistic model of the fluid *and* of the walls and studied it by molecular dynamics at high and intermediate values of Kn. We have demonstrated that the velocity profiles are very different from those predicted by continuum hydrodynamics and show an unexpected dependence on the density and a distinct dependence on the system size on going from low to high Knudsen numbers. For liquids, an increase in the density results in an increased viscosity and, therefore, in a slowing down of the flow. Strikingly, an opposite tendency is present in the gaseous regime due to an interplay between the interactions of the molecules with the wall and with each other.

In order to elucidate the nature of boundary conditions at a fluid–solid interface, Maxwell [1] has considered two limiting types of the walls: thermal and specular. The latter corresponds to a perfectly smooth surface that would reflect an incident molecule specularly. The former – to a highly uneven or granular surface which would scatter an incident molecule internally and multiply and would eventually reemit it fully thermalized. For the general situation, Maxwell postulated a linear combination of the two kinds of behavior with the weights f for thermal and $(1 - f)$ for specular contributions, respectively. This hypothesis has been applied to viscous slip, thermal slip, etc. [3]. Millikan [21], among others, carried out experiments on viscous slip that attempted to unravel Maxwell's linear combination – the effective value of f for common solid and liquids was found to be usually bigger than 0.8 but the determination of f was indirect.

Our own molecular dynamics simulations [20] identified both kinds of limiting behaviors suggested by Maxwell: the collisions with walls were thermal, when there was sufficient attraction between the wall and fluid molecules, or specular, when there was no attractive component in the wall–fluid interactions. However, the crossover between the two behaviors was found to be much more complex and interesting than the simple linear combination envisioned by Maxwell and used by many workers since.

In our previous publication [20], we presented our molecular dynamics results for the case of a wall with strong attraction to the fluid leading to the formation of layers

of fluid atoms next to the wall. This kind of coating provides an efficient thermalizing effect. Here, we focus on the other limiting case and discuss what happens when
the wall is purely repulsive. In particular, we demonstrate that the nonmonotonic dependence of the velocity profiles on the density remains qualitatively similar to the
attractive wall case. The mechanism for this phenomenon, though still many body in
nature, is, however, different. In addition, there are substantial differences in the profiles
of density, velocity and shear stress.

An atomic level modelling requires that the fluid and solid molecules be treated
on an equal footing. In previous molecular dynamics studies in the Knudsen regime
[22,23], it was only the fluid which had an atomic representation and the boundary
conditions were put in by hand. As a simplest model, we consider a Lennard–Jones
fluid in which two atoms separated by a distance r interact with the potential

$$V_{LJ}(r) = 4\varepsilon \left[\left(\frac{r}{\sigma}\right)^{-12} - \left(\frac{r}{\sigma}\right)^{-6} \right] , \tag{1}$$

where σ is the size of the repulsive core (the potential is truncated at 2.2σ). The
flows take place between two solid planar walls parallel to the xy plane with periodic
boundary conditions imposed along the x and y directions. Couette flow was generated
by moving the walls at constant velocity in opposite directions along the x-axis whereas
Poiseuille flow was induced by an application of uniform acceleration g on all the fluid
molecules in the x-direction.

As in the paper by Thompson and Robbins [24], we construct the wall from two
[001] planes of an fcc lattice. Each wall atom is tethered to a fixed lattice site by
a harmonic spring of spring constant k. We chose $k = 400$ so that the mean-square
displacement about the lattice sites did not exceed the Lindemann criterion for melt-
ing but was close to the value set by this criterion. The wall–fluid interactions were
modelled by another Lennard–Jones potential,

$$V_{wf}(r) = 16\varepsilon \left[\left(\frac{r}{\sigma}\right)^{-12} - A \left(\frac{r}{\sigma}\right)^{-6} \right] \tag{2}$$

with the parameter A varying between 1 and 0, corresponding to attractive and repulsive
walls, respectively. For the narrowest channel that we studied, the fluid atoms were
confined to a space measuring $13.6\sigma \times 5.1\sigma$ in the xy plane and 12.75σ between the
walls. We consider a wall which contains 96 wall atoms each (in each wall plane, the
atomic periodicity is 0.85σ). The geometry used is illustrated in Fig. 1 which shows a
snapshot of the positions of the atoms.

The equations of motion were integrated using a fifth-order Gear predictor-corrector
algorithm (see e.g. [25]) with a step size of 0.005τ, where $\tau = (m\sigma^2/\varepsilon)^{1/2}$ and m is the
atomic (wall and fluid) mass. The fluid and wall temperatures were fixed at $k_B T/\varepsilon = 1.1$.
Here, we present results obtained with the use of Langevin thermostat [26]. We have
checked that the results obtained with the use of the Nose–Hoover thermostat [27,28]
for the attractive wall case were virtually identical and we expect the same to hold for
the repulsive wall as well. The noise was applied in all directions during equilibration
and only in the y-direction during the data-taking phase. The averages were obtained

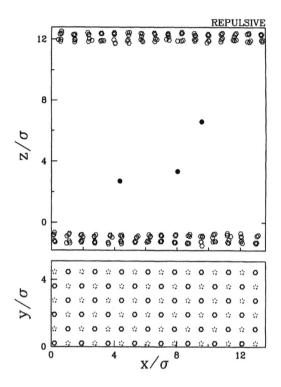

Fig. 1. The snapshot of the 'atoms' of the systems. The black circles represent the fluid atoms. There are three fluid atoms here which illustrates the case of extreme dilution: the central density is $0.004/\sigma^3$. The open hexagons represent the wall atoms. The top panel is the $x - z$ projection whereas the bottom panel is the $x - y$ projection. Here, the hexagons are either solid or dotted to differentiate between atoms from the two wall layers.

over runs which lasted at least 5000 τ, and in the extreme dilution case, for up to 400 000 and 16 000τ for the attractive and repulsive wall cases, respectively. An initial time of at least 400τ was spent for equilibration. The spatial averaging was carried out in slabs of width $\sigma/4$ along the z-axis.

In the dilute gas limit, the Knudsen number, Kn, is defined as λ/L, i.e., as $(\sqrt{2}\pi\rho \sigma^2 L)^{-1}$, where λ is the mean free path, ρ is the fluid number density and L is the characteristic width of the channel. The expression for λ is adopted from the result for the hard core gas. A study of the density profile shows qualitatively different behaviors of the density profiles for the attractive and repulsive walls, as illustrated in Figs. 2 and 3, respectively. In the former case, formation of adsorbed layers form near both walls whereas there is merely a depletion zone, whose thickness is of order σ, alongside the repulsive walls. The first layer at the attractive walls is fairly periodic, as measured by the static structure factor, and the resulting local order corresponds to a square lattice. L is defined as an effective channel width in which the flow takes place (in the attractive case this is the distance between centers of the second layers and in the

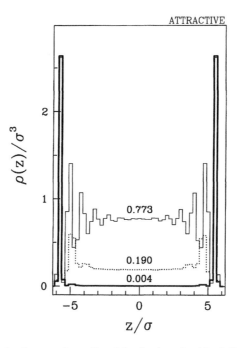

Fig. 2. The density profiles for the attractive wall and for the three densities indicated. The values refer to the density in the central region.

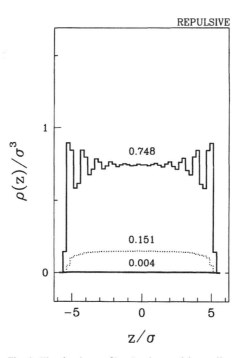

Fig. 3. The density profiles for the repulsive wall.

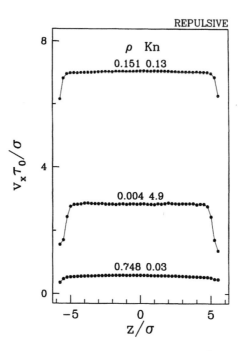

Fig. 4. Velocity profiles of gravity-driven channel flow with $g = 0.01\sigma/\tau^2$ for the smallest width channel with repulsive walls. The values of the interior fluid density and the corresponding Knudsen number are indicated. The inner wall layers are at the edges of the figure.

repulsive case this is the width of a region in which the density is non-zero). In the center of the channel, the fluid density is essentially constant and this value is chosen as ρ in the definition of Kn. Increasing the density to liquid values results, for both kinds of walls, in a buildup of atomic layers further and further into the bulk.

Fig. 4 shows the flow velocity profiles for three different values of ρ (or Kn) for the repulsive wall case. The flow was induced by a small gravitational force with $g = 0.01$ along the x-axis. The profiles correspond to plug flows across the whole density range with some Poiseuille-like parabolicity present only at the highest densities. This is in contrast to the profiles found for the attractive wall case [20]. For the attractive walls, the profiles are plug-flow like only at high dilution whereas they are nicely parabolic at intermediate and high densities. The attractive wall exerts a pull on the fluid molecules leading to virtually no-slip boundary conditions whereas the repulsive wall gives rise to a large slip.

A further understanding of the problem is obtained by studying the shear stress component of the stress tensor. This is shown in Figs. 5 and 6 for the attractive and repulsive walls, respectively. In the attractive case, ρ of 0.004 corresponds to a very small velocity gradient and thence virtually zero stress. At intermediate and high densities, however, the shear stress has an overall slope indicating a non-zero viscosity. In the repulsive case, the behavior is similar but the shear stress is an order

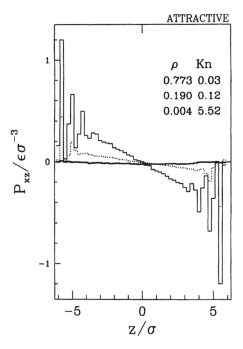

Fig. 5. The shear stress profiles of the gravity-driven channel flow for the smallest width channel with attractive walls. The values of the interior fluid density and the corresponding Knudsen number are indicated and they are ordered top to bottom corresponding to the left-hand side portions of the profiles.

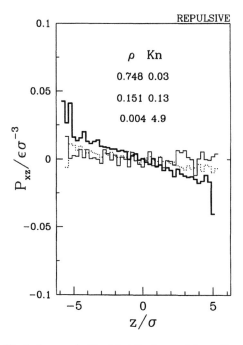

Fig. 6. Same as in Fig. 5 but for the repulsive walls.

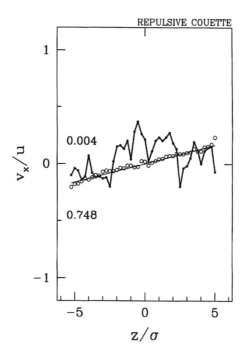

Fig. 7. Velocity profiles of Couette flow with $u = 0.1\sigma/\tau$ for the smallest width channel with repulsive walls. The fluid densities in the central region are indicated.

of magnitude smaller. The decoupling of the fluid flow from the repulsive wall is also seen in Couette flow. This is illustrated in Fig. 7 which shows the velocity profile for the situation in which the top wall moves to the right and the bottom wall moves to the left, both with a speed of u. In the limit of large dilution, the profile is not smooth. The smoothness observed in gravity-driven flow is due to an organizing influence of the gravitational force which acts on all molecules. In Couette flow, the forcing is restricted to the wall area and, in the absence of viscous effects, it does not propagate inward to provide organization.

We now proceed to study the effect of interactions between the fluid particles undergoing gravity-driven channel flow on varying ρ within the channel. In such a flow, the velocity field is the highest in the middle of the channel and is denoted by v_{max}. For the attractive wall, one expects that v_{max} ought to scale as L at small densities and as L^2 at high densities. The low density behavior is implicit in an analytic result for the flowrate derived by Smoluchowski [29,3] and its physical origin is that the driving force is effective in accelerating the fluid particles only during the time spent traversing the channel between collisions with the walls (and the associated adsorbed layers). This time simply scales as the channel width. The high density behavior follows from the Navier–Stokes equation with no-slip boundary conditions. The crossover between the two situations results in a monotonic dependence of v_{max} on ρ – there is a maximum around a characteristic density ρ_K of about $0.19\sigma^{-3}$ above which the velocity profiles

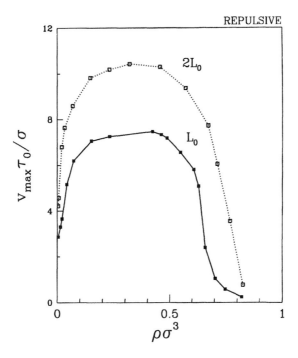

Fig. 8. Plot of the fluid velocity in the middle of the channel in gravity-driven flow with $g = 0.01\tau/\sigma^2$ as a function of fluid density The plot is for two different channel widths. The L_0 in the figure denotes the narrowest channel. $2L_0$ denotes a channel that is effectively twice as wide as the smallest channel.

are clearly parabolic [20]. Above ρ_K, the fluid intermolecular interactions dominate the influence of the wall.

Fig. 8 shows that the repulsive wall also generates a maximum in the dependence of $v_{\max}$ on ρ at ρ_K of about $0.4\sigma^{-3}$. However, the scaling behavior appears different. Doubling the separation between the walls at the highest Knudsen numbers results in increasing $v_{\max}$ only by a factor of about $\frac{3}{2}$. A similar factor characterizes the scaling around ρ_K. A further increase in ρ makes the scaling factor bigger (between 3 and 6) with no evidence for the quadratic law. The low density behavior should be related to the fact that the saturation of the velocity profile with time is not due to the thermalization at the wall but merely due to bulk thermalization by the Langevin noise. Furthermore, the wider the channel, the longer lasting is the accelerating influence of the force. The large density behavior, on the other hand, is modified by the large slip length. A further increase in the wall-to-wall separation is expected to lead to the L^2 law asymptotically.

The most interesting aspect of Fig. 8 is that, as in the attractive wall case, $v_{\max}$ increases with ρ when $\rho < \rho_K$. In the attractive wall case we attributed this phenomenon to an increased scattering which keeps the atoms in the channel longer [20]. This results in two effects: (1) the acceleration time gets extended and (2) the time spent on thermalization at the walls gets reduced. A scattering-induced effective deflection of

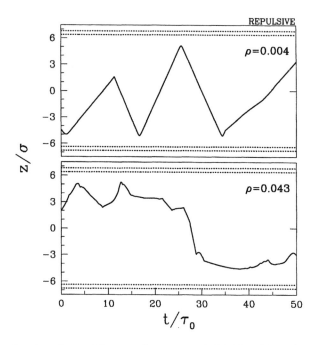

Fig. 9. Plots of the time dependence of the z-coordinate of a typical fluid particle undergoing gravity-driven flow in the narrowest channel. The horizontal dotted lines indicate the z-coordinates of the wall molecules.

ballistic trajectories from the wall is also present in the case of the repulsive wall and is illustrated in Fig. 9. In this case, however, only the first of the above two effects is operational since the specularly reflecting wall has little thermalizing influence. The upper panel of Fig. 9 shows the trajectory of the z-coordinate of a typical atom when three fluid atoms are present in the channel. At the wall, the atom is reflected approximately specularly and moves out of the vicinity of the wall immediately. The bottom panel shows qualitatively different behavior when 27 more atoms are added into the system: each atom hits the wall less frequently.

A useful characterization of the fluid–solid interface is provided by the collision kernel [30,31] – the probability density that a molecule striking the surface with a given velocity reemerges with a specific velocity after a given residence time. We find that, both for the attractive and repulsive walls, the scenarios envisioned by Maxwell hold very well. For the repulsive wall, the residence time is negligible and the reflections are specular. On the other hand, for the attractive walls the distribution of residence times develops a long tail and the residence time is found to be uncorrelated with the outgoing velocity. Strikingly, the velocity distribution of a particle after the collision with the wall is substantially independent of the incoming velocity. The velocity distributions agree with the thermal wall scenario and for v_x and v_z they are given by [2,3,32]

$$\phi_x(v_x) = \sqrt{\frac{m}{2\pi k_B T}} \exp\left(-\frac{m v_x^2}{2 k_B T}\right) , \tag{3}$$

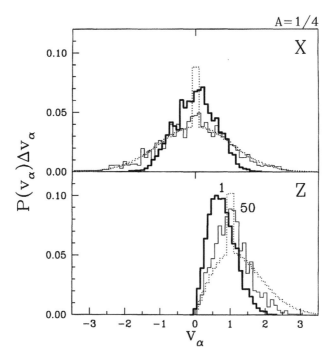

Fig. 10. The thick solid lines (marked additionally by the label 1) represent the probability density distri-butions for the velocity components of the outgoing velocity of a fluid molecule after collision with a wall with the wall–fluid interactions given by Eq. (2) with $A = \frac{1}{4}$. There are no adsorbed fluid atoms on the wall. The upper and lower panels are for the x and z (normal) velocity components, respectively. The statistics are based on about 3000 scattering events with the wall. The dotted lines correspond to the linear combination of to the combined distributions $f\phi_\alpha(v_\alpha) + (1 - f)\delta(v_\alpha - v_\alpha^{sp})$ with $f = 0.9$. Here, ϕ_α is given by Eqs. (3) and (4) and v_α^{sp} denotes the specularly reflected component of the velocity. The selected value of f offers the closest match to the data for both velocity components simultaneously. The thinner solid lines in the main panel (marked by the label 50) represent the velocity distributions obtained for $A = \frac{1}{4}$ when 50 fluid molecules are placed in the narrowest channel ($\rho = 0.066$).

and

$$\phi_z(v_z) = \frac{m}{k_B T} v_z \exp\left(-\frac{mv_z^2}{2k_B T}\right) , \qquad (4)$$

respectively.

Following Ref. [20], we now consider the case in which the wall–fluid potential is given by Eq. (2) with $A = \frac{1}{4}$. This corresponds to the crossover behavior from that of specular reflection to that of a thermal wall. Fig. 10 shows that, in this case, the velocity distributions of the scattered particles clearly deviate from pure thermal behavior and contains a specular component. However, the resulting distributions are more complex than the simple linear combination form.

The behavior of the collision kernel is further enriched on considering the role of the interactions between the fluid molecules as the Knudsen number decreases to a value of 0.27 corresponding to a density of $\rho = 0.066$. The density profile corresponding to this

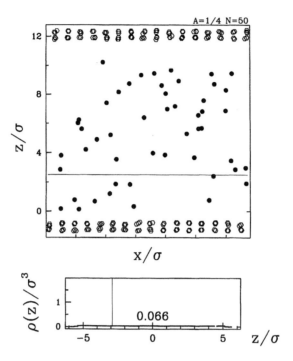

Fig. 11. The top panel shows a snapshot of the atoms, similar to Fig. 1, which depicts the geometry used to determine the effect of other molecules on the effective collision kernel for one molecule. There are 50 fluid molecules here and $A = \frac{1}{4}$. The thin horizontal line corresponds to the monitoring plane. The bottom panel shows the corresponding density profile. The value of the density at the center is indicated.

situation is shown in the bottom panel of Fig. 11 whereas a corresponding snapshot is shown in the top panel of this figure. The distributions are obtained by monitoring molecules incident on the walls at a distance of 1 mean free path (or 3.4σ) from the walls, marked by a line in Fig. 11, during the evolution of the system. The data were obtained based on about 1100 events in which the incoming molecules happened to be moving nearly normally towards one of the walls with a speed between 0.8 and $1.2\sigma/\tau_0$ and then reemerged with velocities distributed as shown in the figure. The influence of other fluid molecules is seen to lead to the distributions more closely approaching the thermal ones. Moving the monitoring plane further away from the wall also results in bringing the distributions closer to the thermal ones.

The nature of the boundary conditions at a fluid–solid interface is thus more complex than those envisioned by Maxwell and molecular dynamics simulations are a useful tool to understand this complexity.

We are indebted to Mark Robbins and Riina Tehver for many valuable discussions. This work was supported by KBN (Grant No. 2P03B-025-13), NASA, and the Petroleum Research Fund administered by the American Chemical Society.

References

[1] J.C. Maxwell, Philos. Trans. R. Soc. Lon. Ser. A 170 (1867) 231.
[2] J.C. Maxwell, Philos. Mag. 19 (1860) 20.
[3] E.H. Kennard, Kinetic Theory of Gases, McGraw-Hill, New York, 1938.
[4] A. Kundt, E. Warburg, Philos. Mag. 50 (1875) 53.
[5] A.K. Timiriazeff, Ann. Phys. 40 (1913) 971.
[6] M. Knudsen, Ann. Phys. 28 (1909) 75.
[7] M. Knudsen, Ann. Phys. 48 (1915) 1113.
[8] W.H. Marlow (Ed.), Aerosols: Aerosol Microphysics I – Particle Interactions, Springer, New York, 1980.
[9] J.H. Seinfeld, S.N. Pandis, Atmospheric Chemistry and Physics, Wiley, New York, 1998.
[10] E.P. Muntz, D. Weaver, D. Campbell (Eds.), Rarified Gas Dynamics, American Institute of Aeronautics and Astronautics, Washington, 1989.
[11] W. Ehrfeld, Elements of Flow and Diffusion Processes in Separation Nozzles, Springer, New York, 1983.
[12] C.-J. Kim, (Ed.), Microelectromechanical systems (MEMS), American Society of Mechanical Engineers, New York, 1997.
[13] C.-M. Ho, Y.-C. Tai, Ann. Rev. Fluid Mech. 30 (1998) 579.
[14] J. Koplik, J.R. Banavar, Comput. Phys. 12 (1998) 424.
[15] J. Koplik, J.R. Banavar, Annu. Rev. Fluid Mech. 27 (1995) 257–292.
[16] S. Sugano, Microcluster Physics, Springer, Berlin, 1991.
[17] B. Bhushan, J.N. Israelachvili, U. Landman, Nature 374 (1995) 607.
[18] I. Bitsanis, T.K. Vanderlick, M. Tirrel, H.T. Davis, J. Chem. Phys. 89 (1988) 3252.
[19] W. Loose, S. Hess, Rheol. Acta 28 (1989) 91.
[20] M. Cieplak, J. Koplik, J.R. Banavar, submitted for publication.
[21] R.A. Millikan, Phys. Rev. 21 (1923) 217.
[22] D.K. Bhattacharya, G.C. Lie, Phys. Rev. A 43 (1991) 761.
[23] D.L. Morris, L. Hannon, A.L. Garcia, Phys. Rev. A 46 (1992) 5279.
[24] P.A. Thompson, M.O. Robbins, Phys. Rev. A 41 (1990) 6830.
[25] M.P. Allen, D.J. Tildesley, Computer Simulation of Liquids, Clarendon Press, Oxford, 1987.
[26] G.S. Grest, K. Kremer, Phys. Rev. A 33 (1986) 3628.
[27] S. Nose, Mol. Phys. 52 (1984) 255.
[28] W.G. Hoover, Phys. Rev. A 31 (1985) 1695.
[29] M. Smoluchowski, Ann. Phys. 33 (1910) 1559.
[30] C. Cercignani, The Boltzmann Equation and its Applications, Springer, Berlin, 1988.
[31] Y. Matsumoto, N. Yamanishi, H. Shobatake, Proceedings of the 19th International Symposium on Rarefied Gas Dynamics, Vol. 995, 1994.
[32] R. Tehver, F. Toigo, J. Koplik, J.R. Banavar, Phys. Rev. E 57 (1998) R17.

ELSEVIER

Physica A 274 (1999) 294–299

www.elsevier.com/locate/physa

Applications of statistical mechanics to natural hazards and landforms

Donald L. Turcotte*

Department of Geological Sciences, Cornell University, 4122 Snee Hall, Ithaca, NY 14853, USA

Abstract

The concept of self-organized criticality was introduced to explain the behavior of the cellular-automata sandpile model. A variety of multiple slider-block and forest-fire models have been introduced which are also said to exhibit self-organized critical behavior. It has been argued that earthquakes, landslides, forest-fires, and extinctions are examples of self-organized criticality in nature. The basic forest-fire model is particularly interesting in terms of its relation to the critical-point behavior of the site-percolation model. In the basic forest-fire model trees are randomly planted on a grid of points, periodically sparks are randomly dropped on the grid and if a spark drops on a tree that tree and the adjacent trees burn in a model fire. In the forest-fire model there is an inverse cascade of trees from small clusters to large clusters, trees are lost primarily from model fires that destroy the largest clusters. This quasi-steady-state cascade gives a power-law frequency–area distribution for both clusters of trees and smaller fires. The site-percolation model is equivalent to the forest-fire model without fires. In this case there is a transient cascade of trees from small to large clusters and a power-law distribution is found only at a critical density of trees. The earth's topography is an example of both statistically self-similar and self-affine fractals. Landforms are also associated with drainage networks, which are statistical fractal trees. A universal feature of drainage networks and other growth networks is side branching. Deterministic space-filling networks with side-branching symmetries are illustrated. It is shown that naturally occurring drainage networks have symmetries similar to diffusion-limited aggregation clusters. © 1999 Elsevier Science B.V. All rights reserved.

1. Introduction

Earthquakes, landslides, and forest and wild fires are natural hazards that are of great concern. The processes that result in these hazards are extremely complex but in each case the frequency–size distributions appear to satisfy power-law (fractal) relations. It

---

* Fax: +1-607-257-4780.
E-mail address: turcotte@geology.geo.cornell.edu (D.L. Turcotte)

0378-4371/99/$ - see front matter © 1999 Elsevier Science B.V. All rights reserved.
PII: S 0378-4371(99)00325-8

can be argued that each of these hazards is an example of self-organized criticality in nature. There is a close relationship between these phenomena and the site percolation model and the behavior can be explained in terms of a self-similar-inverse cascade model.

Landforms are also examples of complex natural phenomena that satisfy power-law (fractal) statistics in a variety of ways. River networks are an essential component of landform evolution on the earth. River networks satisfy self-similar (fractal) statistics including side branching. The behavior is very similar to that of diffusion-limited aggregation.

2. Self-organized criticality

The concept of self-organized criticality was proposed by Bak et al. [1] as an explanation for the behavior of a cellular automata model they developed. In this model a square grid of boxes was considered. At each time step a particle was added to a randomly selected box, when a box contained four particles they were redistributed to neighboring boxes or lost from the grid. Redistributions could lead to further "instabilities" which resulted in an "avalanche" of particle redistributions. Because of this "avalanche" behavior, this was called a "sandpile" model. The frequency–area distribution of model "avalanches" was found to satisfy a power-law (fractal) distribution with a slope near unity.

A second model that can exhibit self-organized critical behavior is the slider-block model [2]. In this model an array of slider blocks are connected to a constant velocity driver plate by puller springs and to each other by connector springs. The blocks exhibit stick–slip behavior due to frictional interactions with the plate across which they are pulled. The frequency–area distribution of the smaller slip events again satisfies a power-law distribution with a slope near unity.

A third model that exhibits self-organized criticality is the forest-fire model [3,4]. In the simplest version of this model, a square grid of sites is considered. At each time step either a tree is planted on a randomly chosen site (if the site is unoccupied) or a spark is dropped on the site. If the spark is dropped on a tree, that tree and all adjacent trees are "burned" in a model "forest fire". The number of sparks dropped divided by the number of steps is the sparking frequency f. The frequency–area distribution of the smaller fires again satisfies a power-law (fractal) distribution with a slope near unity.

3. Earthquakes

Since the concept of self-organized criticality was first introduced, earthquakes have been identified as an example of this phenomena in nature [5]. Earthquakes occur primarily in the brittle upper crust of the earth at depths of less than 20 km. Most earthquakes occur at or near the boundaries of the tectonic plates, but a few earthquakes occur within plate interiors. Stress is added to the crust slowly due to the tectonic motion of the plates and is relieved rapidly in earthquakes (avalanches). The crustal stress oscillates about an equilibrium value.

For over 50 years earthquakes have been recognized as an example of a natural hazard that is accurately approximated by power-law, frequency–area distributions. It is accepted that earthquakes universally obey Gutenberg–Richter scaling, the cumulative number of earthquakes per year in a region with magnitude greater than m, N_{CE}, is related to m by

$$\log N_{CE} = -bm + a , \qquad (1)$$

where the constant b is known as the b-value and has a near-universal value $b = 0.9$. When (1) is expressed in terms of the earthquake rupture area, A_r, instead of earthquake magnitude, this relation becomes a power law [6]

$$N_{CE} \sim A_r^{-b} . \qquad (2)$$

These earthquake frequency–magnitude relations can be used to quantify the earthquake hazard for large earthquakes based on the frequency of occurrence of small earthquakes [7].

4. Forest and wild fires

An obvious application of the forest-fire model is to actual forest and wild fires. Frequency–magnitude distributions for four forest-fire and wildfire data sets from the United States and Australia were given by Malamud et al. [8]. The agreement with power-law (fractal) statistics is quite good. Considering the many complexities of the initiation and propagation of forest fires and wildfires it is remarkable that the frequency–area statistics are very similar under a wide variety of environments. The proximity of combustible material varies widely. The behavior of a particular fire depends strongly on meteorological conditions. Fire-fighting efforts extinguish many fires. Despite these complexities, the application of the frequency–area distributions associated with the forest-fire model appears to be robust.

5. Landslides

Pelletier et al. [9] have considered four data sets for the frequency–area distributions of landslides. For the larger landslides, the frequency–area distributions correlate well with the power-law relation (2). It must be noted, however, that landslides have a depth as well as an area, so any comparison with the strictly two-dimensional sandpile model is only approximate. Nevertheless, all four data sets give reasonably good power-law correlations for the larger landslides and similar slopes are obtained in diverse environments. There are a number of similarities between the sandpile model and the real landslides. Slope instabilities develop slowly and are eliminated in "avalanches". Power-law distributions are obtained.

6. The forest-fire model and the site-percolation mode

The forest-fire model is closely related to the site-percolation model that is known to exhibit critical behavior [10]. If trees are planted on a grid without fires, site-percolation behavior is found. The critical point is reached when a tree cluster spans the grid. This occurs when the ratio of trees N_t to grid size, N_t/G, reaches a critical value of 0.59275. There is an important difference between the forest-fire model and the site-percolation model. The frequency–area distribution of tree clusters in the forest-fire model satisfies the power-law (fractal) relation (2) for the smaller clusters for all values of the sparking frequency f. The frequency–area distribution of tree clusters in the site-percolation model satisfies the power-law relation (2) only in the immediate vicinity of the critical value of N_t/G.

The relationship between the forest-fire model and the site-percolation model illustrates several interesting aspects of "self-organized criticality". The first question is whether the forest-fire model exhibits self-organized critical behavior for all values of the sparking frequency f. The range over which power-law (self-similar) behavior is found is limited except for a tuned value f_s. Is self-similar behavior over the full range of areas required for self-organized critical behavior? If it is, then a parameter must be tuned and there is no difference between self-organized critical behavior and critical behavior.

It is clear that there is an important difference between the behavior of the forest-fire model and the site-percolation model. For large values of f a scaling region is obtained in the forest-fire model, but for small values of N_t/G no scaling region is found for the site-percolation model. This difference in behavior can be explained by an inverse-cascade model [11]. New single-tree clusters are continuously generated by the planting of trees. Tree clusters coalesce to form larger clusters in an inverse cascade that satisfies (2) with unity slope. A significant number of trees are lost only in the fires that destroy the largest clusters and terminate the cascade. Smaller fires sample the distribution of tree clusters but do not significantly affect the number of smaller clusters.

7. River networks, Tokunaga trees, and DLA

Simple branching networks are most easily illustrated by river networks. The merging of small streams gives larger streams, large streams merge to give small rivers, and so forth. But small streams can also merge with larger streams and rivers. The branching statistics of networks can differentiate between alternative models for their formations. The original taxonomy for branching networks was given by Horton [12]. Strahler

[13] introduced a slightly modified system which is now widely used in a variety
of applications. In this classification system a stream with no upstream tributaries is
defined to be a first order ($i = 1$) stream. When two first-order streams combine they
form a second-order ($i = 2$) stream, when two second-order streams combine they form
a third-order ($i = 3$) stream, and so forth. The total number of ith order streams is N_i
and their mean length L_i.

Horton [12] defined the branching ratio $R_N = N_i/N_{I+1}$ and the length–order ratio
$R_L = L_{I+1}/L_i$. Empirically it was recognized that both R_N and R_L are nearly constant
for a range of stream orders for all river networks and this observation constitutes
two of Horton's laws. As Mandelbrot [14] developed fractal concepts he recognized
that Horton's laws define a fractal tree and the validity of Horton's laws implies that
drainage networks are fractal trees.

A major improvement in the quantitative classification of self-similar networks was
introduced by Tokunaga [15]. In order to classify side branching, Tokunaga extended
the Strahler ordering system. A first-order branch joining another first-order branch is
denoted "11" and the number of such branches is N_{11}, a first-order branch joining
a second-order branch is denoted "12" and the number of such branches in N_{12}, a
second-order branch joining a second-order branch is denoted "22" and the number of
such branches is N_{22}, and so forth. The branch numbers N_{ij}, $i \leqslant j$, constitute a square,
upper-triangular matrix.

When considering self-similar (fractal) trees it is convenient to introduce branching
ratios T_{ij}, these are the average number of branches of order i joining each branch
of order j, $i < j$. Again the branching ratios T_{ij} constitute a square, upper-triangular
matrix.

Self-similarity (fractality) requires that $T_{i,j} = T_k$ where T_k is a branching ratio that
depends only on $k = j - i$. A general requirement for self-similarity is [15]

$$T_k = ac^{k-1} \ . \tag{3}$$

This is now a two parameter family of trees and we will define fractal trees in this
class to be Tokunaga trees.

Peckham [16] has determined the branching-ratio matrices for the Kentucky River
basin in Kentucky and the Powder River basin in Wyoming. For the Kentucky River
basin, the bifurcation ratio is $R_b = 4.6$, the length–order ratio is $R_r = 2.5$, and the fractal
dimension is $D = 1.67$; for the Powder River basin the bifurcation ratio is $R_b = 4.7$, the
length–order ratio is $R_r = 2.4$ and the fractal dimension is $D = 1.77$. Both the results for
the Kentucky River basin and the Powder River basin correlate well with (3) taking
$a = 1.2$ and $c = 2.5$.

A statistical model that generates Tokunaga self-similar fractal trees is diffusion
limited aggregation (DLA) introduced by Witten and Sander [17]. Ossadnik [18] has
considered the side-branching statistics of off-lattice DLA clusters. The average bifurcation ratio of the clusters was found to be $R_N = 5.15 \pm 0.05$ and the average
length–order ratio $R_L = 2.86 \pm 0.05$, the corresponding fractal dimension is $D = 1.56$. In
order to analyze the side-branching statistics of DLA clusters Ossadnik [18] utilized the

ramification matrix introduced for DLA by Vannimenus and Viennot [19]. The ramification matrix is entirely equivalent to the branching-ratio matrix introduced by Tokunaga [15]. DLA clusters are found to be Tokunaga self-similar trees which correlate well with (3) taking $a = 1.5$ and $c = 2.7$ discharges. Masek and Turcotte [20] have proposed a modified DLA model for river networks. For comprehensive treatments of the Tokunaga approach the reader is referred to Newman et al. [21].

References

[1] P. Bak, C. Tang, K. Wiesenfeld, Phys. Rev. A 38 (1988) 364.
[2] J.M. Carlson, J.S. Langer, Phys. Rev. A 40 (1989) 6470.
[3] P. Bak, K. Chen, C. Tang, Phys. Lett. A 147 (1992) 297.
[4] B. Drossel, F. Schwabl, Physica A 191 (1992) 47.
[5] P. Bak, C. Tang, J. Geophys. Res. 94 (1989) 15 635.
[6] D.L. Turcotte, Fractals and Chaos in Geology and Geophysics, 2nd Edition, Cambridge University Press, Cambridge, 1997.
[7] D.L. Turcotte, Phys. Earth Planet. Int. 111 (1999) 275.
[8] B.D. Malamud, G. Morein, D.L. Turcotte, Science 281 (1998) 1840.
[9] J.D. Pelletier, B.M. Malamud, T. Blodgett, D.L. Turcotte, Eng. Geol. 48 (1997) 255.
[10] D. Stauffer, A. Aharony, Introduction to Percolation Theory, 2nd Edition, Taylor & Francis, London, 1992.
[11] D.L. Turcotte, B.D. Malamud, G. Morein, W.I. Newman, Physica A 268 (1999) 629.
[12] R.E. Horton, Geol. Soc. Amer. Bull. 56 (1945) 275.
[13] A.N. Strahler, Trans. Amer. Geophys. Union 38 (1957) 913.
[14] B. Mandelbrot, The Fractal Geometry of Nature, Freeman, San Francisco, 1982.
[15] E. Tokunaga, Geog. Rep. Tokyo Metro. Univ. 13 (1978) 1.
[16] S.D. Peckham, Water Resour. Res. 31 (1995) 1023.
[17] T.A. Whitten, L.M. Sander, Phys. Rev. Lett. 47 (1981) 1400.
[18] P. Ossadnik, Phys. Rev. A 45 (1992) 1058.
[19] J. Vannimenus, X.G. Viennot, J. Stat. Phys. 54 (1989) 1529.
[20] J.G. Masek, D.L. Turcotte, Earth Planet. Sci. Lett. 119 (1993) 379.
[21] W.I. Newman, D.L. Turcotte, A.M. Gabrielov, Fractals 5 (1997) 603.

ELSEVIER

Physica A 274 (1999) 300–309

www.elsevier.com/locate/physa

Application of statistical physics to impact fragmentation

Hajime Inaoka[a], Hideki Takayasu[b,c,*]

[a]*Graduate School of Frontier Sciences, University of Tokyo, Tokyo 113-8656, Japan*
[b]*Sony Computer Science Lab., Takanawa Muse Bldg., 3-14-13 Higashi-gotanda, Shinagawa-ku,
Tokyo 141-0022, Japan*
[c]*Earthquake Research Institute, University of Tokyo, Tokyo 113-0032, Japan*

Abstract

Numerical simulation methods of impact fragmentation are reviewed. The well-known power-law distribution of fragment pieces is explained theoretically and its dependence on aspect ratio is derived. © 1999 Elsevier Science B.V. All rights reserved.

1. Introduction

When a glass is dropped on a hard floor it is instantly broken into pieces by the impact. We find several large pieces, many small ones and uncountable fine pieces that are hardly distinguishable by naked eyes. This is a typical example of impact fragmentation phenomenon, and the most interesting feature of this phenomenon is such a wide range distribution of fragment pieces. Largest examples of fragment pieces may be the asteroids that are presumed to be created by a collision of astronomical objects such as small planets or satellites. There are more than 10 000 asteroids having their own names and the largest ones are more than 100 km in diameter. Therefore, the scale of impact fragmentation phenomenon may span over nearly ten orders of magnitude in diameter, from sub-millimeters up to over 100 km.

Impact fracture is a special case of brittle fracture phenomena that occurs in a short time scale whose characteristic speed of failure propagation is of the order of sound velocity. In this paper we are going to review our numerical model of impact fracture and discuss distribution of fragment pieces in view of statistical mechanics.

* Corresponding author. Sony Computer Science Lab., Takanawa Muse Bldg., 3-14-13 Higashi-gotanda, Shinagawa-ku, Tokyo 141-0022, Japan. Fax: 81-3-5448-4273.
E-mail address: takayasu@csl.sony.co.jp (H. Takayasu)

0378-4371/99/$ - see front matter © 1999 Elsevier Science B.V. All rights reserved.
PII: S 0378-4371(99)00426-4

Fracture processes of brittle solid materials have been widely studied because of their importance in various fields of science and technology [1–3]. One of the statistically remarkable features seen in impact fragmentation is a power-law distribution of the fragment mass. For example, Gilvarry and Bergstrom performed experiments of impact fragmentation using glass specimens and reported in their widely known paper [4] that the cumulative distribution of fragment mass $P(\geqslant m)$, the probability that an arbitrarily chosen fragment piece has mass m or larger, follows a power law

$$P(\geqslant m) \propto m^{-\tau}, \quad \tau \sim \tfrac{2}{3}. \tag{1}$$

Eq. (1) generally holds in several decades up to the order of the total mass of fractured object in the case of full three-dimensional specimens like spheres. Recently, Oddershede, Dimon, and Bohr conducted experiments of impact fragmentation and firstly pointed out that the fragment mass distribution depends on the shape of specimen while it is independent of the material [5]. Motivated by their results, Meibom and Balslev managed careful experiments and reported that the fragment mass distribution follows a universal power law accompanied with a nearly flat tail for large mass in cases of plate-like specimens [6]. These works triggered physicists' interest in the universality of fragment mass distribution by impact fragmentation of fragile materials.

Numerical study of fracture phenomena has been developed widely [7,8], however, only two methods are reported on the simulation of the impact fracture phenomenon. One approach is called molecular dynamics approach which is based on solving dynamic equations of particles connected with springs having nonlinear or threshold characteristics [9,10]. The empirical power-law mass distribution can be confirmed by this method, however, such approaches require huge computer facilities to clarify the shape dependence. The other approach is a macroscopic numerical model of impact fragmentation based on a competitive growth process of cracks proposed by the authors [11,12]. The essential point of this method is to extract a possible scale-independent dynamics underlying in real crack developments. As we will see later, the model can simulate the impact fracture phenomenon successfully with even a personal computer and we can clarify the cause of the universality and the shape dependence of the fragment mass distribution. We briefly review the results of the study of impact fragmentation by our numerical model [11].

2. The numerical simulation

The basic idea is based on a competitive growth process that is generally essential for irreversible systems showing fractality [13]. In a system of diffusion-limited aggregation [14,15], a typical fractal growth system, for example, the competition among growing clusters produces a power-law cluster size distribution [16]. Our fracture model is based on a competitive process during crack propagation as follows.

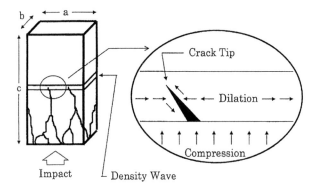

Fig. 1. A schematic figure of our numerical model. A thin compressive density wave causes dilation on the density wave plane, which drives the crack propagation.

The mechanisms of three-dimensional crack propagation in general situations are very complicated, so, we focus on the situation that a rectangular parallelpiped object suffers an impact on one of its sides. It is a reasonable assumption that a plane-like compressive density wave propagates from the hit surface toward the opposite side with a constant velocity v_0 as schematically shown in Fig. 1. We assume that most of the crack propagation occurs only on this density wave forming a so-called failure wave [17]. By this assumption we treat the three-dimensional impact fracture process by considering a time evolution of cracks on the two-dimensional plain of the density wave. We observe the propagation of cracks on a coordinate whose x–y plane is placed on the hit surface of the fractured object and the z-axis is directed toward propagating direction of the wave. The density wave plane is placed at depth $v_0 t$ from the hit surface at time t, and in a small time interval Δt it proceeds $v_0 \Delta t$. In our model time is discretized and crack tip configuration on a layer at depth $v_0 t$ determines the configuration of the next layer at depth $v_0(t + \Delta t)$. We measure the time by using the constant time unit Δt and call it as time step. By setting Δt unity, the time is measured simply by the depth z. The density wave plane is also discretized and the model is defined on a two-dimensional triangular lattice representing the density wave plane.

The dynamics of the crack tips is ruled by stress and strain field in the material. We introduce randomness of material by introducing spatial fluctuation in the elastic moduli. As schematically shown in Fig. 1, the randomness causes random dilation $\theta(x, y; z)$ on the density wave plane when the material undergoes compression by the density wave. This dilation is a driving force of the crack propagation. The deformation of the material on the density wave plane is approximately described by a displacement potential $\phi(x, y; z)$ [11]. That is,

$$\nabla^2 \phi(x, y; z) = \theta(x, y; z) \tag{2}$$

holds. We assume that the boundary condition of $\phi(x,y; z)$ on crack tips as $\phi(x, y; z)=0$, which means that displacement near a crack tip is perpendicular to the crack tip. Solving

Eq. (2) and comparing the magnitude of displacement $\nabla\phi(x, y; z)$ on both sides of the crack, the crack tip is assumed to move toward the place where the displacement is smaller. By this dynamics a larger fragment section tends to expand its area faster than neighboring small fragment sections, where a fragment section is defined by an area on the density wave surrounded by cracks.

By above mechanisms, the cracks can only move and merge on the propagating density wave plane and no bifurcation is allowed and no crack stops spontaneously. As a result the crack density monotonically decreases as the density wave propagates and larger fragments appear near the opposite side of the hit surface of the fractured object. Such decay of crack density is quite natural in view of energy because the crack propagation is dissipative. In our model we assume that the mechanical energy injected by the impact on the hit surface is strong enough for cracks to penetrate the specimen.

As one can see from the above discussion, the dynamics of crack propagation in our model is not directly related to the potential flaw criterion nor Griffith's criterion, which are also considered to be basic concepts of fracture mechanics [18,19]. Potential flaws may control the crack propagation dynamics microscopically with the effect of strain and stress fields. However, we apply the macroscopic dynamics deduced from the knowledge of fractal growth discussed above since we believe that it is the most essential part of the cause of the power-law fragment mass distribution.

Simulations are started from the initial condition that every lattice site belongs to a different fragment section, meaning that the hit surface is completely fractured. Note that, in our model, a region surrounded by cracks on the density wave plane is a section of a three-dimensional fragment and the total number of the sites belonging to each fragment section throughout simulation time steps defines the fragment volume or mass.

The simulations were performed with periodic boundary condition to discuss the universality of the fragment mass distribution. The randomness of the dilation field $\theta(x, y; z)$ is given by uniform random numbers. Examples of the fragments produced by this model are shown in Fig. 2. The cross section parallel to the density wave, Fig. 2(a), shows a configuration of fragment sections at a certain depth. There appear fragment sections having nearly the same size. Contrasted with Fig. 2(a), the cross section perpendicular to the density wave, Fig. 2(b), shows many fine fragments near the initial plane on the top and a few large ones located away from the initial plane indicating that the system contains fragments of various size. The cumulative fragment mass distribution in the model is shown in Fig. 3 in log–log scale. The dots are clearly on a straight line in the mass range from 10 to 10^5 indicating that the cumulative fragment mass distribution follows a power law

$$P(\geqslant m) \propto m^{-\tau}, \quad \tau = 0.66. \tag{3}$$

The value of the exponent $\tau = 0.66$ is very close to $\frac{2}{3}$, and the value is consistent with the values observed in experimental situations.

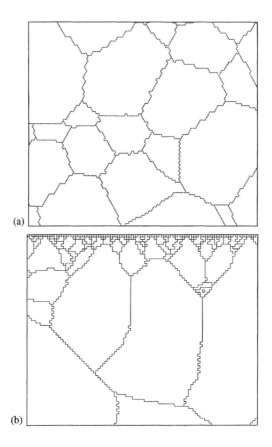

(a)

(b)

Fig. 2. (a) A snapshot of the system of size 256×256. A part of size 100×100 at 30 time steps is presented. The solid lines show cracks. (b) A section perpendicular to (a) is presented. The size is 100 lattice size and 80 time steps. The density wave propagates from the top to the bottom in the figure.

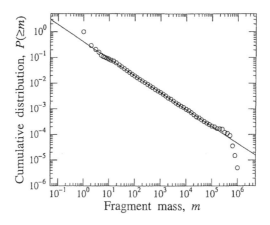

Fig. 3. The cumulative distribution of the fragment mass by the model in log–log scale. The solid line shows a power-law distribution $P(\geqslant m) \propto m^{-0.66}$.

In the model, the nature of the material is reflected only through the dilation field $\theta(x, y; z)$. By changing the randomness of the dilation field $\theta(x, y; z)$ from the uniform distribution to a Weibull distribution with the exponent 0.2, for example, the effect of the distribution of randomness on the fragment mass distribution is observed. In spite of large fluctuation of $\theta(x, y; z)$ compared to that of the model with uniform random distribution, no significant change is observed in the fragment mass distribution. It implies that the impact fracture process of a three-dimensional object-like spheres produces a cumulative power-law fragment mass distribution with exponent $\tau = \frac{2}{3}$ not depending on the detailed information of material of a fractured object.

We discuss the cause of the universality in terms of the competitive process between fragment sections. Let us consider the time evolution of a locally largest fragment section named S whose area is s. The total amount of dilation in S, Θ, at a certain moment $t = z$ is defined as

$$\Theta \equiv \sum_{S} \theta(x, y; z), \tag{4}$$

where the sum runs over the sites which belong to the fragment section S. Let us introduce a flux density field by gradient of the displacement potential $\nabla\phi(x, y, z)$, which is identical to the displacement field. In our model each site in a fragment section generates flux of magnitude $\theta(x, y; z)$, and it propagates toward the crack tips edging the fragment section. Thus, the flux field transmits the information about the size of the fragment section to the cracks, which, in turn, governs the competitive process among the fragment sections. Since the dilation field $\theta(x, y; z)$ is generated by a random number or an independent, stochastic variable θ, Θ is a stochastic variable which consists of the sum of $\theta(x, y; z)$ over s sites. From the central limit theorem, for large s, Θ follows a Gaussian distribution whose expectation is $\langle\theta\rangle \cdot s$ regardless of the distribution of θ. So, the function form of θ is not so important. Considering that the fluctuation of Θ increases with $s^{1/2}$ and is smaller compared to the expectation, we can say that Θ is roughly proportional to s when s is large.

The total dilation Θ in the fragment section S generates flux of amount Θ, which flows toward the crack tips edging S. The flux density near the crack tips surrounding S is roughly proportional to $s^{1/2}$ because Θ is proportional to s and the total length of the surrounding crack is roughly proportional to $s^{1/2}$. In the model, the cracks move by one site at a time step toward the place having smaller magnitude of the displacement. By this rule, the locally largest fragment section almost certainly expands its radius by one site at a time step because the magnitude of the displacement at the edge of the fragment section is a monotonically increasing function of s. Using a monotonically increasing function of the displacement $F(\nabla\phi)$, the dynamics of the cracks such as the cracks move by one site at a time step toward the place having a smaller value of $F(\nabla\phi)$ can be introduced. Even in this case, the situation that the locally largest fragment section can expand its radius is not changed. So the generalization of the model in this direction results in the same competitive dynamics of the fragments to the original model.

Thus, the exponent of the fragment mass distribution seems to have robust universality with respect to the randomness of the material and the mechanisms of the crack propagation.

3. Dependence on the shape

Meibom and Balslev reported that fragmentation of a plate-shaped specimen produced a power-law distribution with a flat tail whose transition point m_t was determined by the thickness of the specimen [6]. Such plate-shaped specimens can be simulated by changing the boundary condition from the periodic condition to the bounded condition where the boundaries of the whole system are treated as straight cracks. By performing simulations with various system sizes and time steps, we observe the fragment mass distribution of plate-shaped specimen by our model [12]. We measure the dimensions of the system in x, y, and z directions by the lattice unit and denote them by a, b, and c, respectively. Note that c is set to be the dimension in the direction parallel to the direction of the density wave propagation and it also means the simulation time steps. In the following discussion we assume $a \geqslant b$ without losing generality.

The simulation results are categorized in several cases. Those results are summarized in Fig. 4.

(1) The case $a > c > b$. The distribution shows a power law with a flat tail connected at transition mass m_t like the ones observed in experiments conducted by Meibom et al. The power-law exponent τ for smaller mass range is estimated to be close to the universal value $\frac{2}{3}$, while that of larger mass range looks very close to zero. The flat tail is followed by a cutoff at m_c. It is observed that m_t and m_c are approximately described by b^3 and $b^2 c$, respectively. These results are consistent with the experimental results [6].

(2) The case $c > a > b$. When c is not so large compared to a, the transition m_t and the cutoff m_c are described by b^3 and $b^2 c$, respectively. When c is very large, the position of the cutoff m_c is described by abc showing dependence on the dimension a. However, the volume abc means the total volume of the specimen meaning that the actual fragmentation process finishes leaving one big fragment piece. So, this case is not physically important.

(3) The case $a > b > c$. The flat tail of the fragment mass distribution vanishes. It follows a power law with a sudden cutoff. The exponent τ for the power-law part of the distribution is again close to $\tau = \frac{2}{3}$. The position of the cutoff is approximately described by c^3. The position is determined only by the dimension c and is not influenced by the aspect ratios of a and b to c as long as $a > b > c$ holds.

In the experiments reported so far, for example, the experiments conducted by Odder-shede et al. [5] or Meibom et al. [6], the impact onto a specimen is loaded by throwing the specimen onto a hard floor or by a hit with a hammer. So, the impact loading by those methods may not be well controlled. For example, when a plate-shaped specimen is thrown onto a floor, it is most likely that one of the edges of the specimen firstly

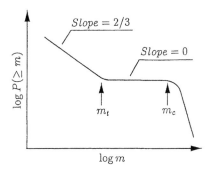

	m_t	m_c
$a > b > c$	—	c^3
$a > c > b$	b^3	b^2c
$c > a > b$	b^3	b^2c
$c \gg a > b$	b^3	abc

Fig. 4. A schematic figure of the positions of the crossover m_t and the cutoff m_c and a table of m_t and m_c by the model with various combination of the system dimensions.

touches the floor and a failure wave propagates from there. This case may be similar to the case (1). But it is ambiguous and the exact experimental condition is unknown. Our result suggests the importance of carefully controlled impact to clarify the behavior of the fragment mass distribution by experiments.

The behavior of the model can be explained intuitively as follows: According to the simulation results [11], the radius of the largest fragment sections in the model increases nearly proportionally to time steps when the competition is effective. So, the cracks making the perimeter of the largest fragment section collide with one of the system boundaries at time step around $t = \min(a, b, c)$, where $\min(i, j, k)$ represents the minimum value in i, j, and k. In the cases of $a > c > b$, the crack collides with the flank side first. Thereafter, most of the flux $\nabla \phi$ propagates towards the edges of the system and the competition among the fragment sections becomes ineffective. As a result, nearly all the fragments which are larger than m_t stop competition. This causes the flat part in the graph of the cumulative fragment mass distribution. Therefore, the position of the crossover m_t is approximately proportional to $\min(a, b, c)^3 = b^3$. The size of the fragments produced by the ineffective competition should be proportional to $c \times \min(a, b, c)^2 = b^2c$, which gives the position of the cutoff m_c. On the other hand, in the cases of $a > b > c$, the cracks edging the largest fragment are likely to reach the opposite side of the hit surface without colliding the flank sides. This means that the competition process is effective all the way to the end of the simulation and the crossover does not emerge. However, a cutoff appears around $c \times \min(a, b, c)^2 = c^3$ anyway since c is the shortest dimension in this case. This discussion is consistent with the simulation results.

4. Discussions

In this paper we discussed the numerical results by a three-dimensional model of impact fracture. One of the main characteristics of our fracture model is its competitive process among fragment sections during the process of crack propagation. Such a competitive process is generally important for a growth process to produce a fractal structure and a power-law size distribution. In fact, the growth process of diffusion-limited aggregation [14,15] and the river network formation by water erosion [20,21] are such competitive processes and produce both fractal structures and power-law distributions. One of the authors proposed a simplified model of the competitive process seen in the fractal growth processes [22]. The model reproduces a space-filling, fractal structure similar to a configuration of river basins and power-law size distribution whose exponent depends on a parameter in the competitive dynamics. Our study has shown that a similar competitive process is effective in the process of impact fragmentation, and it produces the universal, power-law fragment mass distribution.

It is obvious that our numerical model of impact fracture is a very much simplified version and we should prepare more realistic generalization to account for impact fracture phenomenon in the real world. A physical feature we miss in our model is the viewpoint based on energetics of crack propagation. For example if the energy input by the impact is not strong enough the cracks might stop in the middle of the specimen leaving a large un-fractured piece. On the contrary if the impact energy is too strong the fragment pieces may scatter with large velocities that may cause secondary impacts among the pieces. These effects may modify the mass distribution of fragment pieces, which are to be clarified in the near future.

Acknowledgements

This work is partly supported by the Japan Society for the Promotion of Science and Japanese Grant-in-Aid for Science Research from the Ministry of Education, Science, Sports and Culture.

References

[1] H. Takayasu, Prog. Theor. Phys. 74 (1985) 1343.
[2] F. Kun, H.J. Herrmann, Phys. Rev. E 59 (1999) 2623.
[3] Y. Hayakawa, Phys. Rev. B 53 (1996) 1428.
[4] M. Ausloos, Solid State Commun. 59 (1986) 401.
[5] J.F. Knott, Fundamentals of Fracture Mechanics, Butterworth, London, 1973.
[6] J.A. Zukas, T. Nicholas, H.F. Swift, L.B. Greszczuk, D.R. Curran, Impact Dynamics, Wiley, New York, 1982.
[7] T.Z. Blazynski (Ed.), Materials at High Strain Rates, Elsevier, London, 1987.
[8] J.J. Gilvarry, B.H. Bergstrom, J. Appl. Phys. 32 (1961) 400.
[9] L. Oddershede, P. Dimon, J. Bohr, Phys. Rev. Lett. 71 (1993) 3107.
[10] A. Meibom, I. Balslev, Phys. Rev. Lett. 76 (1996) 2492.

[11] H. Inaoka, H. Takayasu, Physica A 229 (1996) 5.
[12] H. Inaoka, E. Toyosawa, H. Takayasu, Phys. Rev. Lett. 78 (1997) 3455.
[13] T. Vicsek, Fractal Growth Phenomena, World Scientific, Singapore, 1989.
[14] T.A. Witten, L.M. Sander, Phys. Rev. Lett. 47 (1981) 1400.
[15] T.A. Witten, L.M. Sander, Phys. Rev. B 27 (1983) 5686.
[16] P. Meakin, Phys. Rev. B 30 (1984) 4207.
[17] S.J. Bless, N.S. Brar, G. Kanel, Z. Rosenberg, J. Am. Ceram. Soc. 75 (1992) 1002.
[18] A.A. Griffith, Philos. Trans. Roy. Soc., London, Ser. A 221 (1920) 163.
[19] J.J. Gilvarry, J. Appl. Phys. 32 (1961) 391.
[20] H. Takayasu, H. Inaoka, Phys. Rev. Lett. 68 (1992) 966.
[21] H. Inaoka, H. Takayasu, Phys. Rev. E 47 (1993) 899.
[22] H. Inaoka, Fractals 1 (1993) 977.

ELSEVIER

Physica A 274 (1999) 310–319

www.elsevier.com/locate/physa

Equations of granular materials

S.F. Edwards

Cavendish Laboratory, University of Cambridge, Madingley Road, Cambridge, CB3 0HE, UK

1. What can we do with a powder?

(1) Choose it (Fig. 1)
(2) Pour it

Surface statistics : $h(x, y; t)$,

$P(h(\mathbf{r}); t)$ or $P([h])$,

$$p(H) = \int P(h(\mathbf{r}))\delta(h(\mathbf{r}) - H)\mathscr{D}h(\mathbf{r}) .$$

(3) Shake it

ρ, structure internally .

'Thermodynamics' of powders .

Miscibility .

(4) Stress it
In this article, we study deposition, i.e. problem 2.

2. Deposition of granular materials

- Particles are deposited uniformly.
- Weak flux: no correlations between incoming particles.
- Particles settle gently: no cooperative reorganisation.

What are the surface fluctuations? Consider the *Edwards–Wilkinson* model:

$$\frac{\partial h}{\partial t} = v\nabla^2 h + \eta(r, t) ,$$

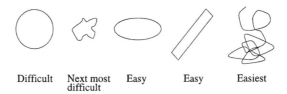

Difficult   Next most   Easy   Easy   Easiest
                difficult

Fig. 1. Possible geometries and their associated difficulties.

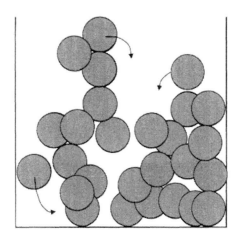

Fig. 2. Random particle falling.

where $v = \kappa a^4$, κ is an average rate of landings per unit time per unit area, and a is particle radius.

$$\langle \eta(\mathbf{r}, t) \eta(\mathbf{r}_0, t_0) \rangle = 2\delta(\mathbf{r} - \mathbf{r}_0)\delta(t - t_0) .$$

It will be found that

$$\langle h^2 \rangle = \frac{a^2}{4\pi} \ln \frac{L}{\pi a} ,$$

where L is the bin's size.

Particles fall at random in space and time. They can roll over (Fig. 2).

In the case of no overhang, we have $z = z(x, y, t)$. If the landings/unit time unit area $= \Omega$, then the average increase in height is Ωh.

$$\frac{\partial z}{\partial t} = \Omega h + \Omega a^4 \nabla^2 z + \xi ,$$

where a is the sphere radius and ξ a random variable.

$$\langle \xi(xyt)\xi(x'y't') \rangle \sim h^2 e^{-r^2/4a^2} \delta(t - t') ,$$

$$r^2 = (x - x')^2 + (y - y')^2 .$$

This is all very simple, and reasonably adequate.

Mathematical theorem:

if $\partial Z/\partial t = \xi$ and ξ is random, i.e.

$$\langle \xi(t)\xi(t') \rangle = h\delta(t - t') \,,$$

then the probability of finding $Z = z$ at t is

$$\frac{\partial P}{\partial t} - h\frac{\partial^2 P}{\partial z^2} = 0 \,.$$

More generally, if

$$\frac{\partial Z}{\partial t} + AZ = \xi \,,$$

$$\frac{\partial P}{\partial t} - \frac{\partial}{\partial z}\left(h\frac{\partial}{\partial z} + Az\right)P = 0 \,.$$

Finally, if Z is $Z(\mathbf{r}, t)$ and conditions are homogeneous, use Fourier variables:

$$Z_{\mathbf{k}}(t) = \int e^{2\pi i \mathbf{k} \cdot \mathbf{r}} Z(\mathbf{r}, t)\, d\mathbf{r} \,,$$

$$\frac{\partial P}{\partial t} - \sum_{\mathbf{k}} \frac{\partial}{\partial z_{\mathbf{k}}}\left(h_{\mathbf{k}}\frac{\partial}{\partial z_{-\mathbf{k}}} + A_{\mathbf{k}}z_{\mathbf{k}}\right)P = 0 \,.$$

In our problem, $h_{\mathbf{k}}$ is independent of $\mathbf{k}$, but it need not be.

Dynamics of fluctuations

$$(z(t + t') - z(t))^2 = \frac{h^2}{4\pi a^4}\log(1 + \Omega t' a^2) \,.$$

The time to add one layer is approximately $(\Omega a^2)^{-1}$, $t' \ll (\Omega a^2)^{-1}$

$$(z - z') = \frac{h^2\Omega t'}{4\pi a^2} \,,$$

i.e. z moves in a random walk.

The curiosity is that

$$\langle z(x, y, t)z(x + x', y + y', t + t')\rangle = \frac{h^2}{8\pi^2 a^4}\int \frac{d\mathbf{k}\, d\mathbf{j}}{k^2 + j^2}e^{ik(x-x') + ij(y-y')}$$
$$\times \exp[-(k^2 + j^2)a^2(1 + \mathbf{r}t'a^2)] \,.$$

This diverges unless cut-off at small $(\mathbf{k}, \mathbf{j})$, i.e. surface fluctuations depend on the size of the box, L:

$$\langle z^2 \rangle = \frac{h^2}{4\pi a^4}\log\left(\frac{L}{\pi a}\right) \,.$$

There is no escape from this, but it is difficult to even measure it.

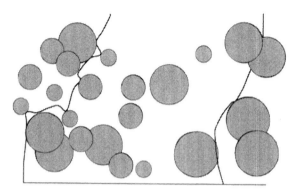

Fig. 3.

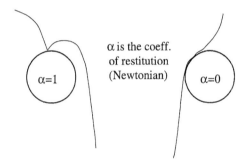

Fig. 4.

3. Deposition of mixed powders

Several different cases (Fig. 3)

Miscibility? Rods/spheres are extremes.

A particular limiting case: two spheres, one radius ≪ other.

Key quantities are *Péclet* numbers. If the diffusion tensor is D_{ij}, c the average velocity, and a the particle radius, then

$$\text{Radial Pec:} \quad \text{Pe}_r = 2ac/D_{11}, \tag{1}$$

$$\text{Axial Pec:} \quad \text{Pe}_a = 2ac/D_{33}. \tag{2}$$

Standard approach:

Write down a Boltzmann equation (Ehrenfest Woodwind) and solve, this involves much algebra. However, a surprisingly simple law emerges: the percolation velocity $\propto (1 - \alpha)^{1/4}$. Clearly, $\alpha = 1$ is of special interest (Figs. 4 and 5); we can derive an

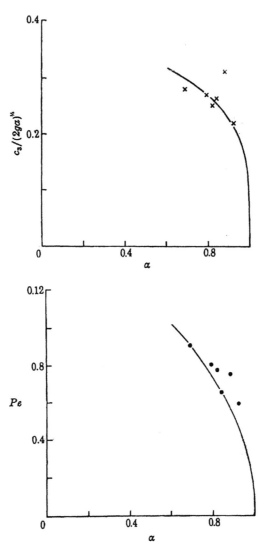

Fig. 5. Graphs showing the Percolation velocity (top) and Péclet numbers (bottom) against the coefficient of restitution, α.

equation for the average descent $\langle z \rangle$.

$$\frac{\partial^2 \langle z \rangle}{\partial t^2} = g - \frac{\partial}{\partial t} \langle z \rangle (2g\mu^2 z)^{1/2}$$

and for large times z is the same as $\langle z \rangle$

$$\langle z \rangle = \left(\frac{9}{8} \frac{g}{\mu^2} \right)^{1/3} t^{2/3}$$

(μ=constant related to packing fraction, i.e. the environment)

   $\mu = 0$ when they cannot get through.

Note comparison with diffusion along a random walk

$$\bar{R}^2 \sim S, \quad \bar{S}^2 \sim t,$$

$$R^2 \sim t^{1/2}, \quad \sqrt{R^2} \sim t^{1/4}.$$

4. A list of $\langle z \rangle$ as a function of t

- Free fall: $\frac{1}{2}gt^2$.
- Driven woodwind: t.
- Driven elastic woodwind: $t^{2/3}$.
- Undriven Einstein: $t^{1/2}$.
- Undriven Einstein, in a random tube: $t^{1/4}$.

(There are many more complex problems solved by Sinai, but only in one dimension $(\log t)^{1/4}$.)

5. A class of non-linear Stochastic equations

Can be vector, e.g. velocity $u_i(\mathbf{r}, t)$ or height of powder $h(x, y; t)$, etc.

$$\frac{\partial h}{\partial t} + v\nabla^2 h + Mhh = \eta,$$

e.g. *KPZ* is (physically one or two) two dimensions

$$\frac{\partial h}{\partial t} + v\nabla^2 h + (\nabla h)^2 = \eta(x, y, t),$$

$$\langle \eta(x, y, t) | \eta(x', y', t') \rangle = D_0(x - x', y - y', t - t'),$$
$$\sim D_0 \delta(x - x')\delta(y - y')\delta(t - t')$$

or

$$\frac{\partial h_k}{\partial t} + v(k)h_k + \sum_{l,j} ljh_l jh_j \delta(k + j + l) = \eta_k.$$

Hydrodynamics:

$$\frac{\partial u_k^i}{\partial t} + vk^2 u_k^i + \sum_{l,m,j,k} M_{kjl}^{ilm} u_k^l u_j^m = \eta_k^i,$$

$$M = k^i O_k^{lm} \delta(k + j + l),$$

where O is the Oseen tensor.

Navier–Stokes:

$$\frac{\partial \mathbf{u}}{\partial t} - \nu\nabla^2\mathbf{u} + (\mathbf{u}\cdot\nabla)\mathbf{u} - \nabla p = \eta\,,$$

$$\nabla\cdot\mathbf{u} = 0 \quad \text{fixes } p\,,$$

where the ∇^2 above is a three-dimensional representation.

KPZ:

$$\frac{\partial h}{\partial t} - \nu\nabla^2 h + (\nabla h)^2 = \eta\,,$$

where $\nabla^2 = \partial^2/\partial x^2 + \partial^2/\partial y^2$ in the *KPZ* equation above. (Also studied in 1D where *NS* is called Burgers equation, and *KPZ* is soluble as a statistical problem.)

Expect:

$$\langle u_\mathbf{k}(t)u_{-\mathbf{k}}(0)\rangle = \phi_\mathbf{k} F(kt^\beta)\,, \tag{3}$$

$$\langle h_\mathbf{k}(t)h_{-\mathbf{k}}(0)\rangle = \phi_\mathbf{k} F(kt^\gamma)\,. \tag{4}$$

Dimensional analysis gives $\beta = \frac{3}{2}$, $\gamma \simeq \frac{2}{3}$.

From

$$\frac{\partial h}{\partial t} + vh + Mhh = \eta\,, \tag{5}$$

we can derive:

(1) an equation for $P(h;t)$

$$\frac{\partial P}{\partial t} + \sum \frac{\partial}{\partial h}(-vh - Mhh + \eta)P = 0\,, \tag{6}$$

$\langle P\rangle_\eta$ satisfies (write P for $\langle P\rangle$ from now on)

$$\frac{\partial P}{\partial t} - \sum_\mathbf{k}\left(\frac{\partial^2}{\partial h^2}D_0 + \frac{\partial}{\partial h}\left(vh + \sum_{j,l}Mhh\right)\right)P = 0 \tag{7}$$

or

(2) an equation for $P([h])$ the probability of whole history

$$(\dot{h} + vh + Mhh - \eta)P = 0\,. \tag{8}$$

Consider the first: If Mhh were a random variable, it would simply enhance D_0, i.e.

$$\frac{\partial^2}{\partial h^2}D_0 \to \frac{\partial^2}{\partial h_\mathbf{k}\partial h_{-\mathbf{k}}}(D_0 + O(Mhh)^2)\,. \tag{9}$$

Following Peierls [1], we can derive a Boltzmann equation for $\phi = \langle h_\mathbf{k}h_{-\mathbf{k}}\rangle$ (for Peierls, ϕ is the number of phonons, for Navier–Stokes, $\phi = \langle u_\mathbf{k}u_{-\mathbf{k}}\rangle$, which is the energy in mode $\mathbf{k}$).

If $\omega_\mathbf{k}$ is the (approximate) lifetime of mode $\mathbf{k}$, the Boltzmann equation is, for steady state,

$$v\phi + \int\frac{MM\phi\phi}{\sum\omega} - \int\frac{MM\phi\phi}{\sum\omega} = D_0 \tag{10}$$

and for the time-dependent equation

$$\frac{\partial \Phi}{\partial t} + v\Phi + \int \frac{MM\phi\Phi}{\sum \omega} - \int \frac{MM\Phi\Phi}{\sum \omega} = 0 \tag{11}$$

with the boundary condition:

$$\left.\frac{\partial \Phi}{\partial t}\right|_{t=0} = D_0 \,. \tag{12}$$

6. Self-consistent expansion

$$\frac{\partial P}{\partial t} + \sum_q \frac{\partial}{\partial h_{-q}} \left(D_{0q} \frac{\partial}{\partial h_q} + v_q h_{-q} + \sum_{l,m} M_{qlm} h_l h_m \right) P = 0 \,, \tag{13}$$

$$\frac{\partial P}{\partial t} + \sum_q \frac{\partial}{\partial h_{-q}} \left\{ (D_{0q} + D_{1q}) \frac{\partial}{\partial h_q} + v_q h_{-q} \right\} P = 0 \,, \tag{14}$$

$$\frac{\partial P}{\partial t} + \sum_q \frac{\partial}{\partial h_{-q}} \left(D_q \frac{\partial}{\partial h_q} + \omega_q h_{-q} \right) P = L_0 P = 0 \,. \tag{15}$$

7. Expansion of correlation functions

$$v_q\phi_q - D_{0q} = \int d^d j d^d l \frac{M_{qjl} M_{qjl} \phi_j \phi_l}{\omega_q + \omega_j + \omega_l} \delta(q + j + l)$$

$$+ \int d^d j d^d l \frac{M_{qjl} M_{jql} \phi_q \phi_l}{\omega_q + \omega_j + \omega_l} \delta(q + j + l)$$

$$+ \int d^d j d^d l \frac{M_{qjl} M_{ljq} \phi_q \phi_j}{\omega_q + \omega_j + \omega_l} \delta(q + j + l) \,. \tag{16}$$

$$v_q - \omega_q = 2 \sum_{l,m} M_{qlm} \frac{M_{lmq} \phi_m + M_{mlq} \phi_l}{\omega_l + \omega_m} \left[\frac{\omega_l + \omega_m - \omega_k}{\omega_l + \omega_m + \omega_k} \right] \tag{17}$$

ω_k is a fit to $\Phi \sim \phi_k e^{-\omega_k t}$, i.e.

$$\omega_k^{-1} = \int_0^\infty \Phi(t) \, dt / \Phi_k(0) \tag{18}$$

(or alternative definitions).

8. Asymptotic power-law solution

$$D_0 - vq^2\phi_q - I_1(q) + I_2(q) = 0 , \tag{19}$$

$$\omega_q - vq^2 - J(q) = 0 . \tag{20}$$

For $|q| < q_0$:

$$\phi_q = Aq^{-\Gamma}, \quad I_1(\Gamma) = I_2(\Gamma) \tag{21}$$

for *KPZ*, $\Gamma \sim 2.6$. For *NS*, $\Gamma = \frac{11}{3}$ (Kolmogoroff).

$$\omega_q = Bq^\mu , \tag{22}$$

where $\mu = \frac{2}{3}$ (Kolmogoroff).

9. Dynamical equations via probability of history

(Eq. motion) $P = 0 .$

After algebra, we find

$$\Phi(t) = \langle h(t)h(0)\rangle \tag{23}$$

to satisfy

$$\frac{\partial\Phi}{\partial t} + v\Phi + \int K_1\Phi\phi - \int K_2\Phi\Phi = 0 , \tag{24}$$

$$\left.\frac{\partial\Phi}{\partial t}\right|_{t=0} = D_0 \tag{25}$$

which at $t = 0$ gives previous

$$v\phi + \int K_1\phi\phi - \int K_2\phi\phi = D_0 \tag{26}$$

which has solution $e^{-kt^{(1/\mu)}}$.

For the Navier–Stokes compressed expression, $e^{-kt^{3/2}}$. Crudely speaking,

$$v\phi + \int K\phi = s , \tag{27}$$

either has solution to

$$\int K\phi = 0 \quad (KPZ) \tag{28}$$

and that ϕ is modified for v and s, *or* there is no solution to

$$\int K\phi = 0 \quad (KPZ) , \tag{29}$$

but K has an inverse, so

$$\phi = K^{-1}s \tag{30}$$

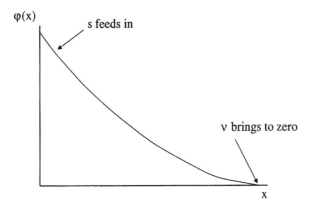

Fig. 6.

and is modified at large K to cover v. For Navier–Stokes, this is the case, but input near $k=0$ (Kolmogoroff) is pathological. Another problem is point probability, $P(h(r)=H)$ is *not* Gaussian. $P = e^{-H^\gamma c}$; fluids: $\langle (u(r) - u(0))^2 \rangle \sim r^{2/3}$ but $P(u(r) - u(0) = U) \neq e^{-U^2/r^{2/3}}$.

10. Model of the Boltzmann equation

$$v(x)\phi(x) + \left[\lambda \int_0^\infty \frac{\phi(y)}{x+y}\, dy - \phi(x) \right] = S(x), \tag{31}$$

$$\int_0^\infty \frac{y^{-\alpha}\, dy}{x+y} = \frac{\pi}{\sin \pi\alpha} x^{-\alpha} \quad (0 < \alpha < 1) \tag{32}$$

which suggests a Stieljes transform (Double Laplace)

$$\phi(y) = \int_0^1 F(\beta)x^{-\beta}\, d\beta. \tag{33}$$

If $v(x) = x^\gamma$ then

$$F(\beta - \gamma) + \frac{\lambda\pi}{\sin \pi\beta}F(\beta) - F(\beta) = S(\beta), \tag{34}$$

where the first term in the above equation is handled by the boundary condition, but the rest is non-trivial.

Expect the solution to look like as shown in Fig. 6.

References

[1] R.E. Peierls, Quantum Theory of Solids, Clarendon Press, Oxford, 1955.

ELSEVIER

Physica A 274 (1999) 320–324

www.elsevier.com/locate/physa

Phase separation in a weak first-order phase transition

H. Arkin[a], T. Çelik[a,*], B.A. Berg[b], H. Meyer-Ortmanns[c]

[a]*Department of Physics, Hacettepe University, 06532 Beytepe, Ankara, Turkey*
[b]*Physics Dept., Florida State University, Tallahassee, FL, USA*
[c]*Fachbereich Physik, Universität Wuppertal, Germany*

Abstract

Recent results on the phase transition in scalar field theory indicate that the strength of the transition is not a minor quantitative detail of the transition, but of considerable impact concerning the phase conversion mechanism. To see which phase conversion mechanism takes place depending on the strength of transition, we have simulated the $q = 5$ state Potts model in two dimensions under an external magnetic field. Our results indicate that the phase conversion mechanism changes from nucleation-to-spinodal decomposition as the strength of the transition is weakened by increasing the external field. © 1999 Elsevier Science B.V. All rights reserved.

1. Introduction

In condensed matter systems with first-order phase transitions, a quenched cooling from a high to a low-temperature phase is expected to proceed via nucleation or spinodal decomposition. The size of the nucleation region should increase with the strength of the order of transition. Vice versa, for a very weak first-order transition, it is more likely to end up in the spinodal region.

If the energy barrier is rather small as in very weak first-order phase transitions, at the critical temperature T_c, the system will be in a mixture state. Then the question as to whether a first-order phase transition is "strong" or "very weak" boils down to how well localized is the metastable state at T_c when the system is quenched to a temperature below T_c [1]. We are interested in the way in which the quenched system approaches equilibrium. The detailed dynamics will depend, besides on the amount of quench, on the relative fraction of the total volume occupied by each phase.

---

* Corresponding author.
E-mail address: celik@thep1.phys.hun.edu.tr (T. Çelik)

0378-4371/99/$ - see front matter © 1999 Elsevier Science B.V. All rights reserved.
PII: S0378-4371(99)00306-4

To study the question of phase conversion as a function of the strength of the first-order transition and in an off-equilibrium cooling process, it is sufficient to consider even simpler spin models in two dimensions which allow for easy graphical presentation of the phase regions. We choose the 5-state Potts model in two dimension which is known to possess a first-order phase transition for vanishing external magnetic field, quite similar as the 3-state Potts model in three dimensions. The transition gets weakened in an external field h. Thus we can tune the strength of the transition via h until the transition line ends in a critical endpoint for some critical magnetic field strength h_c.

2. The model

We have simulated a two-dimensional $q=5$ state Potts model under external magnetic field using a heath bath algorithm. The Hamiltonian of the two-dimensional Potts model is given by [2]

$$-\beta H = K \sum_{\langle i,j \rangle} \delta_{\sigma_i,\sigma_j} + h \sum_i \delta_{\sigma_i,1} .$$

Here $\beta = 1/kT$, $K = J/kT$, k and T are the Boltzmann constant and the temperature, respectively, and J is the magnetic interaction between spins σ_i and σ_j, which can take values $1, 2, \ldots, q$ for the q-state Potts model and $h = H/kT$ with h the external field along the orientation 1. The two-dimensional Potts model undergoes a second-order phase transition for 2, 3 and 4 states, and a first-order phase transition for higher number of states.

3. Results and discussions

The first-order phase transition of the 5-state Potts model is further weakened by increasing the external field. For our purpose of studying the phase mixing and phase conversion, the two-dimensional 5-state Potts model with an external field strength of $h = 0.005$ seems to be a good candidate with a very weak first-order phase transition.

Here, we will refer to the phase consisting of the spins of a chosen orientation (the orientation of the external field) as the "ordered" phase and the rest of the volume as the "disordered" phase.

First, we have done thermalization runs with an external field of $h = 0.005$ and 0 on 256×256 square lattices with a heath-bath algorithm. In Fig. 1 we show how the volume fraction of the total number of spins oriented along the field increases with time (the lower curve) after a totally disordered start, which shows a behavior like the time evolution of magnetization per spin. The upper curve is the volume fraction of the rest of the spins oriented along the other directions. It is seen that at the end of 5000 iterations after a totally disordered start, half of the spins are oriented along the

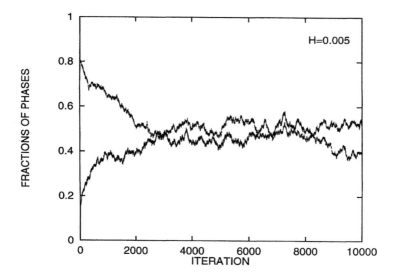

Fig. 1. Time evolutions of the volume fractions of the ordered (the increasing curve) and the disordered (the decreasing one) phase during thermalization runs after a totally disordered start.

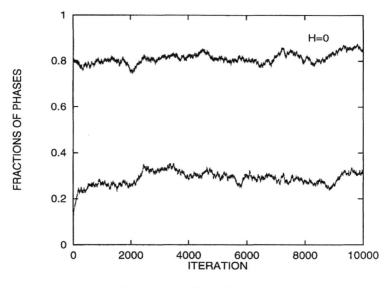

Fig. 2. Same as Fig. 1 for $h = 0$.

field (ordered phase) whereas the other half is filled with a mixture of spins oriented along the other four directions making the disordered phase.

Fig. 2 is the same as Fig. 1 for $h=0$, obviously one does not see any phase mixing at all. A disordered start survives for very long times on a lattice which is chosen big enough not to have spurious mixing arising from the finite size of the system.

Next, we take the system for $h = 0.005$ and 0, both thermalized at corresponding T_c where the specific heat has a peak, for 5000 iterations and quench them to a lower temperature by $\Delta K = 0.5$. The system with $h = 0.005$ starts from a totally mixed phase at T_c when it is quenched, while the pure 5-state Potts model spin system with no external field starts from a well localized disordered state. They both are quenched by same amount of temperature. We then have analyzed their approach to equilibrium, diagnosed the evolution of clusters and displayed their snapshots of the time evolution after the quench in Fig. 3(a) for $h = 0$ and in Fig. 3(b) for $h = 0.005$. The dark color indicates the ordered clusters of the same orientation and the white background is the disordered phase consisting of spins of other orientations.

From the snapshots given in Fig. 3(a), the increase of the ordered regions is such that the smaller clusters disappear while the larger ones grow and coalesce together to build larger domains. As pointed out in the introduction, when the system is quenched from T_c with a quench that crosses and ends in the nucleation region, the time evolution towards equilibrium proceeds via nucleation to the extent that homogeneous nucleation theory is applicable. Pictures for the quenched spin system with a very weak first-order phase transition, displayed in Fig. 3(b), do not exhibit any sign of nucleation, but are different. They show a rather peculiar dynamics indicating spinodal decomposition. A quench in temperature most likely ends up in the spinodal region, the ordered and disordered phases seem to diffuse into each other. The two phases compete, the finger-like structures grow thicker as the system relaxes and eventually the ordered phase takes over. The apparent dynamics here is a large domain coarsening and is much faster than the relatively slow nucleation dynamics delayed by supercooling until bubbles grow around nucleation centers that are created by fluctuations in the metastable region.

In order to estimate the late stage time dependence of the growth of ordered domains in case of the spinodal decomposition like the one depicted in Fig. 3(b) by snapshots, we made fits to the domain size $R(t)$ calculated from the maximum cluster volume [3] and to the radius of gyration of the maximum size cluster data from ten different simulations on 256×256 lattices. The three parameter fits to the data give the power law behavior $R(t) \sim t^{0.51 \pm 0.01}$ for maximum cluster volume data and $\sim t^{0.50 \pm 0.05}$ for the radius of gyration data, which are in good agreement with the expected value 0.5 for spinodal decomposition in systems with a nonconserved order parameter [4] which applies to our simulation.

To summarize our results: For a strong first-order transition nucleation seems to be the preferred way of phase conversion after a quench, for a weak transition it is spinodal decomposition. It would be interesting to see whether spinodal decomposition remains the preferred mechanism as a result of a quenched cooling process even if the external field h exceeds the critical value h_c so that the spin model has a crossover phenomenon with analytic behavior of the free energy everywhere in phase space rather than a true phase transition.

Acknowledgements

This project is partially supported by TÜBİTAK through project TBAG-1699.

References

[1] P.M. Chaikin, T.C. Lubensky, Principles of Condensed Matter Physics, Cambridge University Press, Cambridge, 1995.
[2] F.Y. Wu, Rev. Mod. Phys. 54 (1982) 235.
[3] D. Chowdhury, D. Stauffer, Z. Phys. B. 60 (1985) 249.
[4] A. Sadiq, K. Binder, J. Stat. Phys. 35 (1984) 517.

ELSEVIER

Physica A 274 (1999) 325–332

www.elsevier.com/locate/physa

Fractal-type relations and extensions suitable for systems of evolving polycrystalline microstructures

A. Gadomski *

Department of Theoretical Physics, Institute of Mathematics and Physics, University of Technology and Agriculture, Al. Kaliskiego 7/402, PL-85796 Bydgoszcz, Poland

Abstract

We report on two possible physically justified situations concerning the evolution of polycrystalline microstructures. The first is an observation that a polycrystalline microstructure formation process, considered as a random walk in the space of the crystallites' sizes, can be equivalent to the anomalous kinetic problem in a continuum percolation space. The second, in turn, regards some extension of the random walk process while occurring on a fractal substrate and/or when a single grain boundary can presumably be treated as a random walk trajectory, i.e. a situation quite acceptable when dealing with, e.g. quasicrystals. One may find possible applications of the modeling offered in the areas of biophysics and physical metallurgy. ⓒ 1999 Elsevier Science B.V. All rights reserved.

PACS: 64.60.-i; 05.70.Fh

Keywords: Polycrystalline structure formation kinetics; Diffusion

1. Introduction

Since the appearance of pioneering works of Fortuin and Kasteleyn [1–4], followed thoroughly by more recent studies [5,6], it is accepted that the Ising and percolation systems have very much in common. Moreover, this statement may even be extended when dealing with others, say, Ising-like systems, e.g. the Potts models, or its practical realization in biophysics, named the Pink model of clustering [7–9].

In this short paper, we are going to show that two kinetic processes, being apparently different at a first look, i.e. the formation of polycrystalline microstructures

---

* Fax: +48-52-3408643.
E-mail address: agad@tower.atr.bydgoszcz.pl (A. Gadomski)

0378-4371/99/$ - see front matter ⓒ 1999 Elsevier Science B.V. All rights reserved.
PII: S 0378-4371(99)00310-6

(represented by an inhomogeneous diffusion type model in the space of crystallites sizes [10–16]) and a random walk realized in a continuum (Swiss cheese) percolation matrix near criticality (a position space), seem to be practically equivalent, because of possessing very similar kinetic characteristics. Thus, in Section 2, a presentation of the reasoning, supporting the above stated observation, is given. In Section 3, in turn, some extension of scaling relations characterizing the same process of polycrystaline microstructure formation, quite important in physical metallurgy is presented, in terms of its non-Euclidean geometrical characteristics. Section 4 contains closing remarks.

2. On some equivalence between the formation of polycrystals and a kinetic random walk on the continuum Swiss cheese-like percolation matrix

In a series of papers (see, [10–13] and references therein), we regarded the formation of grains containing (i.e., space filling) systems in a d-dimensional space, where an analytical model of (inhomogeneous) diffusion was used and where the process was grain boundary curvature driven, rather, than with an expected ability to obey the 1st Fick's law. (Therefore, we called sometimes the process to be of *explicitly* non-Fickian character, abbreviated here as the NF-process.) In another study (see, [14–16] and references therein) doing the same, but for some, e.g. recrystallizing assemblies (which do not obey the condition of space filling), we used a diffusion-type model, with an explict requirement of the 1st Fick's law to be fulfilled, contrary to the NF-case mentioned [10–13]. It explored a well-known continuity equation of the form

$$\frac{\partial}{\partial t} f(v,t) = -\frac{\partial}{\partial v} J(v,t), \quad v \in [0,\infty),$$
(1)

where $f(v,t)$ is a distribution of crystallites of volume v (at time t) in the system of total hypervolume $V^{(d)}$, and the local flux $J(v,t)$ is straightforwardly obtainable from the 1st Fick's law. It was asssumd that the diffusion function involved in the flux is a power function of v, cf. [10–16], though an inverse power dependence upon t, like $\sim$ $(const.+t)^{-h}$ (h – a dimensionless competition [17,18] dynamic parameter; cf. [10–13] for details) was considered as well. (We will, however, take $h=0$ in our present study [14–16], which means, that the growth process passes smoothly and within a moderate range of temperature values through a set of thermodynamic quasiequilibria; see [19 –21], too. We called sometimes that process [14–16] to be of purely Fickian character, abbreviated here as the F-process.) Surprisingly, solving both the above introduced systems for the same initial (delta-Dirac) and θ-Dirichlet boundary conditions (it is usually a standard procedure of evaluating the first statistical moments of $f(v,t)$, but one may look into [10–16] for necessary details), one gets the same asymptotic result for the average radius of the crystallite, r_{av}, namely (for $t \gg 1$) [10–16]

$$r_{av}(t) \sim t^{1/d+1},$$
(2)

where d stands for the dimension of the space ($d = 1, 2, 3, \ldots$), irrespective of whether the system is space filling (grain boundary curvature driven) or not. (Note that the standard diffusional scaling form is recovered exclusively when $d = 1$.) It was, however, not a physical paradox, since we knew that the systems of first type (NF-systems) do conserve their $V^{(d)}$ (e.g., in the normal grain–growth or bubbles containing processes [22,13]) while the other ones (F-systems), like certain gelling systems containing microcrystalline regions, do not; see examples listed in [14–16].

Another observation that comes out from these studies would be that the quantity r_{av} can be interpreted as an averaged distance travelled by the boundary of an "averaged" grain from its center of mass (inertia). The process of grain boundary wandering in a random fashion is thought to be realized in the environment of other (neigboring, etc.) grains that stand for randomly distributed obstacles perturbing the walk of the boundary. This scenario resembles very much a random walk of a tracer on the percolation Swiss cheese-like continuum space, both for $d = 2, 3$.

It can be learned from classical literature sources [23–25] that for $t \gg 1$

$$r_{rw}(t) \sim t^{k_p}, \tag{3}$$

where $k_p = v/2v + \mu - \beta$, and v, μ, β represent the standard critical exponents of the percolation system at criticality, when the incipient infinite cluster (IIC) is formed for the first time. If the cluster is finite, but really the largest taken from the population of finite clusters (a FL-cluster), the numerator in k_p (see also the right-hand side of (3) and specification of k_p thereafter), must be replaced by $v - (\beta/2)$. Certainly, r_{rw} stands for the mean distance travelled by the walker on the percolation space [23–25].

It can be anticipated as another evidence that the polycrystalline and percolation systems are dynamically equivalent. It is indeed so, since asymptotic relations (2) and (3) describe almost the same temporal behavior, since the exponent $k_p \approx 1/d + 1$. It is summarized by the following

Polycrystal NF, F	Continuum percolation IIC/FL	
$2d$: $r_{av} \sim t^{1/3}$	$r_{rw} \sim t^{0.3302}$	$r_{rw} \sim t^{0.3483}$
$3d$: $r_{av} \sim t^{1/4}$	$r_{rw} \sim t^{0.2414}$	$r_{rw} \sim t^{0.3103}$

where all the symbols used can be found in the text written in this section, and the data for continuum percolation have been taken from [23–25]. Moreover, note that for an FL-object, which mimics a space available for a random walk realization of a test particle, the listed exponents differ more markedly from the values of $\frac{1}{3}$ or $\frac{1}{4}$, for a $2d$ or $3d$ case, respectively. It is simply due to the finiteness of the system under consideration.

3. An analysis of the kinetics of stochastic grain growth: some extension

A couple of years ago an interesting study on the grain growth [26–29] kinetics was published [22], in which both, a crucial role of deterministic as well as stochastic character of the process was emphasized. In the former, the most explored mechanism proposed for the driving force appears to be a capillary force, connected *via* the Gibbs–Thompson relationship, with a magnitude of the grain boundary curvature [10–13,30,31]. In the latter, in turn, a random walk (RW) concept is usually offered [19], which directly leads to the presumption that there exists a certain analogy between the RW (typically, approximated by the standard diffusion equation for a joint probability of finding a "walker" at a certain time-position point) of a particle in real space and the motion of grains in the space of their sizes [14–16,30,31].

The deterministic mechanism for the growing process mentioned always predicts

$$R(t) \sim t^{1/2} , \tag{4}$$

for the long times limit, which is also very characteristic of the so-called (normal) diffusion-controlled processes of any kind. Note that $R(t)$ is, very often, referred to as the linear grain size (or radius) taken at time t.

The stochastic character of the process was considered quite naturally by the authors of [22]. Namely, they really performed a RW realization for some stochastically changing ensemble of grains constituting a three-dimensional polycrystal, and they found out that the mean increase Δv in the volume of a certain number of growing crystallites reads [22]

$$\Delta v \propto t^{1/2} , \tag{5}$$

i.e., qualitatively the same relation as that written above as (4) has been obtained. There appeared, however, an important qualitative difference, since volume elements are simply related, by a cubic relationship (but, in the Euclidean space, exclusively), to the linear characteristics such as e.g., the mean radius r_{av} of the crystallite. Thus, being in close agreement with this elementary geometrical knowledge and utilizing (5), one is able to provide the asymptotic scaling formula valid for the "stochastic side" of the process under study, namely

$$r_{av}(t) \sim t^{1/6} , \tag{6}$$

which was also estimated by Chen and confirmed specifically by Monte Carlo simulations of Potts-like systems [9], mostly in some correspondence to metallurgical phenomena manifesting strong physi- and chemisorption as well as other retardation effects [10–13,26–31]. Another example may invoke polymeric quasicrystalline assemblies, in which the macromolecular adsorption is very much suspected to play a crucial role in inhibiting the growth. Thus, some experimental biophysical microdomain systems composed of multilamellar lipid vesicles, characterized by small fractional dimensionalities, can probably stand for well-chosen candidates, too [32–34].

Let us point out two useful but rather rarely mentioned extensions of the approach proposed above (and thoroughly presented in [22]). Both of them concern with some 'refinement' of its stochastic part.

Thus, the first extension that can be taken into account is just a simple geometrical observation that it would be useful to consider not only Euclidean objects (crystallites) but also some non-Euclidean, say fractal (see [26–29]), quasicrystalline domains, or simply the evolution of domain systems in a fractal space of dimension d_f; see, also [10–13]. It immediately enforces to replace (6) by a more realistic asymptotic scaling rule

$$r_{av}(t) \sim t^{1/2d_f}, \tag{7}$$

where $0 \leqslant d_f \leqslant d$ (d_f is a non-integer), so that, e.g. the limit of the Euclidean dimension $d = 3$ is readily available, and for such (7) must be fully identified with (6).

The second extension towards our "refinement procedure" is probably more subtle [19]. Namely, it makes use of a more established knowledge on the RW process. That knowledge teaches us, in general, that there are three types of RW realizations, performed with the fractal dimension, d_w, of a random path along which the walker travels across the available space [23–25]. When the walk is realized in a certain accelerated way, $d_w < 2$, so as it would be systematically "kicked" (say, enforced to be in hurry) by the physical environment, in which the action takes place. Such a walk is said to be superdiffusive or just a "busy" RW. If, in turn, the walk is steadily caused to rest (e.g., by a chemical reaction), and its realization is anomalously slow, the walk, with $d_w > 2$, is called to be subdiffusive or just a "lazy" RW. (The walk "in between", considered in [22], is typically named the normal or "neutral" RW).

In this context, Eq. (5) is to be modified as

$$\Delta v \propto t^{1/d_w}, \tag{8}$$

so that (7) has to be rewritten as

$$r_{av}(t) \sim t^{1/d_w d_f} \tag{9}$$

which for the standard case ($d_w = 2$) coincides very well with (7), and for $d_f = d = 3$ also with (6).

One can recall many physical grounds for which (9) is a useful approximation. They are juxtaposed e.g., in [26–29], i.e. in a metallurgical context. Let us mention here, for example, Zener (due to pinning) or Mullins (due to thermal grooves) drag forces providing some rationale for introducing RW with $d_w > 2$, because the drag forces are serious kinetic obstacles for the polycrystalline growth. (Also, by noticing that there exists, in the limit of $q \to 0$, a possibility of mapping the q-state Potts model [5–9], i.e. a very useful tool for reproducing by means of the Monte Carlo simulation technique polycrystalline microstructures [10–13], onto a percolation microstructure [1–6,23–25] at criticality, one may hope for a proper use of (8) at least for having $d_w = 2 + \varepsilon$, i.e., $d_w > 2$, where $\varepsilon > 0$, and typically of fractional value, can be completely determined by the critical percolation exponents [23–25,32–35].) Or, the presence of high-angle grain boundaries, or even small grains (fine-grained structure), both with large curvatures, which in turn ensures a vigorous realization of the growing process.

As was proposed by Furukawa and others [14–16], who studied systems of droplets (clusters) in quenched but unstable states, one may eventually carry the whole analysis to get some interpolating scaling formula (see also [10–13]), which is now not

$r_{av}(t) \sim t^{1/3}$ [8], i.e., in a Lifshitz–Slyozov (LS) form, mostly characteristic of the Ostwald ripening [14] but, by combining (4) and (8), in the long times limit, one provides

$$r_{av}(t) \sim t^{\frac{(1/2)+(1/d_w d_f)}{2}} \tag{10}$$

which for $d_w = 2$ (standard RW) and $d_f = d = 3$ recovers just the LS form.

Let us note that result (10) is readily capable of reproducing a huge spectrum of the temporal behaviours characteristic of the normal as well as a-normal (supernormal or abnormal) grain growth as well as recrystallization processes in pure metals (Al) or alloy systems ($\alpha -$ Fe or $\gamma -$ Fe, for example) [26–29]. Invoking the long-living analogy between metallurgical and biophysical [10–13,32–35] systems, it can also serve to elucidate some asymptotic temporal behaviour in micelles or vesicles; cf. [19–21], and references therein. To summarize the efforts of this section, let us state explicitly that we have proposed a self-consistent and simple picture of the polycrystalline structure formation, realized very much in a spirit of earlier deterministic-stochastic studies [22,30,31], but also extended naturally to a variety of anomalous (random walk) and fractality-dependent cases of its realizations [19–21,10,12]. The extensions proposed may serve to elucidate many anomalous kinetic data, frequently reported in the literature [10–13,32–35].

4. Closing remarks

The presented study exemplifies another way of how to support a formal analogy between the polycrystalline as well as percolation systems. In Section 2, it was done by noticing that a kinetic (percolation) exponent, k_p, is of a value being quite comparable to $1/d + 1$ (d – dimension of the space), and some additional possibilities of how to approach this value more closely have also been proposed in [10–13] (h $\neq 0$). A qualitative rationale that stays behind says that a walk of a random tracer on the continuum percolation matrix can be thought of to be, on an average, (nearly) equivalent to a random motion of the boundary of a crystallite. In Section 3, in turn, certain possibilities of extending well-known kinetic characteristics of a polycrystalline (in particular, grains containing) system to non-Euclidean, let us say fractal cases, have been pointed out, mostly based on [10–13,22,30–35]. In particular, note that when looking at the exponent in (10), one might just redefine the fractal dimension of a "grain walk" in the space of grain (crystallite) sizes as

$$d_w^g = d d_w , \tag{11}$$

where d_w^g could presumably be understood as the RW dimension of a path of a randomly walking particle, of fractal dimension d_w, when the particle jumps over the grain boundaries, assuming for the moment that the grain boundary network were artificially frozen, but with a general rule (restriction) superimposed, namely, that the jumps over low-curvature boundaries are energetically favourable. Although the concepts presented here seem to be still waiting for further more formal exploration, it is worth noticing that contrary to the material presented mostly in [1–6], the analogy between polycrystals and (weighted) percolation is explored here on a dynamic level, where certain asymptotic temporal characteristics (Eqs. (2)–(10)) have to be discussed. (For a recent minireview on the percolation as a phase transition problem one is strongly encouraged to consult [35].)

Acknowledgements

A financial support of the Rector of the University of Technology and Agriculture is acknowledged. Special thanks go to M. Ausloos for activating ideas of the end-of-millennium workshop in applications of statistical physics, and to H.E. Stanley for mentioning, during his talk on the final outlook and perspectives of the NATO ARW, May 19–23, 1999, Budapest, pioneering works of Kasteleyn on Ising-type systems. Last but not the least, the author wishes to express his sincere gratitude to his family, especially to Ewa and Magda.

References

[1] C.M. Fortuin, P.W. Kasteleyn, Physica 57 (1972) 536.
[2] M. Sweeney, Phys. Rev. B 27 (1983) 4445.
[3] J. Kertész, D. Stauffer, A. Coniglio, Ann. Isr. Phys. Soc. 5 (1983) 121.
[4] R.B. Potts, Proc. Cambridge Philos. Soc. 48 (1952) 106.
[5] T. Baker, L. Chayes, J. Stat. Phys. 93 (1998) 1.
[6] A. Coniglio, F. di Liberto, G. Monroy, F. Peruggi, Phys. Rev. B 44 (1991) 12 605.
[7] P.S. Sahni, G.S. Grest, M.P. Anderson, D.J. Srolovitz, Phys. Rev. Lett. 50 (1983) 263.
[8] O.G. Mouritsen, Chem. Phys. Lipids 57 (1991) 179.
[9] F. Yaşar, Y. Gündüç, T. Çelik, Phys. Rev. E 58 (1998) 4210.
[10] A. Gadomski, Philos. Mag. Lett. 70 (1994) 335.
[11] A. Gadomski, Chem. Phys. Lett. 258 (1996) 6.
[12] A. Gadomski, Vacuum 50 (1998) 79.
[13] S.K. Kurtz, F.M.A. Carpay, J. Appl. Phys. 51 (1981) 5725,5745.
[14] H. Furukawa, Adv. Phys. 34 (1985) 703.
[15] M. Niemiec, A. Gadomski, J. Łuczka, L. Schimansky-Geier, Physica A 248 (1998) 365.
[16] G. Mazenko, Phys. Rev. B 43 (1991) 8204.
[17] N. Vandewalle, M. Ausloos, J. Phys. A 29 (1996) 309.
[18] N. Vandewalle, M. Ausloos, Phys. Rev. E 54 (1996) 3006.
[19] A. Gadomski, Phys. Rev. E 60 (1999) 1252.
[20] D. Joseph, Phys. Rev. B 58 (1998) 8347.
[21] M. Tomellini, M. Fanfoni, Surf. Sci. Lett. 349 (1996) L191.
[22] K.J. Kurzydłowski, K. Tangri, Scripta Metall. 22 (1988) 785.
[23] A. Aharony, D. Stauffer, in: R.A. Meyers (Ed.), Encyclopedia of Physical Science and Technology, Vol. 10, Academic Press, Orlando, 1987.
[24] A. Bunde, Sh. Havlin (Eds.), Fractal and Disordered Systems, Springer, Heidelberg, 1991, pp. 51–149.
[25] M. Kurzyński, K. Palacz, P. Chełminiak, Proc. Natl. Acad. USA 95 (1998) 11 685.
[26] K.J. Kurzydłowski, B. Ralph, The Quantitative Description of the Microstructure of Materials, CRC Press, Boca Raton, FL, 1995.

[27] C.S. Pande, Acta Metall. 35 (1987) 2671.
[28] C.S. Pande, 36 (1988) 2161.
[29] D. Łazęcki, J. Bystrzycki, A. Chojnacka, K.J. Kurzydłowski, Scripta Metall. 29 (1993) 1055.
[30] I-Wei Chen, Acta Metall. 35 (1987) 1733.
[31] V.E. Fradkov, Philos. Mag. Lett. 58 (1988) 271.
[32] Q. Ye, W.W. van Osdol, R. Biltonen, Biophys. J. 60 (1991) 1002.
[33] J.F. Nagle, S. Tristram-Nagle, H. Takahashi, I. Hatta, Eur. Phys. J. B 1 (1998) 399.
[34] A. Gadomski, Eur. Phys. J. B 9 (1999) 569.
[35] D. Stauffer, Physica A 242 (1997) 1.

ELSEVIER

Physica A 274 (1999) 333–340

www.elsevier.com/locate/physa

On the application of statistical physics for the description of nonequilibrium effects in chemical systems

Jerzy Gorecki[a,b,*], Joanna N. Gorecka[c]

[a]*Institute of Physical Chemistry, Polish Academy of Sciences and College of Science, Kasprzaka 44/52, PL-01-224 Warsaw, Poland*
[b]*ICM, Pawinskiego 5A, PL-02-106 Warsaw, Poland*
[c]*Institute of Physics, Polish Academy of Sciences, Al. Lotnikow 32/46, PL-02-668 Warsaw, Poland*

Abstract

In this paper we show how the methods of statistical physics may be applied to describe the nonequilibrium effects in systems with thermally activated reactions. A special attention is given to systems with multiple reactions, in which such effects are more important then in systems with a single reaction. Considering a model reaction scheme we demonstrate that a simple phenomenology, which associates different temperatures to different reagents, gives an accurate description of the nonequilibrium effects. © 1999 Elsevier Science B.V. All rights reserved.

1. Introduction

The cross section for a chemical reaction depends on the energetic states of inter-acting molecules of reactants. In the case of a thermally activated reaction [1–13], the reaction cross section is an increasing function of the energy of interacting molecules what means that the probability of reactive collision increases with their energy. There-fore, a thermally activated chemical reaction acts as a Maxwellian deamon, which selects the most energetic molecules of reactants and transforms them into products. As the result, the average energy per molecule of products is higher then that per molecule of reactants [14].

---

* Correspondence address: Institute of Physical Chemistry, Polish Academy of Sciences, Kasprzaka 44/52, PL-01-224 Warsaw, Poland.
E-mail address: gorecki@ichf.edu.pl (J. Gorecki)

0378-4371/99/$ - see front matter © 1999 Elsevier Science B.V. All rights reserved.
PII: S 0378-4371(99)00326-X

The reverse situation may also occur; there are processes for which the reaction cross section decreases with the energy of interacting molecules of reactants [15,16] and in such cases the reaction products become "cooler" than reactants.

Now, let us restrict our attention to the thermoneutral chemical reactions, i.e. to the processes in which the reaction heat is equal to zero. For example, the isotope exchange in molecules may be regarded as an example of such reaction. For a thermoneutral reaction which proceeds in an adiabatic system the reactants and products treated as a whole remain in the thermal equilibrium and the system's energy distribution function is Maxwellian. However, if the reaction is thermally activated than the most energetic molecules of reactants are transformed into a product and therefore the average energy per molecule of reactant will be lower then the average energy of the system treated as a whole. As the result the rate constant calculated on the basis of reactant's energy distributions will be different than the one calculated from the energy distribution of the system treated as a whole [1–13].

In majority of theoretical studies on the nonequilibrium effects the line-of-center model for a reactive collision has been used [17]. This model applies to case in which all the molecules of reactants are represented by identical structureless hard spheres and all reactions are bimolecular. Therefore, the state of a molecule is completely described by its chemical identity and the velocity. According to the line-of-center model the cross section for a reactive collision between X and Y particles reads:

$$\sigma_{X,Y}^*(v_X, v_Y) = \begin{cases} \frac{1}{4}s_F d^2(1 - \frac{E_A}{E_C}) & \text{for } E_C \geqslant E_A, \\ 0 & \text{for } E_C < E_A, \end{cases} \tag{1}$$

where E_A is the activation energy, d and m denote the diameter and mass of a sphere and $E_C = \frac{1}{4}(v_X - v_Y)^2 m$ where v_X and v_Y are the velocities of colliding spheres. It may be shown that the cross section σ_{XY}^* may be equivalently represented by the condition that a collision between X and Y is reactive if the kinetic energy of the motion of spheres along the line of centres, calculated in the center-of-mass reference frame exceeds E_A. The introduced steric factor s_F describes the probability that a collision for which the energetic condition is satisfied is reactive. In this way, the influence of all unknown degrees of freedom on the considered reaction is taken into account.

Having the reaction cross section one can easily calculate the rate constants in a system in which molecules are represented by hard spheres. Let us assume that the velocity distribution functions for molecules of X and Y are $f_X(v)$ and $f_Y(v)$, respectively. The reaction rate constant for the process $X + Y \rightarrow products$ can be calculated as

$$k_{X,Y} = \int (\text{the condition that molecules of } X \text{ and } Y \text{ collide})$$
$$\times \sigma_{X,Y}^*(v_x, v_y) f_X(v_X) f_Y(v_Y) \, d(v_x) \, d(v_y). \tag{2}$$

In the case when the distribution functions $f_X(v), f_Y(v)$ are Maxwellians with temperatures T_1 and T_2, the integration for the cross section (1) can be performed analytically and it gives [18]:

$$k_{X,Y} = s_F 4d^2 \left[\frac{\pi k_B(T_1 + T_2)}{2m} \right]^{1/2} \exp\left(-\frac{2E_A}{k_B(T_1 + T_2)} \right) \tag{3}$$

which for $T_1 = T_2$ leads to over 100 years old Arrhenius formula for the rate constant of a thermally activated process [16].

Let us assume that at the beginning the velocity distributions of both X and Y reactants correspond to the same temperature T_0. As it has been said before, as the result of a thermally activated reaction, the most energetic molecules of reactants are transformed into products and than average energy per molecule of reactant will be lower than its initial value $(3/2k_B T_0)$. In order to calculate the rate constant one needs to know how the velocity distributions f_X and f_Y depend on time. Such information can be obtained from the generalized Boltzmann equation, which takes both reactive and nonreactive collisions into account. For example, in a homogeneous system, the equation for time evolution of the distribution functions of reactant X reads:

$$\frac{\partial n_x f_X}{\partial t} = - \int n_x f_X n_y f_Y \sigma^*_{X,Y}(\boldsymbol{v}_X, \boldsymbol{v}_Y) \, d\Omega \, d\boldsymbol{v}_Y |\boldsymbol{v}_X - \boldsymbol{v}_Y|$$

$$+ \int [n_x f'_X(n_x f'_X + n_y f'_Y + n_p f'_{\text{prod}})$$

$$- n_x f_X(n_x f_X + n_y f_Y + n_p f_{\text{prod}})] \sigma_{el} |\boldsymbol{v}_X - \boldsymbol{v}_Y| \, d\boldsymbol{v}_Y \, d\Omega , \tag{4}$$

where n_x, n_y and n_p are the concentrations of X, Y and products, respectively. The first term in (4) describes the decrease in the distribution function due to reactive collisions and the second one describes how it changes in the elastic collisions (σ_{el}). Of course, Eq. (4) should be solved together with the equations for time dependent velocity distribution functions for Y $(f_Y(t))$ and the products $(f_{\text{prod}}(t))$. In practice the procedure is difficult even for the simplest process $A + A \rightarrow products$ and it is still a subject of intensive studies [6–8,10–13]. However, the microscopic simulations of systems with thermally activated reaction indicate [9,10,14,19] that one can approximate nonequilibrium, time-dependent velocity distribution functions by assuming that they are given by Maxwellians with time-dependent temperatures, different from the temperature of the system as a whole. In such case, the distribution function for X is fully described by the density of X molecules $(n_x(t))$ and their temperature $T_x(t)$. The equations describing time evolution of these quantities can be directly obtained from Eq. (4) by calculating the appropriate moments [9,10,18]. It can be shown that for the unidirectional process $A + A \rightarrow products$ the average energy of A particles may be up to 15% lower than the initial one and that the decrease in the rate constant may exceed 20% of its initial value (for the activation energies in the range between 1 and $3k_B T_0$) [9,14] and therefore the transformation of A into products is slower than in an equilibrium system. Such significant deviation from the equilibrium (initial) velocity distribution influences also the other properties of reactant like its viscosity or the diffusion constant [20–23].

The nonequilibrium effects are more pronounced in systems with coupled, thermally activated reactions [24–31]. It is easy to see that if a fast thermally activated reaction with low activation energy E_{A1}

$$A + B \rightarrow products \tag{5}$$

is coupled with another thermally activated reaction:

$$A + C \rightarrow products \tag{6}$$

which is characterized by a high activation energy E_{A2} than the changes in the energy distribution function for A molecules created by process (5) have a large impact on reaction (6) [18,27,28]. The nonequilibrium rate constant of reaction (6) may be estimated as $k_0 \exp(-\varepsilon_{A2}|\Delta T|/T_0)$ where k_0 is the equilibrium rate constant of (6), ε_{A2} is the activation energy in the units related to the initial temperature of the system ($k_B T_0$) and ΔT is the change of temperature of A created by faster reaction (5). On the other hand, if the faster process (5) is inversely activated [15,16] then the average energy of A molecules is increased with respect to its initial value and activated process (6) proceeds faster. Now, the increase in its rate constant is proportional to $\exp(\varepsilon_{A2}|\Delta T|/T_0)$ and for large ε_{A2} the effect is important [30].

In the following, we discuss the influence of nonequilibrium effects on a model chain reaction.

2. The nonequilibrium effects in a model of a chain reaction

In the recently published paper [31] we have demonstrated that the nonequilibrium effects increase the rate of creation of product C at the initial stages of the $A + A \leftrightarrow B + B \leftrightarrow C + C$ reaction. In such case, the first thermally activated reaction delivers a "hot" reactant B which can be easily transformed into C by the next process. Here, we consider the following modification of the previously studied model. A thermoneutral, thermally activated process:

$$A + A \leftrightarrow B + B \tag{7}$$

is coupled with another thermally activated reactions:

$$B + E \leftrightarrow C + E. \tag{8}$$

In the new reaction scheme the role of the nonequilibrium effects is expected to be more important that in the previously considered model as the rate of creation of C at low concentration of B is higher. To ensure the microscopic reversibility of the considered processes we assume that the steric factors and the activation energies for direct and reverse reactions are the same. Let us consider the system, which at the beginning is composed of equilibrated A and E molecules, both at the same initial temperature T_0. As reactions (7) and (8) are both thermoneutral, the system treated as a whole remain at equilibrium with temperature T_0.

$$n_0 f_{M,T_0}(\boldsymbol{v}) = n_A f_A(\boldsymbol{v}, t) + n_B f_B(\boldsymbol{v}, t) + n_C f_C(\boldsymbol{v}, t) + n_E f_E(\boldsymbol{v}, t), \tag{9}$$

where n_0 is the total density of molecules in the system, f_{M,T_0} the Maxwellian characterizing the velocity distribution of molecules in the system as a whole, n_A, n_B, n_C, n_E

the concentrations of molecules A, B, C, E, respectively, and $f_A(v,t)$, $f_B(v,t)$, $f_C(v,t)$, $f_E(v,t)$ the velocity distributions of corresponding reagents. The molecules of E do not undergo any of the reactions and therefore n_E remains constant and $f_E(v,t) = f_{M,T_0}(v)$. Obtaining the distribution functions for system $(7,8)$ directly from the Boltzmann equations is rather difficult, thus we apply an approximate description of the nonequilibrium effects based on the assumption that the velocity distribution functions of reagents which are equilibrated at the moment reactions start retain the Maxwellian form [18,31] with time-dependent temperatures. In the considered system it means that temperatures exist for A and for reagents A and B treated together:

$$n_A f_A(v,t) = n_A f_{M,T_A(t)}(v),$$

$$n_A f_A(v,t) + n_B f_B(v,t) = (n_A(t) + n_B(t)) f_{M,T_{AB}(t)}(v). \tag{10}$$

Now, it is easy to obtain the equations describing the time evolution of $n_A(t)$, $n_B(t)$ and $n_C(t)$ and $T_A(t)$ and $T_{AB}(t)$ as the formulae for the rates of reactions and for the energy exchange have been derived in Ref. [18]. In the scaled variables defined as: $\alpha = n_A/n_0$, $\gamma = n_C/n_0$, $\varepsilon = n_E/n_0$, $\xi_{ab} = T_{AB}/T_0$, $\xi_a = T_A/T_0$, $\varepsilon_{A1} = E_{A1}/k_B T_0$, $\varepsilon_{A2} = E_{A2}/k_B T_0$ and $\tau = 4d^2 g(\pi k_B T_0/m)^{1/2} n_0 t$, one obtains the following set of equations:

$$\frac{\partial \alpha}{\partial \tau} = (\alpha + \beta)^2 s_{F1} \sqrt{\xi_{ab}} \exp\left(-\frac{\varepsilon_{A1}}{\xi_{ab}}\right)$$

$$- 2\alpha(\alpha + \beta) s_{F1} \sqrt{\frac{\xi_a + \xi_{ab}}{2}} \exp\left(-\frac{2\varepsilon_{A1}}{\xi_a + \xi_{ab}}\right), \tag{11a}$$

$$\frac{\partial \gamma}{\partial \tau} = -s_{F2}\varepsilon \exp(-\varepsilon_{A2}) + s_{F2}\varepsilon^2 \exp(-\varepsilon_{A2})$$

$$+ 2s_{F2}(\alpha + \beta)\varepsilon \sqrt{\frac{1 + \xi_{ab}}{2}} \exp\left(-\frac{2\varepsilon_{A2}}{1 + \xi_{ab}}\right)$$

$$- s_{F2}\alpha\varepsilon \sqrt{\frac{\xi_a + 1}{2}} \exp\left(-\frac{2\varepsilon_{A2}}{\xi_a + 1}\right), \tag{11b}$$

$$\frac{3\alpha}{2} \frac{\partial \xi_a}{\partial \tau} = (\alpha + \beta)^2 s_{F1} \sqrt{\xi_{ab}} \exp\left(-\frac{\varepsilon_{A1}}{\xi_{ab}}\right) \left[\frac{3}{2}(\xi_{ab} - \xi_a) + \xi_{ab}\left(\frac{1}{4} + \frac{\varepsilon_{A1}}{2\xi_{ab}}\right)\right]$$

$$- \alpha(\alpha + \beta) s_{F1} \sqrt{\frac{\xi_a + \xi_{ab}}{2}} \exp\left(-\frac{2\varepsilon_{A1}}{\xi_a + \xi_{ab}}\right)$$

$$\times \xi_a \left[\frac{\xi_a}{2(\xi_a + \xi_{ab})} + \frac{2\varepsilon_{A1}\xi_a}{(\xi_a + \xi_{ab})^2}\right]$$

$$- \alpha(\alpha + \beta) s_{F1} \sqrt{\frac{\xi_a + \xi_{ab}}{2}} \exp\left(-\frac{2\varepsilon_{A1}}{(\xi_a + \xi_{ab})}\right)$$

$$\times \left[\frac{3}{2}(\xi_{ab} - \xi_a) + \frac{\xi_{ab}^2}{2(\xi_a + \xi_{ab})} + \frac{2\varepsilon_{A1}\xi_{ab}^2}{(\xi_a + \xi_{ab})^2}\right] + \alpha \sqrt{\frac{1 + \xi_a}{2}}(1 - \xi_a), \tag{11c}$$

$$\frac{3}{2}(\alpha + \beta)\frac{\partial \xi_{ab}}{\partial \tau} = \sqrt{\frac{1 + \xi_{ab}}{2}}(\alpha + \beta)(1 - \xi_{ab}) + s_{F2}\varepsilon(1 - \varepsilon)\exp(-\varepsilon_{A2})$$

$$\times \left[\frac{3}{2}(1 - \xi_{ab}) + \frac{1}{4} + \frac{\varepsilon_{A2}}{2}\right] - 2s_{F2}\varepsilon(\alpha + \beta)\sqrt{\frac{1 + \xi_{ab}}{2}}$$

$$\times \exp\left(-\frac{2\varepsilon_{A2}}{1 + \xi_{ab}}\right)\xi_{ab}\left[\frac{\xi_{ab}}{2(1 + \xi_{ab})} + \frac{2\varepsilon_{A2}\xi_{ab}}{(1 + \xi_{ab})^2}\right]$$

$$+ s_{F2}\alpha\varepsilon\sqrt{\frac{1 + \xi_a}{2}}\exp\left(-\frac{2\varepsilon_{A2}}{(1 + \xi_a)}\right)\left[\frac{3}{2}(\xi_a - \xi_{ab}) + \frac{\xi_a^2}{2(1 + \xi_a)}\right.$$

$$\left. + \frac{2\varepsilon_{A2}\xi_a^2}{(1 + \xi_a)^2}\right] - \varepsilon(\alpha + \beta)\sqrt{\frac{1 + \xi_{ab}}{2}}s_{F2}\left(1 + \frac{2\varepsilon_{A2}}{1 + \xi_{ab}}\right)$$

$$\times \exp\left(-\frac{2\varepsilon_{A2}}{1 + \xi_{ab}}\right)(1 - \xi_{ab}). \tag{11d}$$

Let us note that the influence of system's density on the chemical processes is taken into account by the value of the radial distribution function at the sphere diameter (g), which appears in the definition of the scaled time [32]. The usefulness of such simple phenomenology can be clearly seen if one compares its predictions with computer simulations of the nonequilibrium effects in a system with reactions (7,8) performed at the microscopic level. We have done a simple molecular dynamics simulations using the reactive hard sphere technique [33] as it is described in Ref. [31].

In simulations the molecules of all reagents are represented by structureless hard spheres with the same mass and diameter. Reagents are marked by a chemical identity parameter, which has no direct influence on the mechanical motion and it is modified in collisions regarded as reactive. The simulations have been performed for 125 000 of hard spheres all with the diameter $d = 5$ Å and mass $m = 32$ a.u. The packing fraction of the considered system is $\eta = 0.098$. Reaction (8) has steric factor $s_{F2} = 0.8$ and activation energy $\varepsilon_{A2} = 4$. We have considered two sets of parameters for reaction (7): $s_{F1} = 0.5 \exp(-2)$ and $\varepsilon_{A1} = 0.0$ or $s_{F1} = 0.5$ and $\varepsilon_{A1} = 2.0$. Let us note that if the system is regarded as an equilibrium one than both these sets of parameters lead to the same rate constants. We assume that the initial composition of reagents is: $\alpha(t = 0) = 0.8$, $\varepsilon(t = 0) = 0.2$ and $\beta(t = 0) = \gamma(t = 0) = 0.0$. If the nonequilibrium effects are neglected (which corresponds to the assumption that $\xi_a = \xi_{ab} = 1.0$ in Eqs. (11a) and (11b)) than the time evolution of concentration of C is shown by a thick solid line in Fig. 1. The mid and the upper solid lines represent solutions of Eqs. (11) for $\varepsilon_{A1} = 0.0$ $(s_{F1} = 0.5 \exp(-2))$ and $\varepsilon_{A1} = 2.0$ $(s_{F1} = 0.5)$, respectively. The corresponding dashed lines are concentrations obtained from molecular dynamics simulations.

This result indicates that at the initial stages of the considered reaction the presence of nonequilibrium effects significantly speeds up the "creation" of C particles. As expected, for reaction (7) characterized by $\varepsilon_{A1} = 2.0$, which generates more important nonequilibrium effect, the rate the reagent C appears in the system is much faster than

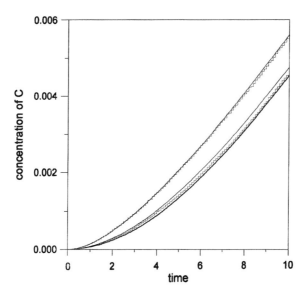

Fig. 1. The concentration of C as a function of time. The comparison between MD simulations (step-like dashed lines), the solution of Eqs. (11) (thin solid lines) and predictions of standard kinetic equations (the thick solid line). The initial condition is: $\alpha(t=0) = 0.8$, $\beta(t=0) = 0.0$, $\gamma(t=0) = 0.0$ and $\varepsilon(t=0) = 0.2$. The upper and lower pairs of dashed and thin solid lines correspond to: ($\varepsilon_{A1} = 0.0$, $s_{F1} = 0.5\exp(-2)$) and ($\varepsilon_{A1} = 2.0$, $s_{F1} = 0.5$), respectively.

for $\varepsilon_{A1} = 0.0$. The predictions based on Eqs. (11) are in very good agreement with microscopic simulation of the system with reactions (7) and (8).

Concluding our paper we would like to focus attention on two aspects of presented results. First, the nonequilibrium effects can have an important influence on the time evolution of thermally activated chain processes, especially in complex nonlinear reaction schemes which are very sensitive to small changes in rate constants and concentrations of reagents. Second, the considered simple phenomenology for the description of the nonequilibrium effects created by a thermally activated reaction is general and it may be applied to any set of reactions for which the cross sections for reactive and nonreactive collisions are known [34]. The comparison between molecular dynamics simulations and phenomenology indicates that the assumption on multiple temperatures characterizing different reagents in a nonequilibrium system can accurately describe the observed influence of the nonequilibrium effects on system's evolution.

References

[1] I. Prigogine, E. Xhrouet, Physica 15 (1949) 913.
[2] K. Takayanagi, Progr. Theoret. Phys. (Kyoto) 6 (1951) 486.
[3] J. Ross, P. Mazur, J. Chem. Phys. 35 (1961) 19.
[4] C.W. Pyun, J. Ross, J. Chem. Phys. 40 (1964) 2572.
[5] R. Kapral, S. Hudson, J. Ross, J. Chem. Phys. 53 (1970) 4387.
[6] B. Shizgal, M. Karplus, J. Chem. Phys. 52 (1970) 4262.

[7] B. Shizgal, M. Karplus, J. Chem. Phys. 54 (1971) 4257,4345.

[8] B. Shizgal, J. Chem. Phys. 55 (1971) 76.

[9] J. Gorecki, B.C. Eu, J. Chem. Phys. 97 (1992) 6695.

[10] A.S. Cukrowski, J. Popielawski, L. Qin, J.S. Dahler, J. Chem. Phys. 97 (1992) 9086.

[11] D.G. Napier, B.D. Shizgal, Phys. Rev. E 52 (1995) 3797.

[12] B.D. Shizgal, D.G. Napier, Physica A 233 (1996) 50.

[13] B. Nowakowski, J. Chem. Phys. 109 (1998) 3443.

[14] J. Gorecki, J. Chem. Phys. 95 (1991) 2041.

[15] A.S. Cukrowski, S. Fritzsche, W. Stiller, Theor. Chem. Acta 73 (1988) 67.

[16] W. Stiller, Arrhenius Equation and Non-Equilibrium Kinetics, Teubner-Texte zur Physik, Band 21, BSF B.G. Teubner, Leipzig, 1989.

[17] B.D. Present, J. Chem. Phys. 31 (1959) 747.

[18] J. Gorecki, J. Chem. Phys. 98 (1993) 7269.

[19] B. Nowakowski, J. Gorecki, Acta Phys. Pol. B 27 (1996) 895.

[20] J. Popielawski, J. Chem. Phys. 83 (1985) 790.

[21] A.S. Cukrowski, J. Popielawski, Chem. Phys. 109 (1986) 215.

[22] B. Nowakowski, J. Popielawski, J. Chem. Phys. 100 (1994) 7602.

[23] B. Nowakowski, A. Lemarchand, J. Chem. Phys. 106 (1997) 3965.

[24] B.D. Shizgal, J.M. Fitzpatrick, Phys. Rev. A 18 (1978) 267.

[25] M.B. Gorensek, M.D. Kostin, J. Chem. Phys. 81 (1984) 1277.

[26] M.B. Gorensek, M.D. Kostin, J. Chem. Phys. 83 (1985) 2280.

[27] J. Gorecki, A.L. Kawczynski, J. Chem. Phys. 96 (1992) 4646.

[28] A.S. Cukrowski, A.L. Kawczynski, J. Popielawski, W. Stiller, R. Schmid, Chem. Phys. 159 (1992) 39.

[29] J. Gorecki, I. Hanazaki, Chem. Phys. 181 (1994) 39.

[30] A.S. Cukrowski, S. Fritzsche, W. Stiller, Chem. Phys. 181 (1994) 7.

[31] J. Gorecki, J.N. Gorecka, Chem. Phys. 240 (1999) 215.

[32] R. Der, S. Fritsche, Chem. Phys. Lett. 121 (1985) 177.

[33] J. Gorecki, J. Gryko, Computer Phys. Commun. 54 (1989) 245.

[34] K. Koura, J. Chem. Phys. 66 (1977) 4078.

ELSEVIER

Physica A 274 (1999) 341–348

www.elsevier.com/locate/physa

Strong motion duration effects on base isolated systems

R. Greco[a,*], G.C. Marano[a], D. Foti[b]

[a]*Istituto di Progettazione, Facoltà di Architettura, Politecnico di Bari, Via Orabona, 4, 70123 Bari, Italy*
[b]*Dipartimento di Ingegneria Strutturale, Facoltà di Ingegneria, Politecnico di Bari, Via Orabona 4, 70123 Bari, Italy*

Abstract

In this paper the non-stationary response covariance of a base isolated building subjected to a seismic excitation modelled through a zero mean Gaussian non-stationary stochastic process, uniformly modulated in the time with a boxcar-shape temporal function, is determined. The results are utilised to give an estimation of the strong motion duration effects on the temporal peak of the root mean square value of the response of the system and on system failure probability for first threshold crossing. © 1999 Elsevier Science B.V. All rights reserved.

PACS: 91.30.P; 07.07.T; 02.50; 05.40

Keywords: Strong motion; Base isolated systems; First threshold crossing

1. Introduction

In the last few years many methods for qualifying earthquake severity have been proposed, with the aim to evaluate their damage potential on structures. Some of these methods considered the seismic peak acceleration, the strong motion duration, the frequency characteristics and the temporal envelope shape of the seismic ground motion. The seismic peak acceleration is the simplest parameter to identify and compare different earthquakes, but it cannot be necessarily correlated to the maximum structural response, especially if the excitation acts on the structure for a very short time and for systems with high fundamental period and a low damping. Arias intensity [1] represents a measure of the seismic excitation which considers at the same time the peak, the

---

* Corresponding author.

duration and the frequency content of the seismic motion. In the following it will be shown that Arias intensity cannot be assumed as a constant measure of the excitation level in a given site. For constant "energy" entered the structure, the root mean square (rms) value of the seismic acceleration or the time duration could influence the system failure probability for first threshold crossing, depending on the system frequency characteristic and the threshold level. This problem is very important for base isolated structures, characterised by high fundamental periods, and, therefore, sensible to the duration of the seismic excitation.

When the seismic motion is represented by a non-stationary stochastic process, the most critic excitation parameters for the structural response are the intensity S_0, the strong motion duration and the "spectral content" described by the spectral density function (PSD) $S(\omega)$. The last represents the Fourier transform of the correlation function of the stochastic process; for stationary random process, in particular, the mean square value is given by the area under a graph of spectral density function against ω.

In literature there are just a few studies on the influence of the excitation duration on the structural response, even if this parameter significantly influences the rms response, the failure probability for the first threshold crossing and the fatigue failure of structures. Zembaty [2] studied this problem related to slender towers (systems with high fundamental periods). The effects of the seismic excitation duration on the response of the isolated system are studied using two different criteria: constant Arias intensity and constant excitation level S_0. In the first case, when the excitation duration increases the excitation rms value decreases; on the contrary, in the second case the mean Arias intensity increases when increases.

2. Description of the study

In this study the seismic acceleration acting on a base isolated building is represented by a non-stationary zero mean Gaussian stochastic process, uniformly time modulated:

$$\ddot{X}_g(t) = V(t)\ddot{X}_{g\text{staz}}(t) . \tag{1}$$

Expression (1) has been obtained enveloping a stochastic stationary process $\ddot{X}_{g\text{staz}}(t)$ with a deterministic time modulation function $V(t)$. The frequency content of the seismic excitation is assumed to be time constant and is described by the Kanai–Tajimi power spectral density (PSD) [3]

$$S_{\text{KT}}(\omega) = S_0 \frac{4\xi_g^2\omega_g^2\omega^2 + \omega^4}{(\omega_g^2 - \omega^2)^2 + 4\xi_g^2\omega_g^2\omega^2} , \tag{2}$$

where ω_g and ξ_g are the ground filter parameters.

In this paper a boxcar temporal modulation function will be considered (Fig. 1):

$$V(t) = \begin{cases} 1 & \text{if } 0 < t < T , \\ 0 & \text{if } t > T . \end{cases} \tag{3}$$

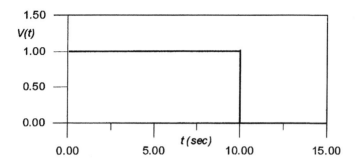

Fig. 1. Modulation function $V(t)$ of the non-stationary excitation process.

In the seismic excitation process (1) T and S_0 represent the two most critical parameters for structural response, when the ground filter characteristics ω_g and ξ_g are assigned. More in detail, S_0 defines the vibration level induced by the earthquake on the structure, while the duration T determines the time-action of the excitation and therefore the mean average number of threshold crossings of the response process. As these two parameters increase, the failure probability of the system increases.

Duration T and S_0 are related through the mean Arias intensity, defined as

$$\langle I_a \rangle = \int_0^{t_0} E[\ddot{x}_g^2(t)]\,\mathrm{d}t = \int_0^{t_0} V^2(t) \int_{-\infty}^{+\infty} S_{\ddot{x}_g^2}(\omega)\,\mathrm{d}\omega\,\mathrm{d}t \; . \tag{4}$$

Vanmarke and Lai [1] defined a stationary equivalent process with duration T and assumed that the total intensity of the earthquake is uniformly distributed along the strong motion duration T. In this case the mean Arias intensity is [4]

$$\langle I_a \rangle = T\sigma_{\ddot{x}_g}^2 \; ,$$

and for the Kanai–Tajimi PSD function

$$\langle I_a \rangle = S_0 T \pi \omega_g \frac{(1 + 4\xi_g^2)}{2\xi_g} \; . \tag{5}$$

3. Covariance response analysis

Fig. 2 shows a simple degree of freedom structure supported on neoprene pads and its dynamical model. The masses m_s and m_b correspond to superstructure and to the base raft, respectively. The stiffness k_s and the damping coefficient c_s define the mechanical properties of the single degree of freedom superstructure. The base isolation system is modelled as a linear viscoelastic system with equivalent stiffness k_b and damping c_b [5]. If $\ddot{x}_g$ is the ground acceleration the equations of the motion are

$$m_s\ddot{x}_s + m_s\ddot{x}_b + c_s\dot{x}_s + k_s x_s = -m_s\ddot{x}_g \; ,$$

$$m_T\ddot{x}_b + m_s\ddot{x}_s + c_b\dot{x}_b + k_b x_b = -m_T\ddot{x}_g \; , \tag{6}$$

where $m_T = m_b + m_s$.

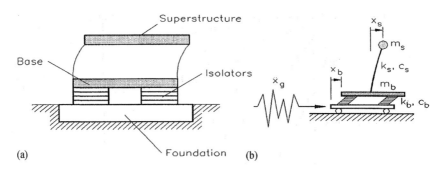

Fig. 2. Single degree of freedom mounted on a laminated rubber bearing.

As the excitation is a stochastic zero mean Gaussian process and the system is linear, the response process will be Gaussian too. Therefore it will be completely characterised by the second-order conjugate means, which are elements of the correlation matrix $R_{xx}(t_1, t_2)$:

$$R_{xx}(t_1, t_2) = E[x(t_1)x^T(t_2)] = \begin{bmatrix} R_{x_b x_b}(t_1, t_2) & R_{x_b x_s}(t_1, t_2) \\ R_{x_s x_b}(t_1, t_2) & R_{x_s x_s}(t_1, t_2) \end{bmatrix}, \tag{7}$$

where $x = \{x_b, x_s\}^T$.

The procedure for evaluating the correlation matrix (7) when the excitation process is represented by Eq. (1) is shown in [6,7]. When $\tau = t_1 - t_2 = 0$, matrix (7) provides the covariance matrix $S_{xx}(t)$:

$$S_{xx}(t) = \begin{bmatrix} \sigma_{x_b}^2(t) & \gamma_{x_b x_s}(t) \\ \gamma_{x_s x_b}(t) & \sigma_{x_s}^2(t) \end{bmatrix}, \tag{8}$$

where the diagonal terms $\sigma_{x_i}^2(t)$ are the variances and the out-diagonal terms $\gamma_{x_i x_j}(t)$ are the covariances relative to the response process of the system.

4. Results: influence of the excitation duration on the system reliability

The effects of the strong motion duration T on the first threshold crossing probability for the base isolated system has been analysed.

If $X(t)$ is the general response process, the structural safety indicates the probability $P_S(t, \xi)$ that the response process $|X(t)|$ does not exhibit any crossing of the threshold limit $\pm \xi$ (double barrier) when $0 < \tau < t$. The system reliability function $P_S(t, \xi)$ is determined in the hypothesis of independent threshold crossings, with a Poisson distribution [8]

$$P_S(t, \xi) = \exp\left\{ -\int_0^t \alpha(t)\,dt \right\} = \exp\left\{ -\int_0^t 2v_{\xi^+}(t)\,dt \right\}$$

$$= \exp\left\{ -\int_0^t 2v_{0^+}(t)\,e^{-\xi^2/2\sigma_x^2(t)}\,dt \right\}, \tag{9}$$

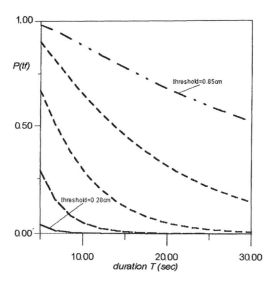

Fig. 3. First passage probability for the base isolator vs. T duration: $\omega_s = 12.5$ rad/s; $\xi_s = 2\%$; $\omega_b = 3.125$ rad/s; $\xi_b = 10\%$; $\omega_g = 25$ rad/s; $\xi_g = 0.6$; $S_0 = 49.19$ cm^2/s^3.

where

$$v_{\xi+}(t) = v_{0+}(t)\exp\left[-\frac{\xi^2}{2\sigma_x^2(t)}\right]$$

and

$$v_{0+}(t) = \frac{1}{2\pi}\frac{\sigma_{\dot{x}}^2(t)}{\sigma_x^2(t)}.$$

In Fig. 3, the values of the first passage probability function $P(t_f)$ for the base isolator displacement evaluated at the end of the observing time t_f are plotted vs. the duration T. It is possible to observe that $P(t_f)$ increases when the threshold level ξ increases; moreover, the reliability function $P(t_f)$ monotonically decreases as the duration T increases.

Fig. 4 represents the variability of the base isolator reliability function $P(t_f)$ when the mean Arias intensity is assumed to be constant. When the threshold level increases $P(t_f)$ shows a qualitative behaviour similar to the one observed for a constant excitation rms level criterion, but in this case the $P(t_f)$ variability vs. duration depends on the threshold level ξ. For a low threshold level, the T increment influences more than the rms excitation reduction the first passage probability for the response process $X_b(t)$. Therefore, $P(t_f)$ decreases when the duration increases, as shown assuming constant rms excitation criterion ($S_0 = $ const.). When the threshold level increases, the rms excitation reduction influences the first threshold crossing probability more than the duration increment: therefore, $P(t_f)$ increases as T increases.

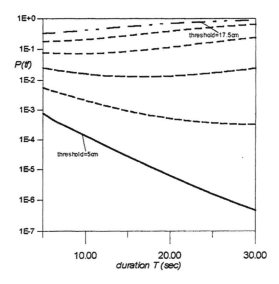

Fig. 4. First passage probability for the base isolator vs. T duration: $\omega_s = 12.5$ rad/s; $\xi_s = 2\%$; $\omega_b = 3.125$ rad/s; $\xi_b = 10\%$; $\omega_g = 25$ rad/s; $\xi_g = 0.6$; $I_a = $ const.

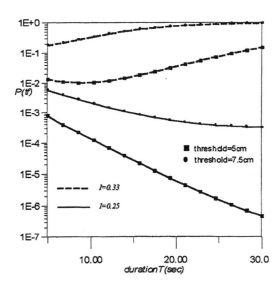

Fig. 5. First passage probability for the base isolator vs. T duration: $\omega_s = 12.5$ rad/s; $\xi_s = 2\%$; $\xi_b = 10\%$; $\omega_s = 25$ rad/s; $\xi_g = 0.6$; $I_a = $ const.

Comparisons of the results for two base isolated systems with different frequency ratio I are shown in Fig. 5; similar results are shown in Fig. 6 for the probability $P(t_f)$ plotted vs. the frequency ratio $I = \omega_b/\omega_s$.

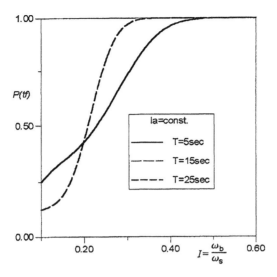

Fig. 6. First passage probability for the base isolator vs. frequency ratio: $\omega_s = 12.5$ rad/s; $\zeta_s = 2\%$; $\zeta_b = 10\%$; $\omega_g = 25$ rad/s; $\zeta_g = 0.6$; $S_0 = 49.19$ cm^2/s^3.

5. Conclusions

In this paper, a two d.o.f. base isolated building subjected to a seismic excitation represented by a non-stationary zero mean Gaussian stochastic process is considered. The covariance analysis has been applied to the system to determine the strong motion duration effects on the temporal peak of the rms value of the system response and on its first passage probability.

The non-stationary covariance solution has been utilised to study the strong motion duration effects on the response of the base isolator. From the results it is only possible to make some general considerations regarding the influence of the seismic excitation duration–intensity on the base isolated buildings response. Arias intensity cannot be assumed as a constant measure of the seismic excitation level in a given site. For constant energy entering the system, the duration T or the rms excitation level could influence the system failure probability for the first threshold crossing, depending on the threshold level and/or the frequency characteristics of the system. Usually, systems characterised by high deformability are more influenced by the mean number of level threshold crossings, that is by the duration T, than by the excitation rms value. On the contrary, these results cannot be extended to systems having the fundamental period in the typical range of the buildings. In fact, in this case the system failure probability for the first threshold crossing is influenced more by the rms of the excitation than by the duration T.

References

[1] E.H. Vanmarcke, S.P. Lai, Strong motion duration and rms amplitude of earthquake records, Bull. Seism. Soc. Am. 70 (1980) 1293–1307.
[2] Z. Zembaty, On the reliability of tower-shaped structures under random excitations, Earthquake Eng. Struct. Dyn. 15 (1987) 761–775.
[3] H. Tagimi, A statistical method of determining the maximum response of a building during earthquake, Proc. 2nd World Conf. on Earthquake Engineering, Tokyo, Japan.
[4] G.W. Housner, Measures of severity of earthquake ground shaking, Proceedings of the US National Conference on Earthquake Engineering, Ann Arbor, Michigan, 1975.
[5] J.M. Kelly, J.M. Eidinger, C.J. Derham, A practical soft story earthquake isolation system, UCB/EEERC-77/270, 1977.
[6] R. Greco, Isolamento alla base di sistemi soggetti a forzanti sismiche stocastiche, Ph.D. Thesis, 1999, Naples, Italy.
[7] H.O. Madsen, S. Krenk, Stationary and transient response statistics, J. Eng. Mech. Div. ASCE V 108 (1982) 622–635.
[8] Y.K. Lin, Probabilistic Theory of Structural Dynamics, Krieger, Huntington, NY, 1976.

ELSEVIER

Physica A 274 (1999) 349–354

www.elsevier.com/locate/physa

Application of the detrended fluctuation analysis (DFA) method for describing cloud breaking

K. Ivanova[a,*], M. Ausloos[b]

[a]*Institute of Electronics, Bulgarian Academy of Sciences, Tzarigrazdsko Chaussee 72, 1784 Sofia, Bulgaria*
[b]*SUPRAS and GRASP, Institute of Physics, Sart Tilman, B5, University of Liège, B-4000 Liège, Belgium*

Abstract

A method to sort out correlations and decorrelations in stratus cloud formation, persistence and breakup is introduced. The detrended fluctuation analysis (DFA) statistical method is applied to microwave radiometer data. The existence of long-range power-law correlations in stratus cloud liquid water content fluctuations is demonstrated over a 2-h period. Moreover using a finite size (time) interval window, a change from Brownian to non-Brownian fluctuation regimes is clearly shown to define the cloud structure changes. Such findings are similar to those found in DNA and financial data sequences when mosaics of persistent and antipersistent patches are present. The occurrence of these statistics in stratus cloud liquid water content suggests the usefulness of similar studies on cloud-resolving model output and for better meteorological predictability. © 1999 Elsevier Science B.V. All rights reserved.

1. Introduction

Substantial progress has been made in the fields of predictive meteorology and climate in the past several decades. In some areas, however, the non-linear processes at work in the atmosphere often produce data series of such complexity that traditional analysis techniques fail to extract meaningful physical information. Better techniques are clearly required. Recently, there have been several reports that long-range power-law correlations [1] can be analyzed in turbulence [2], biological [3] and financial [4] data fluctuations, and more generally in self-organized critical systems [5]. An approach along these lines has recently been reported by two of us [6] for the short time scale

* Corresponding author. Fax: +359-2-975-3201.
E-mail address: kristy@ie.bas.bg, kristy@essc.psu.edu, kristiniv@hotmail.com (K. Ivanova).

0378-4371/99/$ - see front matter © 1999 Elsevier Science B.V. All rights reserved.
PII: S 0378-4371(99)00312-X

variation of financial data. If these methods can be adapted for meteorological and climatological time series, the underlying physical processes could be elucidated in order to offer a path towards improved models of the atmosphere.

The key ingredient to be introduced in this study is based on fractal geometry ideas and phase transitions in non-equilibrium states, i.e., bearing upon scale invariance [7,8], Brownian motion [14,9], a.s.o. phenomenology. Within the content of the *detrended fluctuation analysis* (DFA) method [10] we sort out correlations and decorrelations in the stratus cloud liquid water and radiance data. The DFA method has demonstrated its usefulness in investigations of long-range power-law correlations, e.g. in meteorology [11,12], DNA sequences [3] and foreign currency exchange [4,13]. Our current findings on the transition that occurs from thick uniform stratus formations to broken stratus clouds to clear sky is statistically similar to the mosaic of coding and non-coding *patches* in DNA [3].

2. DFA technique

The DFA technique consists in dividing a random variable sequence $y(n)$ of length N into N/t non-overlapping boxes, each containing t points. Then, the local trend, assumed to be linear in this investigation, $z(n) = an + b$ in each box is computed using a linear least-squares fit to the data points in that box. The detrended fluctuation function $F(t)$ is then calculated as follows:

$$F^2(t) = \frac{1}{t} \sum_{n=3Dkt+1}^{(k+1)t} [y(n) - z(n)]^2, \quad k = 0, 1, 2, \ldots, \left(\frac{N}{t} - 1\right). \tag{1}$$

Averaging $F^2(t)$ over the N/t intervals gives the fluctuations $\langle F^2(t) \rangle$ as a function of t. If the $y(n)$ data are short-range correlated variables, the behavior is expected to be a power law

$$\langle F^2(t) \rangle \sim t^{2\alpha} . \tag{2}$$

By analogy with DNA sequences, an α exponent less (greater) than $\frac{1}{2}$ indicates the so-called non-coding (coding) regions. In financial markets, economic conditions following political events imply rebounds or turnover in the α behavior. A small value of α indicates antipersistence [14] of correlations.

Among the DFA method main advantages, over techniques like Fourier transforms or wavelets, let us notice that (i) local and large-scale trends are avoided, and (ii) local correlations can be easily probed. Moreover the local value of the α exponent indicates what type of physics underlies the phenomenon. Also the time derivative of α can usually be correlated to "entropy" production [4].

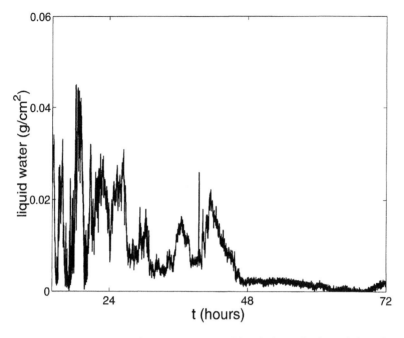

Fig. 1. Evolution of stratus cloud liquid water path measured in Oklahoma for the period April 3–5, 1998 from radiometry data [15].

3. Application to liquid water content in stratus cloud

Consider the liquid water content in stratus clouds as retrieved from microwave radiometer measurements obtained in Oklahoma between 1996 and 1999 [15]. We have studied several cases [16]. The one chosen here is the most representative one for the cloud evolution from thick stratus through patches to clear sky. Special care was taken to choose only data for analysis when no other cloud existed above the stratus deck. This constraint was important because these are remote sensing measurements and one should be sure that the analysed signal represents only the stratus cloud. We chose the period 14:00 UTC April 3 to 24:00 UTC April 5, 1998, when a rather sharp transition to clear sky occured on April 5, 1998. The time evolution of the liquid water content, over a 58 h time period, consisting of 10381 data points, is shown in Fig. 1.

In order to probe the existence of *locally correlated and decorrelated sequences*, we constructed a so-called observation box (a probe) of 6 h size placed at the beginning of the data, and we calculated α for the data in that box. The 6 h size choice is somewhat arbitrary, but the box should be larger than 2 h in order to avoid finite size effects. Then, we move this box by 60 points (20 min) toward the right along the signal sequence and again calculate α. Iterating this procedure for the data sequence, a "local measurement" is obtained for the degree of "local long-range correlations". The local α_6 exponent is found to be constant within the time interval from 2 to 25 min.

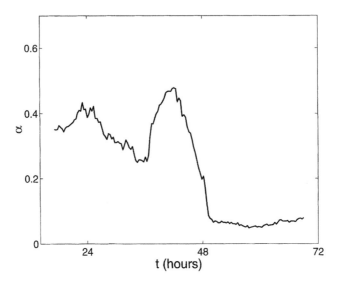

Fig. 2. The evolution of the local value of α estimated with the DFA technique for boxes of size 6 h, which slide to the right with a step of 20 min; see $\alpha_6 = 0.34$ for the cloudy and $\alpha_6 = 0.06$ for the clear sky period.

The results for this local α_6 as shown in Fig. 2 indicate two well-defined regions of scaling with different values of α_6. The first region corresponds to the first two days when thick stratus clouds existed. The average value of the local scaling exponent over this period is $\alpha_6 = 0.34 \pm 0.02$ with a sharp drop to $\alpha_6 = 0.1$ for the clear sky day. Note that α is a measure of the most characteristic dimension of a multifractal process [17,18]. Such a sharp variation of α is similar to that found by Liu et al. [19] in economic time series.

The value of α, which is close to 0.3, indicates a very large antipersistency, thus a set of fluctuations tending to induce a greater stability of the system, in contrast to a persistence of the system fluctuations which would drag the system out of equilibrium.

The appearance of broken clouds and clear sky following a period of thick stratus is hereby obviously interpreted as a non-equilibrium transition or a sort of bifurcation or fracture process in other languages. Notice that the drop of α from 0.34 toward a new lower value at 0.1 indicates a greater antipersistence of the prevailing meteorological process. This implies a specific dynamics to be usefully inserted as ingredients of models.

An $\alpha = 1/2$ value was observed [4] to indicate a period of relative economic calm in financial fluctuations. We emphasize that stable periods can occur for α values which do not correspond to the Brownian $\frac{1}{2}$ value. Moreover, from a fundamental point of view, it seems that the variations of α are as important as its value itself [4]. From the perspective of modeling and better predictability, α values significantly different from $\frac{1}{2}$ are to be preferred.

4. Conclusion

Thus, we have demonstrated that stratus clouds break apart when the fluctuations in the liquid water content of inner columns of stratus clouds become Brownian like. The non-equilibrium nature of the cloud structure and its associated column liquid water path [20] should receive some further thought. As is known from other work, it would be of interest for predictability models to examine whether the long-range fluctuations belong to a Levy-like, rather than Gaussian distribution as in self-organized criticality models [5], and is another phenomenon of Abelian nature, like rain fall [21] or stock market price [22,23], DNA structure [10] or heterogeneous sytems rupture [24]. If the answer is positive, the predictability range of liquid water paths in stratus clouds would be enormously extended and better models could be imagined.

In summary, long-range power-law correlations and anticorrelations have been shown to occur in stratus cloud systems. Moreover, we have quantified that specific sequences appear when the cloud system starts to break apart and undergo a transition to clear sky. It seems that these features can be associated with fundamental and local physical parameters, allowing them to be included in models of predictability and clouds.

Acknowledgements

Authors are grateful to E.E. Clothiaux, T.P. Ackerman from Pennsylvania State University Dept. of Meteorology for providing the data for this analysis. MA thanks an ARC (94-99/174) grant from the Ministery of Higher Education through the Research Council of the University of Liège. KI and MA both thank the NATO ARW organizers for financial support and the fruitful atmosphere they provided during the ARW.

References

[1] R.N. Mantegna, H.E. Stanley, Nature 376 (1995) 46.
[2] S. Ghashghaie, W. Breymann, J. Peinke, P. Talkner, Y. Dodge, Nature 381 (1996) 767.
[3] H.E. Stanley et al., Physica A 200 (1996) 4.
[4] N. Vandewalle, M. Ausloos, Physica A 246 (1997) 454.
[5] P. Bak, K. Chen, M. Creutz, Nature 342 (1989) 780.
[6] K. Ivanova, M. Ausloos, Physica A 265 (1999) 279.
[7] H.E. Stanley, Phase Transitions and Critical Phenomena, Oxford Univ. Press, Oxford, 1971.
[8] D. Sornette, Physics Rep. 297 (1998) 239.
[9] E.F. Fama, J. Fin. 45 (1990) 1089.
[10] C.-K. Peng, S.V. Buldyrev, S. Havlin, M. Simons, H.E. Stanley, A.L. Goldberger, Phys. Rev. E 49 (1994) 1685.
[11] E. Koscielny-Bunde, A. Bunde, S. Havlin, H.E. Roman, Y. Goldreich, H.-J. Schellnhuber, Phys. Rev. Lett. 81 (1998) 729.
[12] E. Koscielny-Bunde, A. Bunde, S. Havlin, Y. Goldreich, Physica A 231 (1993) 393.
[13] N. Vandewalle, M. Ausloos, Int. J. Comput. Anticipat. Systems 1 (1998) 342.
[14] P.S. Addison, Fractals and Chaos, Institute of Physics, Bristol, 1997.
[15] G.M. Stokes, S.E. Schwartz, Bull. Amer. Meteor. Soc. 75 (1994) 1201.
[16] K. Ivanova et al., submitted for publication.

[17] K. Ivanova, M. Ausloos, Eur. Phys. J. B 8 (1999) 665.
[18] K. Ivanova, T.P. Ackerman, Phys. Rev. E 59 (1999) 2778.
[19] Y. Liu, P. Cizeau, M. Mayer, C.-K. Peng, H.E. Stanley, Physica A 245 (1997) 437.
[20] S.S. Seker, O. Cerezci, J. Phys. D 32 (1999) 552.
[21] S.T.R. Pinho, R.F.S. Andrade, Physica A 255 (1998) 483.
[22] N. Vandewalle, M. Ausloos, Eur. J. Phys. B 4 (1998) 139.
[23] N. Vandewalle, Ph. Boveroux, A. Minguet, M. Ausloos, Physica A 255 (1998) 201.
[24] J.C. Anifrani, C. Le Floch, D. Sornette, B. Souillard, J. Phys. I (France) 5 (1995) 631.

ELSEVIER

Physica A 274 (1999) 355–360

www.elsevier.com/locate/physa

Mean-field-type equations for spread of epidemics: the 'small world' model

Adam Kleczkowski[a,*], Bryan T. Grenfell[b]

[a] *Department of Plant Sciences, University of Cambridge, Cambridge, CB2 3EA, UK*
[b] *Department of Zoology, University of Cambridge, Cambridge, CB2 3EJ, UK*

Abstract

In the paper we study a cellular automata (CA) model of epidemic dynamics. The effects of local spatial correlations on a temporal (aggregated) spread of single epidemics are studied, as a function of increasing proportion of global contacts ('small world' model). We conjecture that even in the presence of high local correlations, the aggregated (mean-field-type) models can be quite successful, if the contact rate is treated as a free parameter. The dependence of the (estimated) contact rate on the mixing parameter can be understood in terms of a simple probabilistic model. The contact rate reflects not only a microscopic and epidemiological situation, but also a complicated social pattern, including short- and long-range contacts as well as a possibly hierarchical structure of human society. © 1999 Elsevier Science B.V. All rights reserved.

PACS: 89.90+n; 82.20.D; 64.60.Ak

Keywords: Mean-field-type equations; Epidemics; 'Small-world' model

In recent years, there has been an increased interest in modelling interactions between agents located on networks with a mixture of local and global interactions. When the agents communicate with their nearest neighbours only, the progress of a disease or an innovation is slow. This is caused by high levels of correlations introduced by local interactions, leading to most of the contacts being effectively 'lost'. However, the agents can often make contacts spanning large distances on the network. In a world increasingly dominated by global travel, ideas or exotic diseases can spread very rapidly across countries and continents [1]. Watts and Strogatz [2] analysed a model for disordered networks and shown that they behave like a 'small world' model even for relatively small values of a disorder parameter [3]. Thus, the introduction of even

* Corresponding author. Fax: +44-1223 333 953.
E-mail address: ak133@cus.cam.ac.uk (A. Kleczkowski)

0378-4371/99/$ - see front matter © 1999 Elsevier Science B.V. All rights reserved.
PII: S 0378-4371(99)00393-3

a few long-range interactions causes the epidemics to proceed very fast throughout the population. A similar model was earlier analysed by Boccara and Cheong [4,5], where the individuals were allowed to swap places on a two-dimensional network.

An application of spatial models is, however, limited since they usually require a detailed knowledge of microscopic interactions. Thus, in practice, when the data are often limited to some aggregated global description [1,6,7], simpler methods need to be explored. In this paper we examine the dynamics of a network model motivated by the spread of measles [8–10], one of the best examples for studying a spread of an airborne, childhood viral infection [11]. Using arguments similar to Ref. [12] we examine the initiation and dynamics of clusters in a slightly modified version of Boccara and Cheong model [4]. A mean-field susceptible, exposed, infected and recovered (SEIR) model [13] is formulated as a set of ordinary differential equations, with a single parameter incorporating the spatial structure. A simple theory allows us to relate this parameter to the proportion of global interactions.

In our model, N individuals occupy nodes of a square, two-dimensional lattice, with periodic boundary conditions. The topology is interpreted in terms of an extended — socio-geographical — space rather than a purely geographical location and distance. This reflects a situation where individuals belong to more or less fixed social structures, only occasionally changed.

Each node is assigned one of three states: susceptible or healthy (S), infected and infectious (I) and recovered or immune (R). We concentrate on a single epidemic initiated by a few infective agents immersed either into a fully susceptible population or into a population containing a mixture of susceptible and recovered nodes ($S(0)=0.35$ or $S(0) = 0.99$, $I(0) = 0.01$, $R(0) = 1 - S - I$), thus ignoring births and introductions from outside. During a (short) time Δt, infection is passed to each susceptible agent among z nearest neighbors ($z = 4$, as for interactions limited to a family, or 16, as for an average class size) with a probability $p\Delta t$ per contact, independent of other interactions. In our case, $p = 0.9$ for $z = 4$ and $p = 0.225$ for $z = 16$, so that the product pz is constant. Infected agents recover at a rate $p_r\Delta t$ (corresponding to an infectious period $1/p_r$), and the system is updated synchronously with a small step Δt of one week, reflecting a basic infectious period of 1 week for measles. In addition, we introduce mixing by allowing two agents to exchange their positions on the lattice at any time. The number of changes in each time step is given by a Poisson variable with a mean (and variance) xN. The start and end positions are chosen in random for each move independently. This rule represents social structures better than the one of Refs. [4,14,15], where mixing is allowed when one of the nodes is empty (recovered) only.

The model was simulated 100 times. For varying initial conditions and stochastic realisations, see Fig. 1(a)–(e). For small values of mixing x, the infection progress is limited and the initial infection level quickly decreases (Fig. 1(a)). This corresponds to a 'large world' case in which contacts are local only. As the agents start to mix, the disease spreads more rapidly (Fig. 1(b)). Eventually, if all individuals can move, (Fig. 1(c)), the solution saturates at the mean-field asymptote. Then, the variables

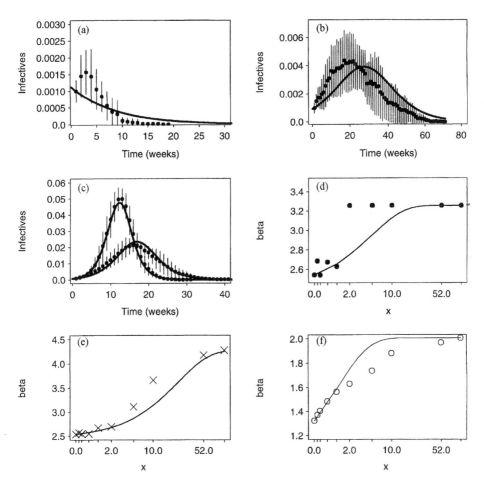

Fig. 1. (a)–(c) Single epidemic waves started by 10 infectives in the mixed ($S(0) = 0.35N$) population sized 100 by 100, $z = 4$, $p = 0.9$, and mixing probabilities x of 0 (a), 5/52 (b), and 10/52 and 1 (c), corresponding to 0, 960, 1920, and 10000 sites being exchanged on average during each step (week). The values represent the mean averaged over 10 runs, and the bars represent standard deviations. Solid lines show the approximate analytical solution (4)–(6), with the 'effective' parameter β obtained by fitting the solution to the data. (d)–(f) The 'effective' parameter β as a function of the mixing probability x, for (d) $z = 4$ and $p = 0.9$, $S(0) = 0.35N$, (e) $z = 16$ and $p = 0.225$, $S(0) = 0.35$ and (f) $z = 16$ and $p = 0.225$, $S(0) = 0.99$. The solid lines correspond to the predictions by model (2). Note the logarithmic scale for x in (d)–(f).

satisfy a set of differential mean-field equations (MF)

$$\frac{dS}{dt} = -\beta SI \ , \tag{1}$$

$$\frac{dI}{dt} = \beta SI - gI \ , \tag{2}$$

$$R = 1 - (S + I) \ . \tag{3}$$

In practice, the only parameter that can be easily estimated independently is g (related to p_r by the formula $p_r = 1 - \exp(-g\Delta t)$). The strength p and range of interactions z as well as mixing x are difficult to estimate independently. However, we can use the solutions to Eqs. (1)–(3) to find the combined force of infection β. Although (1)–(3) are valid in the limit of $x \to \infty$ only, the question arises whether we can use them satisfactorily to describe the dynamics for smaller values of x. We use an approximate analytical solution to (1)–(3) under the assumption that $I(t)$ is not large [16]

$$I(t) \simeq \frac{\alpha^2}{2((\beta S(0))/g)^2 S(0)} \operatorname{sech}\left(\frac{\alpha g t}{2} - \phi\right) \tag{4}$$

with

$$\alpha = \sqrt{\left(\frac{\beta S(0)}{g} - 1\right)^2 + 2\frac{\beta S(0)}{g}\frac{\beta I(0)}{g}}, \tag{5}$$

$$\phi = \tanh^{-1}\left(\left(\frac{\beta S(0)}{g} - 1\right)\bigg/\alpha\right). \tag{6}$$

The approximation works well, even for small x, as shown in Fig. 1(a)–(c). The MF equations are fitted by a nonlinear least-squares method to the mean output from the simulations (as if the latter represented aggregated data). This yields an 'effective' value of β which is then used to predict the shape, and the result plotted in Fig. 1. For small values of mixing, the 'effective' β is small (Fig. 1(d)–(f)), but as x increases, it quickly reaches a saturation point. The asymptotic value of β differs from what we would expect from the MF, i.e. $\beta = pz = 3.6$. Since we are here comparing a spatially extended stochastic system with its deterministic equivalent, such differences in the parameters can be expected [17]. For similar densities of susceptibles, the larger the neighbourhood, the faster the spread, even if probability of individual contacts is smaller. For a completely susceptible population, the mean-field tends to overestimate the spread of infection by neglecting spatial correlations that exist even in the presence of very high mixing. This needs to be corrected by a smaller value of β. It is also possible that in this case, approximations (4)–(6) do not work properly.

The dynamics of the infection can be understood in terms of the clusters evolution by arguments similar to those in Ref. [18]. For the case with no mixing, the infection spreads in the form of expanding foci centered around initial infected agents. Inside the moving wave, an averaged infective mainly contacts other infectives and this leads to a small value of the effective contact rate $\beta = \beta_0$ (non-mixing case). When an infective is removed from the wave by the mixing process, it is often placed in contact with susceptibles outside the focus of the epidemic. As most of its neighbours are now susceptible the contribution to the infection process is much higher. The averaging over areas with different densities of susceptibles and infectives gives the mean-field contribution (i.e. proportional to $S(t)I(t)$). As the mixing is governed by a spatial Poisson process, we can calculate the average number of neighbourhoods in which the infective is placed in the susceptible surrounding as $1 - \exp(-xz p_r^{-1} S(0))$ where $zS(0)$ is an averaged number of susceptibles in the interaction neighbourhood outside the spreading foci, and the average lifetime of a single infective (and therefore the probability that a new focus survives) is p_r^{-1}. This yields a new 'effective' contact parameter

$$\beta_{\text{eff}} \equiv \beta_0 + (1 - \exp(-xz p_r^{-1} S(0)))(\beta_\infty - \beta_0) \tag{7}$$

with two parameters describing the contact rate without mixing (β_0) and with a full mixing (β_∞). Fig. 1(d)–(f) shows the predictions of the model (7) with the parameters estimated by taking the extreme values of β from the fitted β for $x = 0$ and $x = 100/52$.

Although there are differences between the observed values of β and the predictions of the model, especially for the moderate mixing, there is a good qualitative agreement. The model not only predicts a sigmoidal shape for $S(0) = 0.35$ but also a more monotonic change for $S(0) = 0.99$. We also note that substantial levels of mixing are necessary to bring the system close to the MF equations with β similar to the value expected by a simple analogy ($\beta = pz$), contrary to the observation of Watts and Strogatz. However, MF equations with an 'effective' contact rate can be used across a much larger range of mixing. The contact rate as observed in the data is therefore as much a function of microscopic local interactions as of the large-scale social mixing patterns [19].

The next theoretical challenge is to predict the exact values of the asymptotic contact parameters, β_0 and β_∞. For small values of mixing the dynamics are essentially of a percolating system ($I(t) \sim t$), and thus differ from the MF formula ($I(t) \sim \exp(\beta t)$, t small). A comparison of the two limits might lead to other forms of the MF contact term βSI, as predicted by moment-closure methods. For moderate to large mixing, the system behaves like a reaction-diffusion system in an infinitely dimensional space [4,20,21]. However, until the mixing is very large, the dynamics are dominated by the reactions. Linking different paradigms describing the same spatially extended system can lead to a more satisfactory description which can be easily applied for practical purposes. The next step towards a realistic model of measles dynamics is to incorporate births and imports dynamics and to include seasonality in the contact process [22].

Acknowledgements

We thank the Royal Society, London, BBSRC, King's College, Cambridge, the British Council, (A.K.) and the Wellcome Trust (B.T.G.) for their support. Special thanks to the organisers of the Nato ARW, Budapest, 1999.

References

[1] A.D. Cliff, P. Haggett, Atlas of Disease Distributions: Analytic Approaches to Epidemiologic Data, Basil Blackwell, Oxford, 1988.
[2] D. Watts, S. Strogatz, Collective dynamics of 'small-world' networks, Nature 393 (6684) (1998) 440–442.

[3] H. Herzel, How to quantify 'small-world networks'? Fractals — an Interdisciplinary J. Comples Geom. Nature 6 (4) (1998) 301–303.

[4] N. Boccara, K. Cheong, Critical-behavior of a probabilistic-automata network SIS model for the spread of an infectious-disease in a population of moving individuals, J. Phys. A: Math. Gen. 26 (15) (1993) 3707–3717.

[5] N. Boccara, E. Goles, S. Martinez, P. Picco, Cellular Automata and Cooperative Phenomena, Kluwer Academic Publishers, Dordrecht, 1993.

[6] The Registrar General's Weekly Return for England and Wales: Births and Deaths, Infectious Diseases, Weather, Her Majesty's Stationary Office, London, 1948–1968.

[7] B. Finkenstadt, B. Grenfell, Empirical determinants of measles metapopulation dynamics in England and Wales, Proc. Roy. Soc. B 265 (1392) (1998) 211–220.

[8] C. Rhodes, R. Anderson, A scaling analysis of measles epidemics in a small population, Philos. Trans. Roy. Soc. London Ser. B 351 (1348) (1996) 1679–1688.

[9] C. Rhodes, R. Anderson, Persistence and dynamics in lattice models of epidemic sprea, J. Theor. Biol. 180 (2) (1996) 125–133.

[10] C. Rhodes, H. Jensen, R. Anderson, On the critical behaviour of simple epidemics, Proc. Roy. Soc. B 264 (1388) (1997) 1639–1646.

[11] B. Grenfell, Chance and chaos in measles dynamics, J. Roy. Stat. Soc. Ser. B 54 (2) (1992) 383–398.

[12] R. Durrett, S. Levin, The importance of being discrete (and spatial), Theor. Popul. Biol. 46 (3) (1994) 363–394.

[13] R.M. Anderson, R.M. May, Infectious Diseases of Humans, Oxford University Press, Oxford, 1991.

[14] N. Boccara, K. Cheong, Automata network SIR models for the spread of infectious-diseases in populations of moving individuals, J. Phys. A: Math. Gen. 25 (9) (1992) 2447–2461.

[15] N. Boccara, K.O. Cheong, M. Oram, A probabilistic-automata network epidemic model with births and deaths exhibiting cyclic behavior, J. Phys. A: Math. Gen. 27 (5) (1994) 1585–1597.

[16] J.D. Murray, Mathematical Biology, Springer, Berlin, 1989.

[17] G. Gibson, C. Gilligan, A. Kleczkowski, Predicting variability in biological control of a plant-pathogen system using stochastic models, Proc. Roy. Soc. B 266 (1999) 1743–1753.

[18] R. Durrett, S.A. Levin, Stochastic spatial models — a users guide to ecological applications, Philos. Trans. Roy. Soc. London Ser. B 343 (1305) (1994) 329–350.

[19] B. Grenfell, B. Bolker, Cities and villages: infection hierarchies in a measles metapopulation, Ecol. Lett. 1 (1) (1998) 63–70.

[20] J.L. Cardy, Critical exponents of directed self-avoiding walks, J. Phys. A: Math. Gen. 16 (11) (1983) L709–712.

[21] R. Bidaux, N. Boccara, H. Chate, Order of the transition versus space dimension in a family of cellular automata, Phys. Rev. A 39 (6) (1989) 3094–3105.

[22] M. Keeling, D. Rand, A. Morris, Correlation models for childhood epidemics, Proc. Roy. Soc. B 264 (1385) (1997) 1149–1156.

ELSEVIER

Physica A 274 (1999) 361–366

www.elsevier.com/locate/physa

Gaussian-logarithmic distribution of relaxation times in relaxor materials

R. Skulski

Chair of Materials Science, Faculty of Technics, University of Silesia, 41-200 Sosnowiec, ul. Sniezna 2, 41-200 Sosnowiec, Poland

Abstract

It is shown that the Gaussian-logarithmic distribution of relaxation times and its broadening in low temperatures can lead to Vogel–Fulcher relation for frequency of maximum of dielectric permittivity though the temperature dependency of mean relaxation time is of Arrhenius type. © 1999 Elsevier Science B.V. All rights reserved.

1. Introduction

Relaxors are materials in which the dispersion is usually observed at relatively low frequencies. The best-known example is lead magnoniobate PMN-Pb($Mg_{1/3}Nb_{2/3}$)O_3. The physics of relaxation phenomena in these materials is very interesting. Such phenomena are observed in relaxors at RF and lower frequencies. Similar anomalous slowing-down of relaxation phenomena takes place in glass forming liquids and in spin glasses.

As a rule the temperature dependency of relaxation time is of the Arrhenius type [1–4]

$$\frac{1}{\tau} = \frac{1}{\tau_0} \exp\left(\frac{-E_0}{kT}\right) \ . \tag{1}$$

The other possible relation is the Vogel–Fulcher (V–F) relation

$$\frac{1}{\tau} = \frac{1}{\tau_0} \exp\left(\frac{-E_0}{k(T - T_f)}\right) \ , \tag{2}$$

where τ_0, E_0, and T_f are constants. At temperature $T = T_f$ relaxation time τ would stream to infinity (*freezing*). Some authors conclude that V–F relation for temperature

---

E-mail address: skulski@us.edu.pl (R. Skulski)

maximum $\varepsilon'(\omega, T)$ observed in relaxors is a consequence of V–F relation for relaxation times. However, Tagantsev in [4] has shown that the V–F relationship for the positions of the temperature maximum for the real and imaginary parts of the dielectric permittivity can be obtained as a direct consequence of the gradual broadening of the spectrum with decreasing temperature and such a V–F-type relationship does not necessarily imply freezing in the system. On the other hand in the following work by Glazounov and Tagantsev [5] it has been shown that the application of the model presented in [4] to two well-known relaxors PMN and PST leads to the conclusion that the freezing in these materials is "true" freezing, i.e. relaxation times really undergoes V–F relation.

In this work numerical calculations for PMN and SBN are presented which confirm the suggestion that V–F relation between T_m and ω may only be a consequence of the gradual widening of Gaussian-logarithmic distribution of relaxation times while the dependency of mean relaxation time on temperature is Arrhenius type. For this purpose the special method of estimation of the width of Gaussian distribution of relaxation times is proposed. Then the results of calculations for two model relaxors PMN and SBN are presented.

2. The method of estimation of the width of Gaussian-logarithmic distribution of relaxation times

Classic Debay's theory of the dispersion is valid for materials with single relaxation time. In real materials such assumption is in general not true. In such cases we must take into consideration the continuous distribution of relaxation times. Analytical calculations in such cases are possible (for example [6]) only when we can calculate the integral

$$\varepsilon(T) = \int_{-\infty}^{\tau(T)} \varepsilon(T, \tau) y(\tau) \, d\tau . \tag{3}$$

The distribution of relaxation times which seems to be physically reasonable is a Gaussian (in logarithmic scale) distribution. Below, it will be shown how it is possible to estimate the width of Gaussian-logarithmic distribution.

Let us assume that the Gaussian (in logarithmic scale) distribution of relaxation times takes place and it can be described by the function

$$y(z, \sigma) = g(z, \sigma) = \frac{1}{\sqrt{2\pi}\sigma} \exp\left(-\frac{(z - \log(\tau))^2}{2\sigma^2} \right) , \tag{4}$$

where z is the independent variable of integrating and σ the width of the distribution.

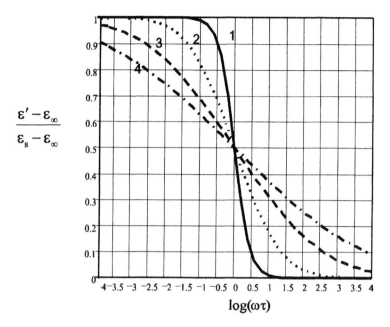

$$\frac{\varepsilon' - \varepsilon_\infty}{\varepsilon_s - \varepsilon_\infty}$$

Fig. 1. Normalized plots of $(\varepsilon' - \varepsilon_\infty)/(\varepsilon_s - \varepsilon_\infty)$ vs. $\log(\omega\tau_0)$ for various widths of Gaussian distribution $\sigma = 0.1, 1, 2, 3$ (in logarithmic scale).

For a given relaxation time τ the dielectric permittivity can be calculated from the well-known Debay's formula

$$\varepsilon(n, \tau) = \varepsilon_\infty + \frac{(\varepsilon_s - \varepsilon_\infty)}{1 - i\omega(n)\tau}, \tag{5}$$

where $\omega(n) = 2\pi \times 10^n$.

It is well known that the plot $(\varepsilon' - \varepsilon_\infty)/(\varepsilon_s - \varepsilon_\infty)$ vs. $\log(\omega\tau_0)$ for all materials should be the same (of course if we assume the single relaxation time). The distribution of relaxation times leads to deviation from such model plot. It is possible to estimate the width of Gaussian distribution using such normalized plots. For this purpose calculations of $(\varepsilon' - \varepsilon_\infty)/(\varepsilon_s - \varepsilon_\infty)$ for various $\log(\tau)$ have been done. It has been stated that the shape of the normalized plot depends only on the width of Gaussian distribution σ (i.e. does not depend on the other parameters in the formula). Results are presented in Fig. 1.

Now it is possible to find the relation between the value of σ and the shape of normalized plot. For this purpose in Fig. 1 the horizontal line has been plotted. The value 0.3 has been chosen arbitrarily but the trials have shown that this value is not critical. Intersection of this line with plots gives us the relation between σ and the value of $\log(\omega\tau)$ for which $(\varepsilon' - \varepsilon_\infty)/(\varepsilon_s - \varepsilon_\infty) = 0.1$. Obtained in such a way the relation can be expressed as

$$\log(\omega\tau) = a + b\sigma . \tag{6}$$

where $a = 0.36$ and $b = 1.2$.

Analyzing the experimental data in such a manner it is possible to obtain the dependencies $\tau_0(T)$ and $\sigma(T)$ for real materials. The procedure described above has been successfully applied to model relaxors PMN and SBN which is described below.

3. Results for PMN-Pb(Mg$_{1/3}$Nb$_{2/3}$)O$_3$

The experimental data for lead magnoniobate (PMN) at various temperatures have been taken from Ref. [7] and the diagram in Fig. 1 has been constructed using this data. It has been assumed that $\varepsilon_\infty = 800$, while the ε_s fulfills the generalized Curie–Weiss law in the form

$$\frac{1}{\varepsilon_s} = A + B|T - T_m|^\gamma , \tag{7}$$

where $A = 5.94 \times 10^{-5}$, $B = 3.44 \times 10^{-8}$ (K^{-1}), and $T_m = 250$ K, $\gamma = 1.8$.

At every temperature the frequency has been found at which $(\varepsilon' - \varepsilon_\infty)/(\varepsilon_s - \varepsilon_\infty) = 0.5$. The relaxation time $\tau(T)$ is the reciprocal of the value obtained for ω. The dependency obtained in such a way is well described by Eq. (1) with $\tau_0 = 0.363 \cdot 10^{-27}$ s, $E_0 = 1.09$ eV (i.e. the dependency is Arrhenius type).

The $\sigma(T)$ dependency found in the manner described in the previous section can be expressed as

$$\sigma(T) = \frac{\sigma_\infty}{(T - T_\sigma)^\xi} , \tag{8}$$

where $\sigma_\infty = 2.69 \times 10^3$, $T_\sigma = 194$ K, and $\zeta = 1.61$.

It means that at the temperature 194 K the width of Gaussian curve streams to infinity, which means that at this temperature $\sigma \to \infty$ and the dispersion curve becomes a line. As a result we can obtain $\varepsilon'(T)$ dependencies for various frequencies which are presented in Fig. 2. From plots in Fig. 2 we can find the relation between T_m and ω. The results are presented in Fig. 3. The dependency presented in this figure can be described by the equation

$$\omega = \omega_0 \exp\left(\frac{-E_0}{k(T_m - T_f)}\right) . \tag{9}$$

Simple transformations lead to

$$T_m = T_f + \frac{E_0}{k \ln(\omega_0/\omega_m)} , \tag{10}$$

where $E_0 = 1.152 \times 10^{-20}$ J, $T_f = 195$ K, and $\omega_0 = 2\pi \times 5.46 \times 10^{10}$ Hz.

These values obtained in a fully numerical way are very similar to that obtained experimentally by Viehland [1], which are also presented in Fig. 3.

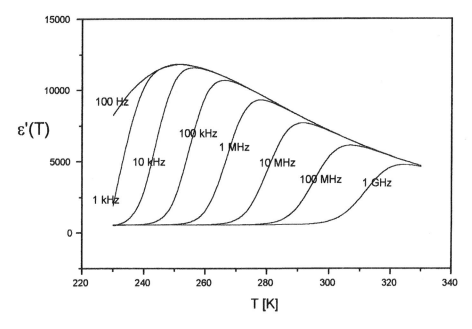

Fig. 2. Numerically obtained dependencies $\varepsilon'(T)$ for PMN, for various frequencies.

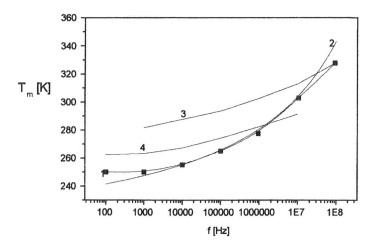

Fig. 3. Obtained as a result of calculations described in the text and experimental relation between T_m and frequency f; (1) Data obtained from Fig. 2. (2) results of fitting to Eq. (5); (3) experimental data from Ref. [7]; (4) experimental data from Ref. [1].

4. Results for SBN-Ba$_{1.25}$Sr$_{3.75}$Nb$_{10}$O$_{30}$-SBN-75 (%Sr)

SBN is a relaxor with the so-called not fully filled tetragonal tungsten bronze structure. The experimental data have been taken from Refs. [8,9] and from the author's own investigations. The plots $(\varepsilon' - \varepsilon_\infty)/(\varepsilon_s - \varepsilon_\infty)$ vs. frequency have been constructed in the manner described above. Next, the frequencies at which $(\varepsilon' - \varepsilon_\infty)/(\varepsilon_s - \varepsilon_\infty)$ is equal to 0.5 and the dependency $\sigma(T)$ were found. Like in PMN such obtained dependency is also Arrhenius type (Eq. (1)) with $\tau_0 = 1.95 \times 10^{-7}$ s or as $E_0 = 3.54 \times 10^{-20}$ J.

Similar analysis as described above for PMN gives us $\sigma(T)$ which can now be described by the equation

$$\sigma(T) = 16.76 - 0.087T + 1.148 \times 10^{-4}T^2 . \tag{11}$$

A comparison with Eq. (8) leads us to the conclusion that this dependency is different than in the case of PMN. However, the physical reasons for such dependency are not clear and Eq. (11) should be rather treated as a result of numerical fitting only. Also in this case the relation between temperature of the maximum dielectric permittivity and the measuring frequency is V–F type in spite of the fact that the relation for relaxation times is Arrhenius type.

5. Conclusions

Specific dispersion behaviours of model relaxors PMN and SBN are the results of superposition for the following reasons:

- dependency of static dielectric permittivity given by Eq. (7) (i.e. generalized Curie–Weiss law);
- dependency of mean relaxation time on temperature given by Eq. (1) (i.e. Arrhenius type);
- Gaussian-logarithmic distribution of relaxation times whose width become wider at lower temperatures.

References

[1] D. Viehland, S.J. Jang et al., Philos. Mag. B 64 (1991) 335.
[2] E. Courtens, Phys. Rev. B 33 (1986) 2975.
[3] Z. Kutnjak, C. Filipic et al., Phys. Rev. Lett. 70 (1993) 4015.
[4] A.K. Tagantsev, Phys. Rev. Lett. 72 (1994) 1100.
[5] A.E. Glazounov, A.K. Tagantsev, Appl. Phys. Lett. 73 (1998) 856.
[6] H. Fröhlich, Theory of Dielectrics, Oxford University Press, Oxford, 1958.
[7] T. Tsurumi, K. Sejima et al., Jpn. J. Appl. Phys. 33 (1994) 1959.
[8] A.M. Glass, J. Appl. Phys. 40 (1969) 4699.
[9] A. Rost, O. Kersten, Ferroelektrizität, Martin-Luther-Uiversitt, Halle (Saale), 1981, p. 78.

ELSEVIER

Physica A 274 (1999) 367–373

www.elsevier.com/locate/physa

The effect of vibrational degrees of freedom on the phase transition in a 2D Ising model

S.V. Stroganova[a], M.I. Vasilevskiy[a,b,*], O.V. Vikhrova[a]

[a]*Faculty of Physics, N.I.Lobachevskii Nizhni Novgorod University, 603600 Nizhni Novgorod, Russia*
[b]*Departamento de Fisica, Universidade do Minho, Campus de Gualtar, Largo do paco,
Braga 4709, Codox, Portugal*

Abstract

The aim of this work is to show how important small atomic vibrations can be for the equilibrium state of a binary alloy or a lattice gas of adatoms on a surface. Using the Monte-Carlo procedure and performing direct lattice dynamics calculations, the total (configurational plus vibrational) free energy of the system is obtained. The calculations show that, if the difference in the bond stiffness between like and unlike atoms is large, it is possible to introduce a renormalised interaction constant and to treat the system as Ising-like. For this case, the effects of lattice dynamics parameters on the order–disorder transition temperature are discussed. ⓒ 1999 Elsevier Science B.V. All rights reserved.

1. Introduction

Usually in the calculation of the thermodynamical potentials of a binary alloy or a lattice gas atomic vibrations are not taken into account. The Ising model, with its well-known properties [1], has been applied to the description of order–disorder phenomena in these systems for a long time. This model assumes that atoms are at rest upon fixed lattice sites and are not allowed to vibrate. Consequently, only the configurational free energy is considered. However, there is experimental evidence that, in some cases, the vibrational free energy can be as important as the configurational one. It has been known for a long time that the entropies of mixing for concentrated solid and liquid solutions are larger than those predicted from configurational entropy alone [2]. More recently [3], the vibrational entropy (S_V) of Ni_3Al was measured in two states of chemical order: as a disordered f.c.c. solid solution, and as the equilibrium

---

* Corresponding author. Fax: +351-53-678-981.
E-mail address: mikhail@fisica.uminho.pt (M.I. Vasilevskiy)

0378-4371/99/$ - see front matter ⓒ 1999 Elsevier Science B.V. All rights reserved.
PII: S 0378-4371(99)00315-5

LI_2 ordered structure. Using three independent methods, it was obtained that, at high temperatures, S_V of disordered Ni_3Al is about $0.3k/$atom greater than for the ordered alloy. This value is comparable to the configurational entropy. Another striking fact showing the importance of the vibrational degrees of freedom for atomic thermal equilibrium was pointed out in Ref. [4]. The equilibrium concentration of deuterium on a Si surface was found to exceed 15 times (at 700 K) that of hydrogen. This was explained solely by the difference in the surface vibration frequencies of D and H adatoms.

From the theoretical viewpoint, including lattice vibrations in the calculation of the partition function is rather complicated because it is necessary to know the complete phonon spectrum. The latter changes substantially if there is a short-range order in the alloy [5]. These changes in the phonon spectrum lead to the difference in the vibrational free energy between random, ordered and clustered alloys (except for the case of purely isotopic disorder). The dependence of the vibrational entropy and free energy of two-component one-dimensional chains on the short-range-order parameter (r) in the high-temperature harmonic approximation was studied in [6] and it was found that ΔS_V depends linearly on r. In [7,8], this result (generalised for a 3D simple cubic lattice) was used to introduce a renormalised Ising interaction constant taking account of the atomic vibrations. From this it was concluded that the temperature scale 'shrinks' in comparison with the situation without vibrations, i.e. the critical temperature should decrease. However, the consideration of [7,8] is correct only for high temperatures compared to the Debye temperature, since it makes use of the following approximation for the vibrational entropy difference between two different alloy states (I and II):

$$\Delta S_V \equiv S_V^{\mathrm{I}} - S_V^{\mathrm{II}} = -\frac{1}{2}k\ln\frac{\prod \gamma_{ij}^{\mathrm{I}}}{\prod \gamma_{ij}^{\mathrm{II}}}, \tag{1}$$

where γ_{ij} are the force constants between nearest neighbours and k is the Boltzman constant hereafter put equal to unity.

In this work, we treat the problem by performing direct lattice dynamics calculations, which provide the vibrational free energy for any temperature. Explicitly, we study a 2D binary alloy $A_{0.5}B_{0.5}$ on square lattice, which is described by the following Hamiltonian:

$$H = -2E_m \sum_{(i,j)} c_i c_j + \frac{M}{2}\sum_i (1 - \varepsilon c_i)\dot{u}_i^2 + \frac{\gamma}{2}\sum_{(i,j)}(1 - \varepsilon_\gamma|c_i - c_j|)(u_i - u_j)^2 \tag{2}$$

where E_m is the mixing energy, $c_j = 0$ or 1 for atoms A and B, respectively, u_i are (small) atomic displacements, $M = M_A$, $\varepsilon = 1 - M_B/M_A$ (M_A, M_B-atomic masses), $\gamma_{AA} = \gamma_{BB} = \gamma$ and $\gamma_{AB} = \gamma(1 - \varepsilon_\gamma)$ are the force constants. The first term in (2) corresponds to the standard Ising model, while the other two represent lattice vibration's energy in the harmonic approximation.

In the next section we explain the numerical procedure used to tackle the problem. Some calculated results are presented and discussed in Section 3, while Section 4 serves for concluding remarks.

2. Numerical procedure

The probabilities of different microstates of the system are determined by

$$P\{c_i\} = \exp\left[-\frac{H_1 + H_2}{T}\right], \tag{3}$$

where H_1 and H_2 stand for the 'pure' Ising and phonon terms in (2), respectively. To decouple the vibrational and configurational degrees of freedom in (3), we make use of the fact that the vibrational density of states (DS) is a self-average quantity [9]. This means that, for a sufficiently large system, thermodynamical quantities, which can be expressed in terms of the DS, have reliable values, i.e. their fluctuations are small. We consider H_2 as a function of just $c = \langle c_i \rangle$ (0.5 in our case) and the short-range-order parameter, thus assuming that this part of the total energy does not depend on micro-realizations $\{c_i\}$. The idea is then to find an effective short-range-order parameter (or, equivalently, an effective mixing energy $J^*(T) \neq E_m$), which would minimize the total (vibrational plus configurational) free energy. The procedure consists of the following steps:

(i) A variable mixing energy J is introduced. Using the standard Metropolis technique one calculates the configurational free energy per atom $F_c(J, T)$ for the 'pure' Ising model with the mixing energy J at tempearature T. Note that $f_c = F_c(J, T)/T$ depends only on (J/T); so, these functions for different T can be obtained by rescaling a single curve. Of course, $F_c(J)$ has a minimum at $J = E_m$ for any T.

(ii) For a given T and different J, an ensemble (of a few tens) of alloy crystallites is generated. For each crystallite, one constructs the dynamical matrix and calculates the vibrational DS $g(\omega)$. The DS is averaged over the ensemble of alloy realizations for each J (see [5] for details of the lattice dynamics calculations). Then the vibrational free energy per atom is calculated according to

$$F_v(J, T) = T \int_0^{\omega_{max}} d\omega\, g(\omega) \ln\left(2\sinh\left(\frac{\hbar\omega}{2T}\right)\right), \tag{4}$$

where $\omega_{max} = 4\max\{\gamma_{ij}\}/M$ is another independent (of E_m) parameter of the problem.

(iii) The total free energy $(F_c + F_v)$ is considered as a function of J. In general, it has a minimum at some $J^*(T) \neq E_m$. This way one determines the effective mixing energy. If it eventually does not depend much on the temperature, system (2) can approximately be considered as Ising-like with the renormalised interaction constant. If this is not the case, one should collect the minimum values of $(F_c + F_v)$ corresponding to different temperatures and calculate the equilibrium total energy as a function of T. Then the total specific heat should be analysed as a function of T in the usual way and conclusions concerning the phase transitions can be made.

3. Results and discussion

The phonon DS calculated for alloys with different short-range order is shown in Fig. 1. We used for simplicity $M_A = M_B$ ($\varepsilon = 0$). As seen from Fig. 1, clustering and ordering lead to some rearrangement of the alloy vibrational modes which has effect on the thermodynamical properties. For $E_m < 0$ and $\varepsilon_\gamma < 0$ (that is, the mixed bond is stiffer than AA or BB bonds), atomic ordering leads to some loss in the vibrational free energy increasing together with γ_{AB}/γ (see Fig. 2). In contrast to ordering, clustering ($E_m > 0$), if $\varepsilon_\gamma > 0$, produces some gain in the vibrational free energy compared to the random alloy. In the case of $\varepsilon_\gamma > 0$ (i.e., $\gamma_{AB} < \gamma$) the behavior is the opposite to that for $\varepsilon_\gamma < 0$.

The free energy difference due to short-range order increases with the increase of the short-range parameter (Fig. 3). At temperatures higher than the mixing energy (and $\hbar\omega_{max}$ since we chose $q = \hbar\omega_{max}/E_m = 0.5$) this dependence is linear, in accordance with the results of Ref. [6] and formula (1). However, this is not so at lower temperatures when high-frequency vibrations are not excited.

The minima of the total free energy plotted as a function of $x = |J|/T$ (see Fig. 4) deviates in x from the 'pure' Ising value of $x_c \approx 0.881$. The deviation is positive or negative in the cases of clustering and ordering ($\gamma_{AB} > \gamma$). This means that thermal equilibrium with respect to the vibrational degrees of freedom enhances clustering and suppresses ordering when the mixed bond is stiffer than that between like atoms. Such a

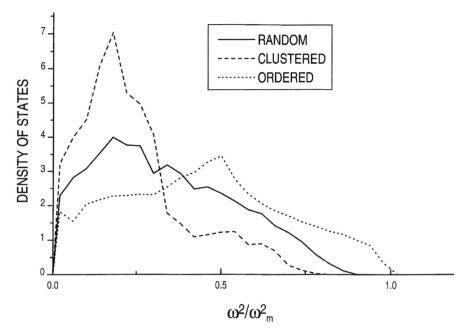

Fig. 1. Calculated phonon density of states for random ($E_m = 0$), clustered ($E_m = 0.85$ T) and partially ordered ($E_m = -0.85$ T) alloys with $M_A = M_B = M$ and $\gamma_{AB}/\gamma = 3$. The spectral variable is scaled by $\omega_m^2 = 4\gamma/M$.

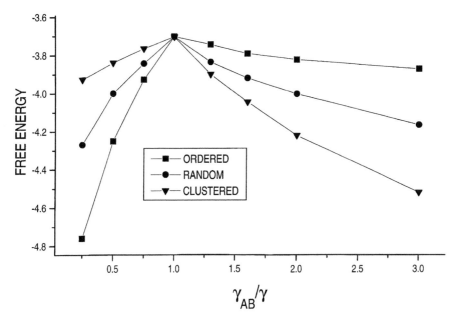

Fig. 2. Vibrational free energy calculated for random ($E_m = 0$), clustered ($E_m = 0.85$ T) and partially ordered ($E_m = -0.85$ T) alloys as a function of the force constant ratio γ_{AB}/γ.

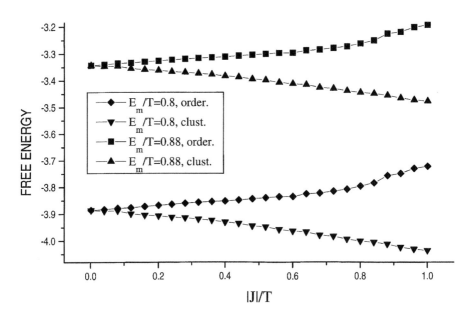

Fig. 3. Vibrational free energy as a function of the variable mixing energy calculated for the cases of clustering and ordereing and two temperatures.

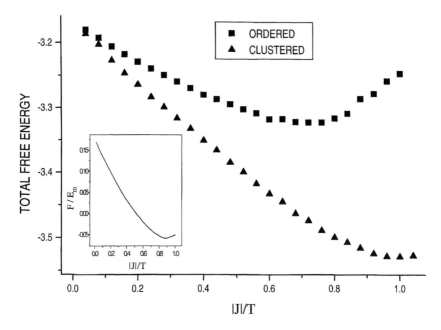

Fig. 4. Total (configurational plus vibrational) free energy calculated as a function of the mixing energy for the cases of ordering and clustering at temperature $T = |E_m|/0.88$. The inset shows the configurational free energy.

situation where $\gamma_{AB} > \gamma$ but $\varepsilon_{AB} < \varepsilon_{AA}, \varepsilon_{BB}$ (ε_{ij} are static pairwise interaction constants) corresponding to clustering should be unusual but not impossible.

For partially ordered alloy ($E_m < 0$), the renormalised interaction constant (corresponding the minimum of the total energy) $|J^*| < |E_m|$. It does not seem to depend significantly on the temperature for $\gamma_{AB}/\gamma \leqslant 2$; so, the system can still be considered as Ising-like. The critical temperature decreases with an increase of γ_{AB}/γ and a decrease of $q = \hbar\omega_{max}/|E_m|$. The former observation is in accordance with the mean-field theory prediction [3,7] (as we have mentioned, this theory is valid for $q \ll 1$). However, it should be pointed out that for a large γ_{AB}/γ the renormalised interaction constant depends on temperature and the situation is more complicated than just a change of the temperature scale.

4. Concluding remarks

In conclusion, we proposed an algorithm for including the vibrational degrees of freedom in the phase equilibrium calculations for an Ising alloy. We have shown that, if the mixed bond is stiffer than the like-atom's one, clustering is enhanced and ordering is suppressed compared to the case without vibrations. If the mixed bond is softer, the situation is the opposite. If the difference in the bond stiffness is large, the system cannot be described as Ising-like, with a renormalised mixing energy. A more detailed study of this case shall be the subject of the future work.

Finally, we would like to point out the practical aspect of these results. Since the vibrational degrees of freedom significantly affect the order–disorder transition for certain alloys, it can sometimes be useful to exclude thermal equilibrium with respect to lattice vibrations in order to obtain a different short-range-order state. For example, by creating a non-equilibrium phonon population one may increase the ordering critical temperature of a surface alloy thus promoting the formation of the ordered phase.

Acknowledgements

MIV wishes to acknowledge FCT/PRAXIS XXI/BCC Fellowship and the financial support from the NATO ARW, May 19–23, 1999, Budapest.

References

[1] H.E. Stanley, Introduction to Phase Transitions and Critical Phenomena, Oxford University Press, London, 1971.

[2] B.A. Oriani, Acta Metall. 4 (1956) 15.

[3] L. Anthony, J.K. Okamoto, B. Fultz, Phys. Rev. Lett. 70 (1993) 1128.

[4] I.P. Ipatova, O.P. Chikalova-Luzina, K. Hess, J. Appl. Phys. 83 (1998) 814.

[5] M.I. Vasilevskiy, O.V. Baranova, S.V. Stroganova, Comp. Phys. Commun. 97 (1996) 199.

[6] J.A.D. Matthew, R.E. Jones, V.M. Dwyer, J. Phys. F 13 (1983) 581.

[7] H. Barker, C. Tuijn, J. Phys. C 19 (1986) 5585.

[8] C. Tuijn, H. Barker, Phys. Stat. Sol. B 155 (1989) 107.

[9] I.M. Lifshitz, S.A. Gredeskul, L.A. Pastur, Introduction to Theory of Disordered Systems, Nauka, Moscow, 1982.

ELSEVIER

Physica A 274 (1999) 374–380

www.elsevier.com/locate/physa

Aging and self-organization of shear bands in granular materials

János Török[a,b], Supriya Krishnamurthy[b], János Kertész[a,*],
Stéphane Roux[c]

[a]*Department of Theoretical Physics, Institute of Physics, Technical University of Budapest,
8 Budafoki út, H-1111 Budapest, Hungary*
[b]*Laboratoire de Physique et Mécanique des Milieux Hétérogènes, Ecole Supérieure de Physique et
Chimie Industrielles de Paris, France*
[c]*Surface du Verre et Interfaces, UMR CNRS/Saint-Gobain, 39 Quai Lucien Lefranc,
93303 Aubervilliers Cedex, France*

Abstract

We introduce a simple lattice model for the formation and evolution of shear bands in granular media. We study this model by numerical simulations and find that due to the localization of the shear band and the consequent inhomogeneous aging of the material, the average density shows very unusual system-size-dependent features. We find qualitatively similar features on studying the same model on a hierarchical lattice for which we are able to derive several analytical results. ©1999 Elsevier Science B.V. All rights reserved.

1. Introduction

The response of granular materials to shearing is much more complex than that of solids. When the external shearing force is very small, the material deforms plastically. However, beyond this limit the material develops what is called a shear band: a region of finite width in which there is a relative motion of particles. Usually, for dense samples the shear band is well defined and stable. For loose materials the shear band wanders through the whole sample affecting different macroscopic properties of the material such as the density, the Coulomb angle and the internal structure.

The macroscopic coarse-grained study of this process has a long history [1–3] while microscopic modeling of the phenomenon has only a sparse literature. Molecular dynamics simulations have concentrated on calculating instabilities [4] but not on the role and structure of shear bands.

* Corresponding author. Fax: +36-1-463-3567.
E-mail address: kertesz@planck.phy.bme.hu (J. Kertész)

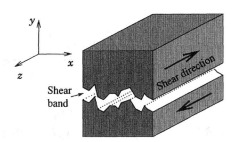

Fig. 1. Schematic picture of the shear process, the shear direction is z. Due to the periodic boundary conditions in this direction, the shear band is parallel to the z-axis. In order to reduce the problem to a two-dimensional one, we average along the z direction.

A number of experiments have been carried out on the shear process in granular materials due to its frequent occurrence in nature and its wide applications in industry [5,6]. Most of the experiments are triaxial tests [6,7] to determine macroscopic properties of materials such as the shear stress or the volumetric strain as a function of the shear strain. Models to explain these experimental results are mainly qualitative in nature with a number of parameters fed in from the experiments [7].

In this paper we will focus mainly on the microscopic details of the shearing process using a simple model. The aim of this model is to describe the formation and evolution of the shear bands in granular materials. The formation of new shear bands follows a global optimization procedure in the spirit of searching for the ground state of the directed polymers in a random potential [8–10]. However, in our case, this potential is not a priori frozen in, it has a self-organized development, reminiscent of the Bak–Sneppen model of evolution [11].

The relevance of the directed polymer model to fracture phenomena was suggested some time ago [12,13]. Here we demonstrate that this model also gives insight into the mechanism of failure formation in a very different context.

In the following section we define the model. We then present some of the results of our simulations on this model as well as a brief review of our analytical results on the hierarchical lattice.

2. Definition of the model

Let us consider a three-dimensional sample under shear where we have periodic boundary conditions in the shear direction (for an experimental realization see, e.g., Ref. [14]). After a given threshold (in stress and/or strain) is reached, the material is not able to support the external load anymore [1,15]. At this point a shear band is formed. The imposed periodic boundary conditions in the shear direction have the effect that the shear band has to be parallel to this direction (see Fig. 1). This facilitates the problem as we can sum up along the shear direction (z) thus reducing the problem to two dimensions.

Due to the geometrical constraints in the packing, the density and Coulomb angle are expected to fluctuate from place to place. At a suitable coarse-grained scale the medium can be described as a continuum, where the density is a random function displaying fluctuations around a mean value. In the spirit of the dilatancy concept [5] we assume that the Coulomb angle is linearly related to the density; the denser the material the more stable it gets. In the rest of this paper we describe the material at every point by only one parameter which will be denoted by ϱ and called either the density or, equivalently, the stability.

Our next task is to define the shear band. We have the following three constraints which define it unequivocally: (a) it is continuous, (b) it spans the sample in the x direction and (c) the sum of the stability along it is minimal among all possible paths satisfying (a) and (b). One can recognize that this is the well-known problem of finding the ground state of a directed polymer in a random potential [8–10].

As already mentioned, all the relative motions of the particles take place within the narrow confines of the shear band. It is known that granular materials are very sensitive to small fluctuations and motions [16–18], thus small movements can induce large changes in the local density. We model this very complex density change along the shear band by choosing random densities independent of their previous values.

In order to be able to simulate the above model we discretized it on a square lattice. We consider here square samples with system size $N \times N$ with N varying from 32 to 512. Every site is assigned a "density", initially a random number uniformly distributed between 0 and 1. The shear band is identified as the path ("directed polymer") explained above and the density at every site along this path is changed by choosing a random number from the $(0,1)$ uniform distribution. We model by this rule the density change induced by the relative motion of the particles in the shear band. This process is continued and the developement of system properties (such as the average density) are monitored as a function of the time.

3. Results

We define the average density $\langle \varrho \rangle$ as the mean value of the density of the sites not belonging to the shear band. This definition guarantees that the average density is a monotonically increasing function of time. This is due to the fact that any part of the newly chosen shear band that is different from the previous one must have a lower density than the previous shear band, by the rules of the model.

The steady state is numerically not reachable except for small lattice sizes where it turns out that $\lim_{t \to \infty} \langle \varrho \rangle = 1$. The detailed study of the model on the hierarchical lattice [19] leads us to believe that – though the dynamics of the system gets extremely slow for large systems – the asymptotic density on any finite sample is unity since the system never freezes completely and there is always a finite probability to visit a site which has a stability less than 1.

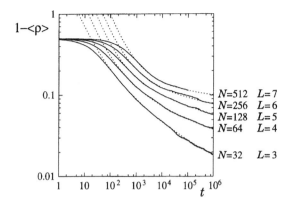

Fig. 2. The difference between the average density and its asymptotic value plotted as the function of time. The fits (dashed lines) are obtained using Eq. (1). The relation between the generation number L and the size of the square lattice N is logarithmic: $\log N \sim L$.

Having found the asymptotic value it is of interest to determine the functional form of the increase of the average density. This is found, very interestingly, to depend on the system size, the convergence to the asymptotic regime getting slower and slower for larger system sizes. From the analytical solution on a hierarchical diamond lattice we obtained the following form for large enough densities:

$$\varrho(t) = 1 - \sum_{i=1}^{L} \frac{a_i}{t^{0.5^i}}, \tag{1}$$

where L is the number of generations of the diamond lattice and a_i are computed coefficients. If i is large enough, a better approximation to the asymptotic form is given by $\sqrt{(\log t)/i}$ instead of $1/t^{0.5^i}$. Surprisingly, this functional form for the approach of the density to the asymptotic value as a function of time also gives a reasonable fit for the numerical results on the square lattice (Fig. 2). While not going into the details of the reasoning leading to the functional form in Eq. (1) [19], it is worth mentioning that the main physics is the inhomogenous aging of the medium. In the process of choosing a shear band, sites are discarded for different reasons. Some because their density was larger than a nearest-neighbor site, others because when summed together with another site, the total was larger and so on. The aging of a particular site (i.e., the increase of its density with time) depends crucially on the stage this site reached before the shear band left it.

In the following, we will try to find further consequences of the slow dynamics of the system.

Let us first define the (Hamming-) distance d which measures the number of different sites between successive minimal paths. The time dependence of this distance can be seen in Fig. 3.

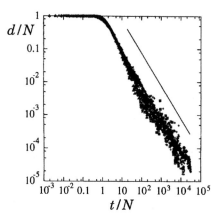

Fig. 3. Log–log plot of the time dependence of the distance d for five different system sizes (from 32 to 512) scaled together. Both the distance and the time scale with the system size. The solid line by the curves has the slope of -1.

It is apparent that in the early regime, below a characteristic time $t_c \simeq 2L$ the distance is nearly constant with $d \approx N$. Thus there is practically no overlap between two successive shear bands. The band wanders through the whole sample inflicting a continuous movement of the material. This mirrors the fact that in loose granular material the shear band is mobile and not localized.

The second regime is characterized by a fast, power law decrease of the distance. This means that larger and larger parts of the sample become quenched and motionless while the shear band becomes more and more localized.

The process of localization can be understood better if we look at the density maps in Fig. 4.

From Fig. 4 one can easily establish the history of the localization. At the beginning there is no structure in the density map therefore the position of the shear band is determined by uncorrelated fluctuations. Thus, we get no correlation between different successive positions of the shear bands: the shear band is delocalized. As time passes and the density increases, the previous path gets more and more favorable (since the mean density of the shear band is fixed by the given uniform distribution). As the path starts to get localized it increases the density only of the neighboring sites and finally finds itself in a very deep "canyon" girdled by very high density regions which makes escape improbable. A big jump can, however, occasionally take place to another part of the sample where the path restarts the above-described procedure.

This quasi-localization of the shear band from time to time is responsible for the very slow increase in the density. As time goes on, it gets harder and harder to make a change. Most of the time, it is the previous path which is the minimum and there is no change in the average density. However, there can be occasional large jumps to a much older path and then the density increases due to small fluctuations in the path as it carves out a "canyon".

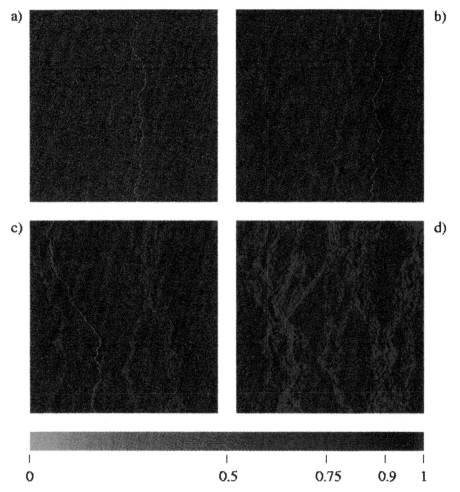

Fig. 4. Snapshots of densities at different times on a system of size 256×256: (a) $t = 1000$, (b) $t = 10\,000$, (c) $t = 100\,000$, (d) $t = 1000\,000$. The color code is indicated at the bottom of the figure. The actual shear band is plotted in green.

4. Conclusion

Though very simple, our model seems to capture the essential features of granular shear and provides at the same time several predictions. The model demonstrates the self-organized mechanism of the shear-induced density increase and of the localization of the shear band in loose granular materials. As the sample ages, very high fluctuations in density appear where we can observe some kind of arching effect: high-density, stable regions protect the loose, less stable ones. Due to the nature of the dynamics nontrivial size effects occur. The heterogeneous structure developing during the time evolution should be experimentally observable.

Acknowledgements

This research was partially supported by OTKA T029985 and T024004.

References

[1] D.M. Wood, Soil Behaviour and Critical State Soil Mechanics, Cambridge University Press, New York, 1990.
[2] R.P. Behringer, J.T. Jenkins (Eds.), Powder and Grains'97, Balkema, Rotterdam, 1997.
[3] H.J. Herrmann, J.P. Hovi, S. Luding (Eds.), Physics of Dry Granular Media, NATO ASI E 350, Kluwer Academic Publishers, Dordrecht, 1998.
[4] M. Sasvári, J. Kertész, D.E. Wolf, Instabilities in sheared granular matter, in: M. Schreckenberg, D.E. Wolf (Eds.), Traffic and Granular Flow'97, Springer, Berlin, 1998, p. 141.
[5] O. Reynolds, Philos. Mag. 20 (1885) 469.
[6] D.M. Wood, M. Budhu, Proceedings of the International Symposium on Soils under Cyclic and Transient Loading, Swansea, 1980.
[7] A. Ahadi, S. Krenk, Non-Associated Plasticity for Soils, Nordic Association for Computational Mechanics, Stockholm, Sweden, 1998.
[8] M. Kadar, G. Parisi, Y.-C. Zhang, Phys. Rev. Lett. 56 (1986) 889.
[9] Y.C. Zhang, T. Halpin-Healy, Phys. Rep. 254 (1995) 215.
[10] A.L. Barabási, H.E. Stanley, Fractal Concepts in Surface Growth, Cambridge University Press, Cambridge, 1995.
[11] P. Bak, K. Sneppen, Phys. Rev. Lett. 71 (1993) 4083.
[12] K.J. Må løy et al., Phys. Rev. Lett. 68 (1992) 213.
[13] J. Kertész, V.K. Horváth, F. Weber, Fractals 1 (1993) 67.
[14] B. Miller, C. O'Hern, R.P. Behringer, Phys. Rev. Lett. 77 (1996) 3110.
[15] J. Török, S. Roux, Heterogeneous Mohr–Coulomb plastic material, Granular Matter, submitted for publication.
[16] F. Radjai, M. Jean, J.-J. Moreau, S. Roux, Phys. Rev. Lett. 77 (1996) 274.
[17] C.T. Veje, D.W. Howell, R.P. Behringer, Phys. Rev. E tentatively scheduled for 59 (1999).
[18] S. Ouaguenouni, J.-N. Roux, Europhys. Lett. 32 (1995) 449.
[19] J. Török, S. Krishnamurthy, J. Kertész, S. Roux, preprint.

ELSEVIER

Physica A 274 (1999) 381–384

www.elsevier.com/locate/physa

Markov and non-Markov processes in complex systems by the dynamical information entropy

R.M. Yulmetyev *, F.M. Gafarov

Department of Theoretical Physics, Kazan State Pedagogical University, Mezhlauk Street 1, 420021 Kazan, Russia

Abstract

We consider the Markov and non-Markov processes in complex systems by the dynamical information Shannon entropy (DISE) method. The influence and important role of the two mutually dependent channels of entropy alternation (creation or generation of correlation) and anti-correlation (destroying or annihilation of correlation) have been discussed. The developed method has been used for the analysis of the complex systems of various natures: slow neutron scattering in liquid cesium, psychology (short-time numeral and pattern human memory and effect of stress on the dynamical taping-test), random dynamics of RR-intervals in human ECG (problem of diagnosis of various disease of the human cardio-vascular systems), chaotic dynamics of the parameters of financial markets and ecological systems. © 1999 Elsevier Science B.V. All rights reserved.

PACS: 05.40.+j; 05.60.+w

Keywords: Complex systems; Nonlinear phenomena; Markov and non-Markov processes; Dynamical information entropy; Creation and annihilation of correlation; Application to psychology; Cardiology; Finance and ecology

1. The objective of this report is to describe the dynamical behavior of a complex systems of various nature using notions from dynamical information Shannon entropy [1,2]. The complexity, nonlinearity and nonstationarity of physical, chemical, biological, physiological and financial systems have been of main interest in the last few years (see, for example, [3–12]). Our main idea that is that information (Shannon) entropy of the random processes is abundantly supplied with the qualitative and quantitative data on the object under investigation.

2. Let us consider the time chaotic evolution of the dynamical variable $A(t)$, statistical average over the distribution $\langle A(t) \rangle$ and fluctuation $\delta A(t) = A(t) - \langle A(t) \rangle$. For the

---

* Corresponding author. Fax: +7-8432-32-4269.
E-mail addresses: rmy@dtp.ksu.ras.ru (R.M. Yulmetyev), gfm@dtp.ksu.ras.ru (F.M. Gafarov)

0378-4371/99/$ - see front matter © 1999 Elsevier Science B.V. All rights reserved.
PII: S 0378-4371(99)00332-5

physical systems the variable $A(t)$ and fluctuation $\delta A(t)$ obey the Liouville equation of motion, this makes possible the use of the technique of Zwanzig-Mori's non-Markov kinetic equations [1,2,12].

Now, let us take up further the normalized time correlation function (TCF)

$$a(t) = \frac{\langle \delta A^*(0) \delta A(t) \rangle}{\langle |\delta A(0)|^2 \rangle} \ ,$$

$$A(0) = 1, \quad \lim_{t \to \infty} a(t) = 0 \ . \tag{1}$$

It is convenient to introduce the relaxation (correlation) time as [11]

$$\tau_c = \Re \tilde{a}(0), \quad \tilde{a}(s) = \int_0^\infty dt \, e^{-st} a(t) \ , \tag{2}$$

where $\Re$ is the real part.

In Ref. [11] it was demonstrated, that there exists the possibility of another definition of relaxation time

$$\tau_c = \left\{ \int_0^\infty dt \, t^K \rho(t) \right\}^{1/K} , \tag{3}$$

where K is an integer. The function $\rho(t)$ is the time-dependent probability density associated with the TCF $a(t)$ and normalized by

$$\int_0^\infty dt \, \rho(t) = 1, \quad \rho(t) = C|a(t)|^2, \quad C^{-1} = \int_0^\infty dt \, |a(t)|^2 \ . \tag{4}$$

Let us note that the case with $K = 2$ in Eqs. (4) and (5) has an analogy with the definition of the coherence time in optics [13] and with the density

$$\rho(t) = a(t) \left\{ \int_0^\infty dt \, a(t) \right\}^2$$

from Egelstaff's definition in the theory of slow neutron scattering in condensed matter [14,15].

Therefore, it may be safely suggested that probability function

$$W(t) = |a(t)|^2 \tag{5}$$

is the probability of the state of creation of correlation. Because of the normalizing condition of the total probability

$$\sum_{i=c,a} W_i(t) = 1, \quad W_c = |a(t)|^2 \ , \tag{6}$$

we can introduce the probability of annihilation of correlation

$$W_a(t) = 1 - W_c(t) = 1 - |a(t)|^2 \ . \tag{7}$$

3. Starting from the definition of Shannon information entropy,

$$S = -\sum_{m=1}^n p_m \ln p_m, \quad \sum_{m=1}^n p_m = 1 \ , \tag{8}$$

where p_m is the probability of m's state of a system. One can come to recognize that the dynamical function

$$S(t) = -\sum_{i=c,a} W_i(t) \ln W_i(t) = -|a(t)|^2 \ln |a(t)|^2 - \{1 - |a(t)|^2\} \ln \{1 - |a(t)|^2\}$$

(9)

is the dynamical information entropy of correlation in the system considered.

To assess quantitatively the differences between two dynamical states (creation and annihilation of correlation) we can introduce two partial entropy channels

$$S_{cc}(t) = -|a(t)|^2 \ln |a(t)|^2 \,,$$

(10)

$$S_{ac} = -\{1 - |a(t)|2\} \ln \{1 - |a(t)|^2\} \,.$$

(11)

The introduction of the different dynamical entropy channels will allow us to understand the hidden role of the existence of correlation in the complicated behavior of the system considered.

4. Valuable information on the Markov and non-Markov behavior of the complex systems of various natures can be extracted from a direct measurement and calculation of dynamical entropy of the time correlations.

We are carrying on research by DISE method for the different living systems in psychology (short-time numeral and pattern human memory, stress effects on the neurophysiological activity), medicine (diagnotics of cardiovascular systems by the study of dynamics of RR-intervals of human ECGs), financial systems (dynamics of exchange rate Russian Rbl/USD nearly to a crash on August 17, 1998 in Russia) and ecological systems.

The obtained data testifies that the dynamical Shannon entropy and its frequency spectra serve as a powerful method for the study of the different complex life systems.

Acknowledgements

This work was partially supported by the Russian Humanitar Science Fund, Grant 97-06-08048 and the Competetive Centre of Fundamental Research at Saint-Peterburg University, Grant 97-0-14.0-12.

References

[1] R.M. Yulmetyev, M.Ya. Kleiner, Nonlinear Phenomena Complex Systems 1 (1998) 80.
[2] R.M. Yulmetyev, F.M. Gafarov, Physica A 273 (1999) 416–438.
[3] S. Clausen, G. Helgesen, A.T. Skjeltorp, Int. J. Bifurcation Chaos 8 (1998) 1383.
[4] N. Vanderwalle, M. Ausloos, Int. J. Mod. Phys. C 9 (1998) 711.
[5] N. Vanderwalle, Ph. Boveroux, A. Minguet, M. Ausloos, Physica A 255 (1998) 201.
[6] O. Mishima, H.E. Stanley, Nature 396 (1998) 329.
[7] G.M. Viswanathan, C.K. Peng, H.E. Stanley, Phys. Rev. E 55 (1997) 845.

[8] P.Ch. Ivanov, M.G. Rosenblum, C.K. Peng, J. Mietus, S. Havlin, H.E. Goldberger, Nature 383 (1996) 327.

[9] C.L. Webber, J.P. Zbilut, in: C.H. Dagli, B.R. Fernander, J. Ghosh, R.T.S. Kumara (Eds.), Intelligent Engineering Through Artificial Neural Networks, vol. 4, ASME, New York, 1994, p. 695.

[10] J.J. Zebrowski, W. Poplawska, R. Baranowski, Phys. Rev. E 50 (1994) 4187.

[11] V.Yu. Shurygin, R.M. Yulmetyev, V.V. Vorobjev, Phys. Lett. A 148 (1990) 199.

[12] R.M. Yulmetyev, V.Yu. Shurygin, T.R. Yulmetyev, Physica A 242 (1997) 509.

[13] M. Born, E. Wolf, Principles of Optics, Pergamon Press, Oxford, 1964.

[14] P.A. Egelstaff, Phys. Rev. A 31 (1985) 3802.

[15] P.A. Egelstaff, Z. Phys. Chem. 156 (1988) 311.

List of contributors

List of participants

KEYNOTE LECTURERS:

Preben Alstrøm
Niels Bohr Institute
Hollaendervej 7 st. tv.
DK-1855 Frederiksberg C
DENMARK

Marcel Ausloos
S.U.P.R.A.S.
Institut de Physique B5
Université de Liège
B-4000 Liège
BELGIUM

Marcia C. Barbosa
Instituto de Fisica
Universidade Federal do Rio
   Grande do Sul
P.O. Box 15051
91501-970 Porto Alegre
BRASIL

Françoise Brochard-Wyart
PCC Physico-Chimie-Curie
Institut Curie
11, rue Pierre et Marie Curie
75231 Paris Cedex 05
FRANCE

Marek Cieplak
Institute of Physics
Polish Academy of Sciences
Instytut Fizyki PAN
Al. Lotników 32/46
PL-02-668 Warsaw
POLAND

Pierre-Gilles de Gennes
ESPCI
10, rue Vauquelin 75005 Paris
FRANCE

Sam F. Edwards
Cavendish Laboratory
Madingley Road
Cambridge CB3 0HE
UNITED KINGDOM

Michael E. Fisher
Institute for Physical Sciences and
   Technology
University of Maryland
College Park, MD 20742
USA

Serge Galam
Laboratoire des Milieux
   Desordonnes et Heterogenes (LMDH)
Université Paris 6
Case 86, T13
4 Place Jussieu
75252 Paris Cedex 05
FRANCE

Shlomo Havlin
Department of Physics
Bar-Ilan University
Ramat-Gan
ISRAEL

Peter R. King
Director, Polymer Research Group
BP Research – BP International Limited
Sunbury Research Centre
Chertsey Road
Sunbury-on-Thames
Middlesex TW16 7LN
UNITED KINGDOM

Imre Kondor
Department of the Physics of
   Complex Systems
Eötvös University
Budapest
HUNGARY

Ryszard Kutner
Institute of Experimental Physics
Warsaw University
Pl-00681 Warsaw, Hoża 69
POLAND

Jerzy Łuczka
Department of Theoretical Physics
Silesian University
40-007 Katowice
POLAND

Rosario Mantegna
Dipartmento di Energetica ed
  Applicazioni di Fisica
Palermo University, Palermo
ITALY

Mitsugu Matsushita
Department of Physics
Chuo University
Kasuga, Bunkyo-ku
Tokyo 112
JAPAN 1007

David R. Nelson
Dept. of Physics
Harvard University
Cambridge, MA 02138
USA

Zoltán Noszticzius
Inst. Phys. TU Budapest
Budafoki ut 8, H-1111 Budapest
HUNGARY

Andrzej Pękalski
Institute of Theoretical Physics
University of Wroclaw
pl. M. Borna 9
PL 50-204 Wrocław
POLAND

Zoltán Rácz
Dept. Theor. Phys.
Eötvös University
Budapest
HUNGARY

Arne Skjeltorp
Institutt for Energiteknikk
POB 40
N-2007 Kjeller
NORWAY

Harry L. Swinney
Physics Department and Center
  for Nonlinear Dynamics
University of Texas at Austin
Austin, TX 78712
USA

Hideki Takayasu
Sony Computer Science Laboratory
3-14-13 Higashi-Gotanda, Shinagawa-ku
Tokyo 151
JAPAN

Misako Takayasu
Research for the Future Project
Faculty of Science and Technology
Keio University
Shin-Kawasaki-Mitsui Bldg. West 3F
890-12 Kashimada, Saiwai-ku
Kawasaki-shi, 221
JAPAN

Tamás Tél
Dept. Theor. Phys.
Eötvös University
Budapest
HUNGARY

Donald L. Turcotte
Department of Geological Sciences
Cornell University, Snee Hall
Ithaca, NY 14853-1504
USA

Tamás Vicsek
Dept. Biol. Phys.
Eötvös University
Budapest
HUNGARY

Dietrich E. Wolf
Theor. Physik, FB 10
Gerhard-Mercator-University
D-47048 Duisburg
GERMANY

Marek Wolf
Institute of Theoretical Physics
University of Wroclaw
pl. M. Borna 9
PL-50204 Wrocław
POLAND

ADDITIONAL PARTICIPANTS:

László Borda
Institute of Physics
Technical University of Budapest
H-1111 Budafoki ut 8.
HUNGARY

Vladislav Capek
Faculty of Mathematics and Physics
Institute of Physics of Charles University
Ke Karlovu 5
CZ-121 16 Prague 2
CZECH REPUBLIC

Tarik Çelik
Hacettepe University
Dept. of Physics Engineering
Beytepe 06532, Ankara
TURKEY

Susanne Cheybani
TU Budapest
Institute of Physics
Budafoki ut 8.
H-1111 Budapest
HUNGARY

Dionizy E. Czekaj
University of Silesia
Department of Materials Technology
2, Śnieżna St.
Sosnowiec, 41-200
POLAND

Dora Foti
Dipartimento di Ingegneria Strutturale
Politecnico di Bari
Via Orabona 4
70125 Bari
ITALY

Jose Viana Gomes
Departamento de Fisica
Universidade do Minho
4719 Braga Codex
PORTUGAL

Jerzy Górecki
Institute of Physical Chemistry
Polish Academy of Science
Kasprzaka St. 22/44
Warsaw
POLAND

Bogdan Grabiec
Department of Theoretical Physics
Silesian University
40-007 Katowice
POLAND

Rita Greco
Dipartimento di Ingegneria Strutturale
Politecnico di Bari
Via Orabona 4
70125 Bari
ITALY

Kristinka Ivanova
Institute of Electronics
Bulgarian Academy of Sciences
72 Tzarigradsko Chaussee
Sofia 1784
BULGARIA

Kimmo Kaski
Helsinki University of Technology
Laboratory of Computational Engineering
Miestentie 3
FIN-02150 Espoo
FINLAND

Adam Kleczkowski
King's College and Dept. of Plant Sciences
University of Cambridge
Downing Street
Cambridge CB2 3EA
UNITED KINGDOM

Miroslav Kotrla
Institute of Physics
Academy of Sciences
Na Slovance 2
180 40 Prague 8
CZECH REPUBLIC

László Kullmann
Institute of Physics
Technical University of Budapest
H-1111 Budafoki ut 8.
HUNGARY

Beppe Marano
Dipartimento di Ingegneria Strutturale
Politecnico di Bari
Via Orabona 4
70125 Bari
ITALY

Miroslav Pojsl
Faculty of Mathematics and Physics
Institute of Physics of Charles University
Ke Karlovu 5
CZ-121 16 Prague 2
CZECH REPUBLIC

Ryszard Skulski
University of Silesia
Department of Materials Technology
2, Śnieżna St.
Sosnowiec, 41-200
POLAND

Frantisek Slanina
Institute of Physics
Academy of Sciences
Na Slovance 2
182 21 Prague 8
CZECH REPUBLIC

János Török
Institute of Physics
Technical University of Budapest
H-1111 Budafoki ut 8.
HUNGARY

Balint Toth
TU Budapest
Inst. of Mathematics

Muegyetem rkp. 3.
H-1111 Budapest
HUNGARY

Mikhail Vasilevsky
Departamento de Fisica
Universidade do Minho
4719 Braga Codex
PORTUGAL

Rinat Yulmetyev
Kazan State Pedagogical University
Department of Theoretical Physics
P.O. Box 21, Mezhlauk Street, 1
420021 Kazan
RUSSIA

Printed and bound by CPI Group (UK) Ltd, Croydon, CR0 4YY

08/05/2025

01864929-0001